REASON AND NATURE

REASON AND NATURE

MORRIS R. COHEN

Routledge
Taylor & Francis Group

First published in 1931 by Harcourt, Brace and Company

This edition first published in 2018 by Routledge
2 Park Square, Milton Park, Abingdon, Oxon, OX14 4RN
and by Routledge
711 Third Avenue, New York, NY 10017

Routledge is an imprint of the Taylor & Francis Group, an informa business

© 1931 by Taylor & Francis

Publisher's Note
The publisher has gone to great lengths to ensure the quality of this reprint but points out that some imperfections in the original copies may be apparent.

Disclaimer
The publisher has made every effort to trace copyright holders and welcomes correspondence from those they have been unable to contact.

A Library of Congress record exists under ISBN: 31008779

ISBN 13: 978-1-138-31057-5 (hbk)

ISBN 13: 978-1-138-31058-2 (pbk)

ISBN 13: 978-0-429-45934-4 (ebk)

REASON AND NATURE

AN ESSAY ON THE MEANING OF
SCIENTIFIC METHOD

by

MORRIS R. COHEN

HARCOURT, BRACE AND COMPANY

NEW YORK

Typography by Robert S. Josephy
PRINTED IN THE UNITED STATES OF AMERICA
BY QUINN & BODEN COMPANY, INC., RAHWAY, N. J.

TO MR. JUSTICE OLIVER WENDELL HOLMES

The Courageous Thinker and Loyal Friend

Preface

THE distinctive intellectual traits of Western civilization, or of what is sometimes called the modern mind, have been largely moulded by the appeal to nature against conventional taboos and by the appeal to reason against arbitrary authority. That nature and reason, like warmth and light, may be intimately joined was made evident in the Hellenic ideal of science as a free inquiry into nature and of ethics as concerned with a rational plan for attaining the natural goods of life. Unfortunately for the career of liberal civilization, however, various circumstances have brought about a mutual hostility between these two appeals to what are popularly called the heart and the mind. The appeal to nature is frequently a form of sentimental irrationalism, and the appeal to reason is often a call to suppress nature in the interest of conventional supernaturalism. To understand fully the grounds, causes, and effects of this conflict would involve a thorough survey of contemporary civilization and carry us far into the complexity of the human mind. One of the elements, however, in such a survey is a right understanding of the general bearing or meaning of scientific method, i.e., of the principles of procedure according to which scientific results are obtained and according to which these results are being constantly revised. In developed natural science, reason and nature are happily united.

Such a study of the principles of science I began some twenty years ago. Untoward circumstances have prevented the continuous toil necessary for such a task. Different parts of this work have been written at widely different times and in somewhat different keys. But the favourable reception accorded to those portions of it which have been published from time to time, and the insistence of friends whose judgment compels respect, induce me somewhat reluctantly to publish the present essay rather than wait any longer for the completion of a more satisfactory systematic treatise. I am induced to do this not only by a sense of the inherent importance of the issues I have faced but also by my conviction that the present state of philosophy is especially in need of the method of approach here represented.

There are many indications of a widespread dissatisfaction with the arid state of present-day technical philosophy. Seldom before has the general craving for philosophic light seemed so vast and the offerings of professed philosophers so scant and unsubstantial. Some there are who attribute this poverty to the preoccupation of professional philosophers with the technical problems of epistemology, e.g. how the mind can know a supposedly external world. It has been justly urged that this baffling and strangely fascinating problem has in fact never thrown much light on the nature of any of the great objects of our vital interests, though subjectivist solutions of it have often been used to support traditional views and evaluations. We may grant this without denying that the concentration on the technical problems of epistemology arose out of a well-justified dissatisfaction with the older romantic fashion according to which every philosopher was expected to spin a new view of the universe out of his inner consciousness, or else confirm in a new and strange manner the old familiar views. The critical question, "How do we know?" is a much-needed challenge to those who complacently claim to have solved all the enigmas of existence. Yet the main motive for epistemology is professional. As teachers of philosophy see their colleagues gaining prestige through contributions in special technical fields, they are tempted to take the position: "We too are specialists. We too have a definite technical field of our own, to wit: the nature of knowledge as such." But alas! The ideal of technical competency is not without its snares. A good deal of the unsubstantiality of later scholasticism was no doubt due to the fact that having elaborated a very subtle technical vocabulary, men felt themselves to be distinguished scholars by the mere mastery of such a vocabulary. The change from Latin to the vernacular revealed this emptiness and compelled a greater attention to substantial content. But this gain is now largely frittered away in philosophy and in the related fields of psychology and sociology, in which exercises in technical vocabulary frequently hide the paucity of substantial insight.

Meanwhile, the ancient need of a more or less integrated view of the general panorama of life and existence has shown itself to be too deep to be permanently neglected. New voices have arisen urging philosophy to become again constructive and to cease to lose itself in historical, philologic, and other technical minutiae. There can be little doubt that the philosophic writers of recent days who have most stirred men's imaginations and found the greatest popular response have been of this character, witness James, Bergson, Croce, and Spengler.

Can those who have a sober knowledge of the many failures of previous romantically "constructive" philosophies entertain a naïve and unquestioning hope in these new efforts? This is especially difficult for those who know something about physical science and who cannot agree to banish its solid theoretic achievements as merely practical devices devoid of philosophic value. Doubtless philosophy will long continue to represent interests wider than those of rigidly demonstrative science. Philosophy is primarily a vision and all great philosophers have something in common with the poets and prophets. But while vision, intuition, or wisdom is the substance of any philosophy that is worth while, serious philosophy must also be something more than a poetic image or prophecy. It must, like science, be also vitally concerned with reasoned or logically demonstrable truth. Granted that great truths begin as poetic or prophetic insights, it still remains true that the views of poets and prophets have in fact often proved narrowly one-sided, conflicting, incoherent, and illusory. To introduce order and consistency into our vision, to remove pleasing but illusory plausibilities by contrasting various views with their possible alternatives, and to judge critically all pretended proofs in the light of the most rigorously logical rules of evidence, is the indispensable task of any serious philosophy. The seed which ripens into vision may be a gift of the gods but the labour of cultivating it so that it may bear nourishing fruit is the indispensable function of arduous scientific technique.

If these considerations are in any way sound, philosophy can hope to be genuinely fruitful only by being more scientifically critical or cautious than the recent romantic efforts, and at the same time more daring and substantial than those microscopic philosophies which lose sight of the macrocosm. This can be brought about only by intimately confronting the great classical views of the world at the heart of the great humanistic tradition with the painfully critical methods by which the natural sciences have built up their great cosmic vistas.

In turning to the sciences I emphasize their method rather than their results. For, in an age of scientific expansion, not only are the methods the more permanent features of the sciences, but the supposed results are often merely popularized conventions, utterly misleading to all those who do not know the processes by which they are obtained. The life of science is in exploration and in the weighing of evidence. Dead or detached results lend themselves to the mythology of popular science, and ignorance of method leads to the view of science as a new set of dogmas to be accepted on the authority of a new set of priests called scientists. It is doubtless impossible for any single individual to be a trained investigator in all the different fields upon which philosophy touches. It may also be urged that too great devotion to rigorous scientific evidence may narrow our sympathies and prevent us from dealing justly with those vital interests which science has not yet been able to organize. But the difficulty and the dangers of our task cannot prevent it from being indispensable in any case. All that is absolutely worth while has something of the unattainable about it. No faith can live today in anything but a fool's paradise unless it ventures out into the high and open but biting air of critical reason as natural science does.

Greater regard for the rational methods of natural science must be joined with a more serious concern with the great historic traditions in philosophy. Sound logic and the history of science unite to show that an adequately critical appreciation of the reported facts in any realm depends on a knowledge of the hitherto prevailing views which condition what we shall regard as facts.

Moreover to take adequate cognizance of those views which have in fact proved influential is a powerful protection against the temptation to triumphantly exploit one's own views in a narrowly partial and one-sided manner. The notion that we can dismiss the views of all previous thinkers surely leaves no basis for the hope that our own work will prove of any value to others.

The philosopher whose primary interest is to attain as much truth as possible must put aside as a snare the effort at originality. Indeed, it seems to me that the modern penchant for novelty in philosophy is symptomatic of restlessness or low intellectual vitality. At the same time

PREFACE

I should not make this book public if I did not hope that thoughtful readers will find its ideas timely, worthy of consideration, and capable of extension. Particular attention is directed to the principle of polarity, the principle that opposite categories like identity and difference, rest and motion, individuality and universality, etc., must always be kept together though never identified. Whatever the inadequacies of my formulation and application of this principle, I am sure that by a more persistent use of it many of the traditional controversies of philosophy can be eliminated, or at least shown to be inadequately stated. The principle of polarity calls attention to the fact that the traditional dilemmas, on which people have for a long time taken opposite stands, generally rest on difficulties rather than real contradictions, and that positive gains in philosophy can be made not by simply trying to prove that one side or the other is the truth, but by trying to get at the difficulty and determining in what respect and to what extent each side is justified. This may deprive our results of sweep and popular glamour, but will achieve the more permanent satisfaction of truth.

While speculations about the future of civilization and philosophy must in the nature of the case be very uncertain, all evidence seems to indicate that in the rhythmic alternation of periods of expansion and consolidation which characterize the life of civilization, different sets of categories are emphasized. Thus the mediaeval mind was dominated by the demand for order in our thought as well as in the world at large. Faced with the centrifugal tendencies due to the pressure of population against the bounds of a fixed land economy, and faced with the doubts as to the traditional theology that resulted from contact with Saracen civilization, the mediaeval mind found in the scholastic method of drawing distinctions a way of introducing order and reconciling conflicting views. It did so by trying to assign to everything its just realm. This method of scholasticism, originated by Abelard in his *Sic et Non*, independently embodied in Gratian's great work on Canon Law, and perfected by St. Thomas in his Summa, rests ultimately on the axiom of Aristotelian logic that opposites cannot be identical, illustrated in Kipling's maxim, "East is East and West is West." The wise or good man, on this view, will draw and respect the proper boundaries between them.

Modern thought, i.e. the thought since the middle of the fifteenth century, has been operating in what has in the main been an expanding economy. The opening of new lands and new methods of production have made the idea of movement or growth the dominant one in the internal as well as in the external world. But the idea of change involves some continuity between opposites. The first modern man who saw this both in mathematics and philosophy was the great Cardinal Cusa, whose doctrine of the coincidence of opposites marks the beginning of modern kinetic views of the world which terminate in nineteenth century Hegelianism and evolutionism. Modern thought emphasizes mobility and the principle that everything changes to something else and must indeed become something other than it was in order to exhibit its true character. The nature of things unfolds itself in time.

There are, however, signs that this period of expansion is about to close. Despite the steady progress of science, Europe seems already to have reached the maximum population that it can support, and the opportunities of exploiting other parts of the world by migration or conquest seem steadily decreasing. Within measurable time, then, the whole earth will become fully explored and the limits of its food and other resources may be definitely attained. The dominant values of life must then become somewhat similar to those of mediaeval and Chinese society—greater stress on order than on expansion. Indeed, already the popular demand is for the organization of knowledge rather than for its expansion. But whatever the future may bring, it is clear that any attempt to see the limitation of the characteristically modern emphasis on the "dynamic," the "evolutionary," and the "progressive" must pay greater attention to the principle of polarity as the union of the values of order and stability with the values of change and progess. Because of these considerations I am confident that no matter how inadequate the results of the studies here published, no just criticism, not even the withering refutation of time, will prove the effort itself to be in the wrong direction.

To readers who have a predilection for conventional labels, I offer the following:

I am a rationalist in believing that reason is a genuine and significant phase of nature; but I am an irrationalist in insisting that nature contains more than reason. I am a mystic in holding that all words point to a realm of being deeper and wider than the words themselves. But I reject as vicious obscurantism all efforts to describe the indescribable.

I reject the euthanàsia or suicide of thought involved in all monisms which identify the whole totality of things with matter, mind, or any other element in it. But I also reject the common dualism which conceives *the* mind and *the* external world as confronting each other like two mutually exclusive spatial bodies. I believe in the Aristotelian distinction between matter and form. But I am willing to be called a materialist if that means one who disbelieves in disembodied spirits; and I should refer to spiritists who localize disembodied spirits in space as crypto-materialists. However, I should also call myself an idealist, not in the perverse modern sense which applies that term to nominalists like Berkeley who reject real ideas, but in the Platonic sense according to which ideas, ideals, or abstract universals are the conditions of real existence, and not mere fictions of the human mind.

To those who labour under the necessity of passing judgment on this book in terms of current values, I suggest the following:

The author seems out of touch with everything modern and useful, and yet makes no whole-hearted plea for the old. He believes in chance and spontaneity in physics and law and mechanism in life. He has no respect for *experience, induction, the dynamic, evolution, progress, behaviourism* and *psychoanalysis,* and does not line up with either the orthodox or revolutionary party in politics, morals, or religion, though he writes on these themes. He offers no practical message to the man engaged in the affairs of life, and seems to be satisfied with purely contemplative surveys of existence.

But to the thoughtful reader I can offer as a preliminary only the expression of my profound faith in philosophy itself. The task of philosophy is too complicated to be solved by simple magical formulae. The age of panaceas, nostrums, and philosopher's stones belongs to the adolescence of philosophy. In its maturer period, it must, like science and rational industry, depend upon more modest and workmanlike efforts, though it can never abandon the search for comprehensive vision. We cannot by reasoning achieve absolute certainties as to matters of fact. But we may clarify our minds as to the rational strength of the evidence for our convictions. We shall never overcome the infinite sea of ignorance, and we should remember that if ignorance were bliss, this would be a much happier world than it is. Let not philosophy, therefore, deal in false comforts or gratify those who crave from it confirmation of established prejudices. It has a far higher function,—to give men strength to envisage the truth. Vision is itself a good greater than the perpetual motion without any definite direction which modernists regard as the blessed life. Cosmic vision ennobles the pathetic futility of our daily crucifixions. But all philosophy is blasphemous if it denies to the gods a sense of humour or the gift of laughter.

One who ranges over as wide a field as that involved in the present volume cannot hope to escape detectable errors. I should, however, add that where my views differ from those of well-known authorities it is not always because of my ignorance as to what these authorities maintain. I have taken pains to familiarize myself directly with the views of the great classical philosophers and with the scientific researches on which my generalizations as to scientific methods are based, but I have not tried to escape responsibility for my own judgment.

I have not been at all impressed by the religious and philosophic lessons drawn from science by men like Millikan, Eddington, Coulter, A. H. Compton, E. G. Conklin, and the like. I respect, as every one must, the great achievements of these distinguished workers in their special fields. But scientists do not always carry scientific method into their views of manners, morals, or politics, of justice between nations or social classes, of the reliability of mediums, etc. Neither are they scientific when they make their professional work a springboard from which to jump off into amateurish speculative flights in the fields of religion and philosophy.

I have in the substance and manner of this book tried to reach thoughtful readers irrespective of their previous philosophic studies and have therefore avoided technical terms as far as was consistent with substantial accuracy. Indeed I have used some technical terms of logic in their popular sense, trusting the context to make my meaning clear. It is difficult to know how much of the factual content of science to take for granted. No two readers are exactly alike in this respect, but I have addressed myself to the generally educated public that is not afraid of close and sustained reflection. Some of the chapters, notably the third of Book I, require more effort in that direction than others. But the reader may skip some chapters or change the order of reading them. Though they are all parts of an integral thesis, they were written at different times and have a certain relative independence of each other.

PREFACE

It is the pleasant privilege of the author of a book, no matter what its achievement, to acknowledge his obligations. This permits me to associate this volume with the great names of my teachers, Josiah Royce and William James—unfortunately no longer with us—and with the names of G. H. Palmer, Felix Adler, F. J. E. Woodbridge, C. A. Strong, and W. H. Sheldon, among the living. If I have succeeded in indicating the genuine interplay of practical issues and theoretic philosophy the credit is largely due to encouragement in that direction from Professor Adler. If I have succeeded in maintaining the need for the classical spirit of discrimination in philosophy, the reader is indebted to Professor Woodbridge, whose seminar in Aristotle was my first regular academic course in philosophy. To Professors Woodbridge and Bush, as editors of the *Journal of Philosophy*, I am also indebted for their kind encouragement to publish most of my papers that were read before the American Philosophical Association or the Philosophical Club of New York.

My debt to George Santayana's *Life of Reason* might have been much greater if I had not arrived at my fundamental positions by a road which I imagine to be altogether different from his own. Only after the essentials of this book were worked out in my own mind did I begin to appreciate the profound insight with which naturalism and spirituality are united in the latter parts of the *Life of Reason*. I mention this not for the sake of any vain claims to originality but in the hope that my readers may be led to a greater appreciation of the richest philosophic classic produced on this side of the Atlantic. To Bertrand Russell's *Principles of Mathematics* I owe the greatest of all debts,—it helped me to forge the instruments for acquiring intellectual independence.

Last I must mention that the philosophic studies of which this book is an expression would probably have been impossible if it were not that in dark and dispiriting hours I was helped by the sustaining friendship first of Thomas Davidson, William T. Harris, and Janet E. Ruutz-Rees—would I had something worthier to offer on the altar of their memory!—and later of Arthur S. Meyer, Felix Frankfurter, Dr. A. A. Himwich, and a life-long companion of an understanding heart whose modesty prevents a more explicit reference. I should be derelict if I did not also acknowledge my indebtedness to Mr. and Mrs. Osmond K. Fraenkel and to Mrs. Edward MacDowell, the inspiring leader of the

MacDowell Colony at Peterborough, N. H., for opportunities to do the actual writing of a large part of this book. Mr. Fraenkel has also increased my obligations to him by reading the book in manuscript. My son, Dr. Felix S. Cohen, has been so helpful in suggestions, criticisms, and preparing the manuscript for the press, that I find it difficult to indicate the great extent of my—and the reader's—obligation to him.

In dedicating this book to Justice Holmes I wish to indicate not only my reverence and affection for him personally but also the fact that I have not forgotten the others who ten years ago were associated with him in a venture which heartened me more than I can readily express.

Contents

CONTENTS

CONTENTS

CONTENTS

CONTENTS

CONTENTS

BOOK III. REASON IN SOCIAL SCIENCE

CONTENTS

CONTENTS

Book I

REASON AND THE NATURE OF THINGS

Chapter One

THE INSURGENCE AGAINST REASON [1]

D ESPITE the frequent assertion that ours is the age of science, we are witnessing today a remarkably widespread decline of the prestige of intellect and reason. Though the most successful of our modern sciences, the various branches of mathematics, physics, and experimental biology, have admittedly been built up by intellectual or rational methods, "intellectualistic" and "rationalistic" are popular terms of opprobrium. Even among professed philosophers, the high priests of the sanctuary of reason, faith in rational or demonstrative science is systematically being minimized in the interests of "practical" idealism, vitalism, humanism, intuitionism, and other forms of avowed anti-intellectualism. A striking instance of this is William James' attack, in his *Pluralistic Universe,* on the whole enterprise of intellectual logic, in the interest of Bergsonian intuitionism and Fechner's mythologic speculations about the earth-spirit.

There can be little doubt that this distrust of reason has its roots deep in the dominant temper of our age, an age whose feverish restlessness makes it impatiently out of tune with the slow rhythm of deliberate order. The art, literature, and politics of Europe and of our own country show an ever-growing contempt for ideas and form. The popular philosophies of the day, those which emanate from James, Bergson, Croce, Nietzsche, Chamberlain, Spengler, and others, are certainly at one with the recent novel, drama, music, painting, and sculpture in attaching greater value to novel impressions and vehement expression than to coherency and order. The romantic or "Dionysiac" contempt for prudence and deliberative (so-called bourgeois) morality is a crude expression of the same reaction against scientific or rigorous intellectual procedure,—a reaction which makes our modern illuminati like Bergson and Croce dismiss physical science as devoid of

[1] The substance of the present chapter was printed in the *Journal of Philosophy,* Vol. XXII (1925), p. 113.

any genuine knowledge, or as at best a merely practical device for manipulating dead things. It would preposterously exaggerate the actual influence of philosophy if the outrageously shameless contempt for truth shown in the various forms of recent propaganda were attributed to the systematic scorn heaped by modernistic philosophies on the old ideal of the pursuit of truth for its own sake "in scorn of consequences." Yet this decline of respect for truth in public or national affairs is certainly not devoid of all significant connection with its decline in philosophy and art.

Is it far-fetched to correlate the distrust of intellectual procedure (and consequent revival of all sorts of ancient superstitions) with the growing bigotry, intolerance, and remarkable resurgence of faith in violence? Violence is certainly no new phenomenon in human affairs. But avowed faith in violence has seldom before raised its head so high among educated men, or led them to such contempt for free parliamentary discussion, and to such readiness to suppress contrary opinion by arbitrary force and dictatorial power. From Moscow to the Mediterranean there reigns a pathetic faith in salvation through brutality. Nor is this entirely the result of the war. Even before the war, unabashed faith in violence showed itself in such diverse movements as nationalism and syndicalism and in such incidents as the organized Tory riot in the British House of Commons.

This decline in the popular prestige of the appeal to presumably universal reason is naturally associated with the growing contempt for the ideal of humanity professed in the days of the rationalistic enlightenment by men like Voltaire, Lessing, Diderot, Kant, Condorcet, Thomas Paine and Goethe. The new philosophies of "sacred" national egoism put temporary, local, or racial interests so high as to hold in contempt the common life of humanity.

In thus calling attention to the connection between anti-rationalism and the violent temper of our age, I have no desire to cry: Woe is us, we have fallen on evil days! Nor do I think that all the new popular philosophies can justly be dismissed by simply being thrown together in bad company—no more than their truth can be established by sweeping unhistorical claims to novelty, or by indiscriminate rejection of all previous philosophy. I wish merely to call attention to the fact that

4

the issues of rationalism have all sorts of vital backgrounds and bearings.

Current anti-rationalism endeavours to soften its opposition to rigorous logical procedure by representing modern science as empirical rather than rational. But without reasoning, as the process of drawing logical inferences, there is no science. We may grant this without prejudging the claims of mystic vision, intuition, or higher non-rational illumination. The proper relation between the rational and empirical elements in science will occupy our attention later. Here it is sufficient to note that none of the great founders of modern science felt any opposition between the rational (mathematical or logical) and the empirical (or experimental) elements in their procedure. Literary historians and philosophers, unacquainted with the actual scientific work of men like Harvey, Kepler, Galileo, Descartes, and Newton, have been misled in this respect by Bacon, and by some polemic passages in the more popular works of Galileo and Descartes. But Bacon was certainly not himself a scientist and the great scientific achievements of Galileo and Descartes were thoroughly mathematical and rationalistic.

A just historical perspective of the relation between rationalism and naturalism—at least before the nineteenth century—views them as necessary allies in the war of emancipation from what many like Goethe regarded as the essence of mediaevalism, namely, the view that nature is sin and intellect the devil. The appeal to reason was a favourite weapon against superstitions and needlessly cruel restraints on natural life. The great enemy of rationalism was, therefore, not empiricism, but some form of non-rational authoritarianism, generally supernatural. Thus in religion, rationalism in the form of deism was opposed to special revelation, to miracles, and to special interventions of the deity in behalf of privileged people. In politics and economics, rationalism challenged the established privileges and monopolies to justify themselves at the bar of common human interests. So in science rationalism opposed the traditionally authoritative view of the world, as well as popular credulity in the strange, the marvellous, and the magical. Thus it was the intimate union of rationalism and naturalism which, through men like Kepler, Galileo, Campanella, Spinoza, Grotius, Leibniz, Newton, Beccaria, and the Encyclopaedists, liberated science from the tutelage of theology, overthrew the Inquisition, prepared the way for a notable humanization of our treatment of the criminal, the sick, and the insane, and liberalized civil and international law. The framers of the American Declaration of Independence and the French Declaration of the Rights of Man appealed so confidently to reason that their age may be referred to as the Age of Reason. It is in the subsequent eclipse of the prestige of reason, following the French Revolution and the excesses of the romantic *Naturphilosophie*, that the opposition between empiricism and rationalism came to be prominently emphasized.

The decline of the avowed faith in reason is thus one of the central facts of recent intellectual history. To attribute it to the reaction due to fright at the violence of the French Revolution and to the stirring up of the national passions by the ensuing wars, is but part of the truth. More is needed to explain the persistent effects of that reaction, and the relative weakness of the liberal rationalistic forces, which, under the aegis of natural science, seemed about to prevail in the middle of the nineteenth century. One who owes no allegiance to the democratic dogma may well associate the decline of rationalism with the decline of aristocracy, or to put it more paradoxically: reason lost ground because of the spread of literacy. For, by removing the political and economic restraints which had kept the multitude from the realm of education, the Industrial Revolution and its political consequences let loose a horde of barbarians for the invasion of the fields of intellectual culture that had hitherto been restricted to those elaborately trained or specially gifted. The spread of literacy, without the prolonged discipline on which aristocracies must depend for the maintenance of their powers and privileges, diluted the intellectual life and brought about a flabby popularity not conducive to rigorous reason. Few can talk sensibly and accurately when they raise their voices high to address very large audiences; and newspapers which seek to attract vast multitudes of readers must adopt an admittedly low intellectual level. These and other courtiers of King Demos dare not remind him that despite modern "outlines," "stories," and other pretended short-cuts to omniscience, there is still no easy royal road to real knowledge. Why, indeed, bother about any road at all when the ability of every citizen to settle all sorts of issues (provided he votes with the prevailing plurality) is sanctified by the motto, *vox populi vox dei?*

These reflections may seem offensive, and they have obvious limitations. They certainly do not take account of certain great gains, even in the intellectual realm, brought about by the democratic movement, e.g. increased respect for hitherto undignified facts. Yet they certainly contain a measure of truth. While some rationality is latent in all conscious life, the clear perception and grasp of its principles is involved in great difficulty so that at all times only a few reach the heights. In any case it is unquestionably true that, previous to the nineteenth century, education, according to the Greek model of rational self-control, was the privilege of the few. Even Rousseau thought of education in terms of one who could have a private tutor accompany him all the time. When schools have to teach large classes, and students must be promoted and graduated en masse in order to make room for newcomers, insistence on the attainment of intellectual standards becomes in fact increasingly difficult. The spread of popular education, the bringing of the matter of education down to a level where every one can reach it, has certainly not been directed to emphasize the prolonged discipline which is necessary for the proper exercise of reason. Moreover, increase of knowledge is not necessarily the same as increase of rationality. Indeed the decline of respect for reason may also in part be due to the unprecedented growth of factual knowledge brought about by the amazing discoveries of new peoples, new historic epochs, and new fields of nature—all of which discoveries have been facilitated by improved means of travel, communication, etc. The first result of all this progress has been to produce an impression of bewildering factual diversity and to weaken the sense of rationality or logical connection of nature as a whole. Scepticism and philomathia thus replace philosophia.

Let us, however, pass from speculations as to historical causes and consider the force of the arguments actually advanced against the indispensable rôle of rational thought in the attainment of truth. These may be grouped under four main heads: (1) the psychological, (2) the historical, (3) the empirical, and (4) the kinetic.

7

§ I. THE ARGUMENT OF PSYCHOLOGISM

THE first and the most widespread of these arguments is the psychological one. The most popular version of it may be stated as follows: The old rationalism absurdly overemphasized the power of conscious reasons as motives. For instance, it thought of law and religion as the conscious invention of some legislator or priest. The sentiment that

> Courts for cowards were erected,
> Churches built to please the priests

may not have been typically respectable, but all classes, Tories as well as revolutionists, appealed to the terms of the social contract by which human society was supposed to have been instituted. Not only the state and the family, but even language, was regarded as instituted by conscious convention.[2]

Against this, the romantic movement since Schlegel, Schelling, and Savigny has insisted that human institutions are matters of growth rather than creation, and that the great achievements of life are the result of unconscious spirit rather than conscious deliberation. Even Hegel, despite or because of his extravagant panlogism, has so emphasized the immanent necessary evolution (or dialectic) at the basis of human history, politics, religion, art, and philosophy as to leave nothing to conscious human effort.

It cannot well be denied that the roots of our conscious being are in a dark soil where the light of deliberate reflection does not directly penetrate. But this does not deny the reality or diminish the unique worth of the light. Furthermore, this argument certainly does not justify the romantic moral that we should trust all our "unconscious" promptings. For the latter are often conflicting and sometimes self-destructive, and none of the specifically human values which we call civilization are independent of long and painful conscious effort. Plants that have their

---

[2] See Hobbes, *Opera*, Vol. I, p. 14, Vol. II, p. 90; *Leviathan*, Pt. I, Ch. 4; Condillac, *Sur l'origine des connaissances humaines* (1746), Vol. II, § 1; Rousseau, *Sur l'inégalité*, *Œuvres* (1790), Vol. VII, pp. 80 ff.; Monboddo, *Of the Origin and Progress of Language* (2nd ed., 1774), especially p. viii and Bk. II, Chs. 6 and 7. Both Rousseau and Lord Monboddo are aware of difficulties, yet still speak of language as an invention, just as Democritus, Epicurus, and Lucretius did.

roots in the dark soil depend no less on the rain and the sunshine from above. The romantic effort to enthrone Dionysius and the chthonic deities has always proved worse than sterile when it has meant the banishing of the Olympic gods of the air and the light. Inspiration or Dionysiac frenzy is barren or destructive except when it submits to rational labour. For not all who rave are divinely inspired.

There is doubtless a certain superficial truth in the anti-rational psychology which maintains that reason is cold and lifeless and an irksome restraint on the creative imagination. But the oft-quoted lines,

> Grau . . . ist alle Theorie,
> Und grün des Lebens goldner Baum,

express not the judgment of Goethe, but a dictum of Mephistopheles, whose motive is revealed in a preceding passage:

> Verachte nur Vernunft und Wissenschaft,
> Des Menschen allerhöchste Kraft, . . .
> So hab' ich dich schon unbedingt . . .

We need not minimize the creative imagination as an inspiration or gift of the gods, to appreciate the importance of well-organized rational routine as the necessary mode whereby our inspiration can find effective expression.

In any case we should not reject the practical claims of rational motives without noting that the worship of reason was a potent influence in minimizing or abolishing age-long abuses like slavery, serfdom, and persecution for witchcraft and heresy. There are not wanting signs that a continued eclipse of the old-fashioned rationalism may bring these abuses back in full force.

There is, of course, no necessary connection between the scientific study of psychology and the vogue of irrationalism. To the extent that it is itself scientific, psychology naturally pursues the rational methods common to all true sciences. Indeed, the modern study of the human mind was largely moulded by the individualistic rationalism typified by Descartes, Hobbes, Leibniz, Wolff, and even, in a measure, Locke, whose French followers were certainly not anti-rationalistic. British associationist psychology, to be sure, always remained extremely nominalistic, that is to say, it so reduced all mental life to separate atomic states as to leave no room for any rational connection. But in its concrete applications, such as in its account of economic and political motives, it long retained the naïve rationalism of the eighteenth century. The great impetus to anti-rationalism in psychology seems primarily due to the great romantic writers, from Rousseau and Novalis to Stendhal and Nietzsche, and to philosophers of the school of Schelling, Schleiermacher, and Schopenhauer. These men, with their fresh and daring vision into the intricacies of the human heart, liberated us from the old over-simple conception of human nature as a logically utilitarian machine. No historian seems to me to have as yet done full justice to the magnitude of the contributions to individual and social psychology due to the romantic school. At the same time it is also true that speculative introspective psychology has largely served as a haven of refuge to those who find the harder facts and the closer and more accurate reasoning of modern physical science uncongenial. Thus it comes to pass that a large part of the idealistic philosophy of the nineteenth century is at bottom hostile to strictly rational procedure, so far as the latter insists on laborious methods to check or prevent the facile confusion between the fanciful world of our heart's desire and the more sober world of actual existence. Hard distinctions and necessary rational order are certainly more easily ignored in speculative psychology than in physics.

Nor is this subjective wilfulness confined to idealism. Positivism, reacting violently against the romantic "rational" or soul psychology, and following the directly opposite route, arrives at strikingly similar results. Thus positivism begins with rejecting the fantastic claims for the creative powers of the mind, and naturally falls into the uncritical worship of "sensations" as the deliverances of the real or external world. This, together with the habit of introspection that stops the processes of thought to see what images exist in the mind, leaves no clue to the rational connectedness of the objective world which science so laboriously seeks. Positivism, therefore, lands us at the conclusion that all logical laws or rational connections of nature are mere fictions or creations of the mind. Thus both romantic idealists and positivists banish rigorous reason as a true integral part of the natural world.

This anti-rationalism common to positivism and to certain forms of idealism shows itself in the effort to deny the significance of logic by reducing it to psychology. It is, of course, perfectly obvious that no branch of philosophy can hope to attain substantial truth if it ignores or goes counter to the teachings of psychology. But it ought also to be equally obvious that psychology is only one among a number of sciences, that it must take the results of logic and mathematics, physics and physiology for granted, and that the attempt to make psychology identical with the whole field of science or philosophy can lead only to confusion. This is certainly the case when in the interests either of positivism or of psychologic idealism we deny the distinction between logic and psychology. Mr. Schiller has written a whole volume to catalogue what he regards as the absurdities of the traditional logic. To me the greatest absurdity of all is the fundamental premise which Mr. Schiller shares with thinkers as diverse as Mill and (at times) Bradley, viz. that logic should be a description of the way we actually think. Granted that all thinking goes on in individual minds, it does not follow that a psychologic description of reasoning as a mental event can determine whether the resulting conclusion is true. All sorts of variations in imagery or motive may take place in the minds of those who come to the same conclusion as to the multiplication table, but these considerations are irrelevant to the truth or validity of what is asserted. You may, of course, define logic as the psychology of thought and deny the very existence of fallacious thinking by refusing to call it thought; and you may also define functional psychology to include both a description of what goes on in the mind and an evaluation of the truth or the correctness of arguments. But the distinction between the descriptive and the normative points of view is not thereby avoided, though it may be confused by a violent use of terms. A psychologic description of what goes on in my mind as I deal with an ethical or practical problem will not determine the correctness of the solution arrived at; and psychology can no more include the whole of logic and ethics than it can the whole of technology.

§ II. THE ARGUMENT OF HISTORICISM

SIMILAR to the attempt to dispense with reason by reducing everything to the facts of psychology is the attempt to dispense with reason by reducing everything to the facts of history.

The historical studies of the nineteenth century, as is well known,
were largely stimulated by the romantic and nationalistic reaction
against the cosmopolitanism of the eighteenth century. Hence history
became in large measure a basis for attacks on rationalism. Thus no
argument is more familiar than that the rationalists of the eighteenth
century had a deficient sense of history. They certainly tried to derive
the detailed nature of human institutions, such as government or
religion, from reasons which seemed plausible a priori, but ignored
actual history and the perverse complexity of the facts. We cannot
refute this argument by pointing to the great historical work produced
by such typical representatives of eighteenth-century rationalism as
Voltaire and Gibbon. There really was, in the heyday of the Enlightenment, a great deal of effort to paint the history of mankind on the
basis of a priori considerations. It is no denial of this to show that the
naturalism of the nineteenth century has produced a great deal more
a priori history under the guise of Spencerian evolution, with its sweeping dogma that all peoples and institutions must pass through the same
stages of development from the simple to the complex. But it is instructive in passing to call attention to the latter belief as an illustration of crypto-rationalism, i.e. whenever reason is ostentatiously banished through the front door, it is unavowedly or secretly admitted
through the back door to perform its necessary functions.

But though in the nineteenth as well as in the eighteenth century
thought could not free itself from rationalism and suffered equally from
an undue desire to have the world very simple, there are important
differences between the conscious pursuit of reason and unavowed or
crypto-rationalism. Conscious rationalism starts from the present and is
abstract and universalistic. But crypto-rationalistic historicism prefers to
start from the exotic and always emphasizes difference and particularism.
This can be seen by comparing Gibbon's *Decline and Fall of the Roman
Empire* with any typical nineteenth-century history. The figures in

Gibbon, subjects or rulers, are all men and women, enlightened or corrupt, but all capable of stepping out into our own scene and playing their petty or grandiose rôles over again. The sense of differences, such as those between the Romans of the Augustan and of the Byzantine periods, between Lombards and Saracens, is hardly felt. But in rebelling against this conception which makes history somewhat monotonous even in the hands of a master like Gibbon, popular nineteenth-century historicism has undoubtedly exaggerated human differences to the extent of losing sight of our common fundamental humanity. As a result, we have all sorts of wild inhuman interpretations of the motives of ancient and primitive peoples. Even with regard to modern European peoples the mythology of racial souls has obscured the fact that the origin and development of differences in culture can be rationally studied only on the postulate of a common human nature. The practical result of particularism is that while the old rationalistic historians were devoted to the virtue of toleration and the arduous pursuit of enlightenment, the historicism of the nineteenth century was frequently the handmaiden of nationalistic and sectarian claims.

The fact that literary historians are generally more interested in the concrete picture of the events they portray, while scientific physicists are generally more interested in the laws which physical phenomena illustrate, has given rise in recent times to the view that history is nearer to reality, which is always individual, and that rational or scientific physics is a more or less useful fiction. Critical reflection, however, shows that despite differences of subject-matter, the same type of reason underlies scientific history whether human or natural. Hence any successful attack on the truth-value of reason in physics would be fatal to the claims of history.

That all existing beings, animate or inanimate, are individuals no rationalist needs to deny. But that is in no way inconsistent with the fact pointed out by Aristotle that there is no knowledge or significant assertion with reference to any individual, except in terms of universals or abstractions. I may in a dumb way point to a single object or I may grasp it with my hand, but I cannot mean anything and I cannot even say "this" about it without using an abstract term applicable to other individuals. If I use a proper name, it has meaning only by convention which ultimately involves abstract terms. You may pretend to despise language as much as you please, but if you talk or think at all you cannot consistently deny all significance to that which is expressed in language. The proposition that all reality is individual must not be confused with the proposition that the individual apart from all its abstract determination is the only reality. If what is abstract is unreal, then the detached and characterless individual is the worst of all abstractions, and the most unreal! Moreover, what distinguishes human life and history from merely physical motions is precisely the greater wealth of meaning or significant relations which human objects have in an intelligent view of them. Monuments, coins, and other "remains" are historical only as they point beyond their own present existence. Black marks on paper may mean the right of millions to live or the obligation to give up all their possessions.

So long as this is the case, no historical description or explanation can possibly dispense with abstractions. Consider the description of any historical event whatsoever, e.g. the life or death of Caesar. Can we say that Caesar was rich, profligate, brave, ambitious, or what not, without using abstractions or aspects of life capable of indefinite repetitions? To be sure, the individual man Caesar will never occur again, but all that we know about him depends upon the assumption of certain laws. In this respect human history does not differ from natural history. The geologist as historian of the individual or unique earth of ours is concerned with describing events (like the fusion of certain rocks) which need never occur again any more than some of the revolutions of human history. Indeed, every physicist in the laboratory as well as the engineer engaged in testing a bridge or particular engine is engaged with an individual object. But just as natural history depends upon general physical science, so does human history depend upon and presuppose general knowledge of the laws of human conduct. Take any attempt at historical description or explanation (explanation is but a developed description). We say Caesar was killed because he was ambitious or too generous to his former enemies, and Brutus conspired because he was patriotic or because he was a greedy usurer who could not resist the opportunity to gather the revenues of provinces, etc. Do not these and all other historic explanations assume certain laws or uniformities in human conduct? To be sure, in human history these laws are assumed tacitly, while in natural history they are explicit; and this is true partly because the laws of physics, dealing with phenomena capable of indefinite repetition, are simpler and more readily verified. But when we see a critical historian engaged in determining whether an alleged fact did or did not take place, his weighing of evidence does not differ in method from that employed in natural science.

Indeed, the facts of history as past cannot be directly observed. They are inferred from monuments, documents, or other present objects which, for certain reasons, we judge to be remains of the past. But from present facts we can conclude as to past facts only if we assume certain universal laws connecting them. The fact that the historian is not always aware of his assumptions or that some things are so constantly taken for granted that there seems no point in mentioning them does not deny their logical necessity. Many of these laws are physical, e.g. that coins or monuments cannot grow naturally but must be made by certain physical processes involving the use of certain materials and tools. Other assumed laws deal with social-psychological elements such as the credibility of witnesses, e.g. that men unconsciously and sometimes deliberately exaggerate their own achievements or those of their country or party and disparage those of their enemies.

Thus the establishment of facts in history, as in courts of law, depends upon both popular assumptions and expert scientific knowledge. We should certainly be critical to the claims that certain eclipses (on which ancient chronology is so often based) can take place only in certain intervals, or to the assertion of certain laws as to the physiology of plants and animals (on the basis of which we conclude that certain regions were depopulated or that certain migrations were botanically necessitated). But if so, how much more critical must we be to "facts" of history that are based on assumed laws of sociology and psychology which have never even been definitely formulated!

The contention of historicism generalized appears as the third argument against rationalism.

§ III. THE ARGUMENT OF EMPIRICISM

THE growth of natural science, with its extensive use of mechanical instruments for observation and experiment, has brought about an impression that science distrusts reason and relies rather on observation and experiment. In proof of the latter it is frequently asserted that the Greeks failed to lay the permanent foundations of science precisely because they were too much addicted to reasoning and did not observe with sufficient care and fulness. Though this assertion has often been repeated by scientists and historians who ought to know better it has no claim to truth. The Greeks were very keen observers precisely because they reasoned clearly and boldly; and it was thus that they did lay the foundations of modern science. The contrary impression that men like Democritus indulged in happy but unverified guesses as to the physical world is not well founded. Democritus was not only a great mathematician, but a list of his works, e.g. "On the Cough which Follows Illness," "On Agriculture," "On Geography," shows that he was certainly not an idle speculator. The Greeks could not have created the science of astronomy and introduced the heliocentric hypothesis if they had not combined very careful observations with rigorous mathematics. It required both to discover the size of the earth's diameter and the precession of the equinoxes, with no better instruments to aid them than the astrolabe. Only a marvellous power of fine observation could, without any clinical thermometer, discover the variations of the temperature curve in various fevers (see the Hippocratic writings). Despite the fashion to abuse Aristotle, it certainly required both active thought and fine observation to discover parthenogenesis and other biologic facts recorded in the writings of Aristotle and of his immediate pupils. The union of clear ideas with accurate observation may also be well illustrated by Xenophanes' noting the impress of fishes in the rocks and boldly concluding that these rocks must at one time have been at the bottom of the sea. If despite that, Greek science did not prosper more, other causes than devotion to reason were responsible.

It is important to press this point because not only among the Greeks but at all times is rigorous reasoning, the essence of mathematics, the necessary condition of accurate and fruitful observation and experiment.

If you had watched the most famous and epoch-making experiments of modern times, e.g. Hertz's on electric waves, or Michelson's on the velocity of light, you might have seen all sorts of apparatus but you could not possibly have observed what these men observed unless you had gone through all the reasoning which these men had gone through before setting up their apparatus. The same is true in principle of almost every observation made today under laboratory conditions. Our fundamental units, light waves, electric current and resistance, rates of metabolism, etc., are never visible except to the eye illumined by all sorts of ideas and rigorous deductions from them. Accidental discoveries of which popular histories of science make mention never happen except to those who have previously devoted a great deal of thought to the matter. Observation unillumined by theoretic reason is sterile. Indeed, without a well-reasoned anticipation or hypothesis of what we expect to find there is no definite object to look for, and no test as to what is relevant to our search. Wisdom does not come to those who gape at nature with an empty head. Fruitful observation depends not as Bacon thought upon the absence of bias or anticipatory ideas, but rather on a logical multiplication of them so that having many possibilities in mind we are better prepared to direct our attention to what others have never thought of as within the field of possibility.

If we thus realize how necessary reason is for the undertaking of scientific observations and experiments, we need not stop to note how necessary it is for the proper interpretation of the results of such experiment and observation. Indeed only theorists utterly ignorant of modern scientific procedure and laboratory practice can ignore the fact that actual observations are often rejected as too improbable and almost always "corrected," i.e. replaced by results that are dictated by the theory of the probable error of observation.[3]

---

[3] In recent years there has been an impressive increase of interest in the history of science, both in its totality as a human institution and in its special fields; and students of philosophy have much to learn from the works of Duhem, Merz, Lasswitz, Radl, L. Thorndike, Singer, Sudhoff, and Neubauer. Nevertheless we can well say that philosophically the history of science is the least developed branch of history. The bulk of it is still in the annalistic or memoir form. "In the year 1895 Roentgen discovered X-Rays." Such work is indispensable, but hardly constitutes significant history unless the historian has some coherent ideas as to what facts are important for his purpose.

When our historians of science do reflect on these ideas they are for the most part

§ IV. THE ARGUMENT OF KINETICISM

AS MATHEMATICS and physics were the first sciences to be developed systematically, rationalism has naturally drawn on them for illustrations of rational truth. The great *éclat*, however, of the doctrine of universal evolution which, as a matter of historic fact, originated more in the romantic philosophy of Schelling than in the biologic observations of Darwin, brought it to the foreground as the popular example of scientific truth. Moreover, the progress of the kinetic theory of matter, the gradual elimination of the inert atom, and the consequent abandonment of Maxwell's view that each individual molecule has remained unaltered since the day of creation, have fortified the impression of universal change. Molecules and atoms, like the hills and the forms of species, have lost their traditional eternity. The sight of so much change where formerly we saw only constancy has produced the dizzy romantic generalization that only change is real and that nature contains no constant elements. Despite the widespread character of this assertion, we need not hesitate to characterize it as a snap judgment resting on no proof of logic or fact. Indeed, how could this universal judgment itself ever be proved by changing empirical facts? Change is doubtless a universal aspect of nature, but so is its correlative constancy. Surface is undoubtedly a universal trait of all physical objects; but any one who argued from this universality to the denial of correlative depth or volume would not be more foolish than one who argues from the universality of change to the denial of constancy. We might as well argue from the universality of death to the absence of life. The truth is simply that there is no change except in reference to something constant. An object, for instance, can be said to move only in so far as it remains the same object and changes its distance from a definite point as measured

dominated by the Voltairean version of the philosophy of the Enlightenment, according to which mankind dwelt in utter scientific darkness, ruled by superstition, theology, and metaphysics, until Bacon, Newton, and Locke came and taught men to observe, calculate, and measure. This view has gained popularity because those who now cultivate a science like physics have not as a rule the historic and philologic training, if they have the interest, to read the older scientists who wrote in a different language and used terms no longer in use. But in recent years, thanks to the labour of men like Duhem and others in ancient, mediaeval, and early Renaissance science, we can see more continuity in the history of science and are in a better position to reflect on its relation to the history of philosophy and of civilization in general.

by some definite unit. Abolish the fixed starting point or goal, the definite direction and the constancy which we call the identity of the object, and nothing is left of the fact of motion. Professor Montague has shown this by asking us to imagine what would become of a race if not only the runners, but the judges, the goal, and the whole track itself, were moving without fixed direction.

Logically, the fact of change or motion is nothing more than the correlation between different moments of time (as determined by some clock) and the different spatial positions of an object. To a creature living entirely in a single moment there is no motion at all. It is only to a being to whom different moments of time are formally present together that motion has a meaning. There is, therefore, nothing paradoxical in saying that the meaning of any motion does not itself move, but is rather a timeless fact or phase of nature. The difficulty in grasping and isolating the timeless or significant aspect of nature is from one point of view simply the difficulty of leaving the ordinary practical interests in the personal possession or manipulation of things, and rising to a reflective insight as to what the world contains. But it may remove needless obstacles to the understanding of this if we insist that the principal difficulty of grasping the nature of universals is the tendency to confuse thoughts with images, and thus reify all objects of discourse. This is shown by the naïve query, "Where are these universals?" (which assumes, of course, that universals are particular things in space). But all truths, even of events in a given place and time (e.g. Caesar's crossing the Rubicon in 49 B.C.), are as truths eternal and independent of spatial location. To argue, as James [4] and others have, that the constant rules of logic cannot be true of a world in flux is a confusion as gross as to argue that motion cannot have a constant velocity or a fixed direction, or that one standing still cannot catch a flying ball. Indeed, are not flux, change, and motion themselves concepts? The Bergson-James argument that all concepts give us only fixed cross-sections of the stream of reality rests at bottom on a confusion between concepts and images—a confusion rendered canonical by the old doctrine of concept as generalized pictures or images. If, however, we examine the actual concepts of science, we find that it is precisely by indicating relations or essential transformations between terms that

[4] *Pluralistic Universe*, pp. 252-253.

they make possible synoptic views of comprehensive changes and enable us to grasp the meaning that change and life have for us. There is no conflict between scientific biology and rationalism.

In the sense in which romanticism denotes distrust of reason, all the foregoing arguments are romantic. But we may regard the last as the essentially romantic one because it makes explicit the romantic impatience with the fixed, clear, and orderly routine of reason, as a restraint on the wish to regard the world as entirely responsive to our fancy and heart's desire. Romanticism, however, is not merely negative. It involves an abounding faith in some inner, creative, and unlimited source of illumination or revelation superior to ordinary reason. Before examining this, however, we may make a few comments on the popular sceptical effort to undermine reason in order to make room for the certainties of faith.

§ V. CAN FAITH BE BASED ON SCEPTICISM AS TO REASON?

IF LOVE is blind, if, where our emotional interests are involved, clear thinking is most difficult, it follows that in matters in which we care most, our thoughts must be full of obstinate confusion and prejudice. The inconsistency of men's professions, as well as the folly of their conduct, bears this out. Historically speaking, few human societies have been interested in matters of belief, and few men care much about intellectual consistency for its own sake. Most men have little more inclination to examine the foundations of their beliefs than trees have to bring their roots to the sunlight.

When, however, we meet those whose conduct and professed beliefs are markedly different from ours, our own accustomed standards are challenged, and we become concerned. In our irritated reaction we may turn away in loathing, or try to exterminate the strange abomination. But when a régime of outer toleration of differences is established, men must intellectually adjust themselves in some way. One easy way is to hold on to your own traditional practice and to minimize the importance of these differences. It then becomes "bad form" to argue matters of religion. In the fashionable theory all good men are religious. Let each worship according to the religion of his father, or as his conscience dictates. But the exercise of unqualified freedom of conscience would bring the practitioner into conflict with the established social order. Another adjustment to the diversity of faiths is to despise contrary beliefs as not held by the proper people. Certain views are ruled out of discussion because they are held only by infidels, bolsheviks, stuck-in-the-mud conservatives, reactionaries, etc.

But in a heterogeneous society men cannot altogether escape the challenge to give reasons for the inherited faith within them. Out of the difficulties of meeting this challenge arises the doctrine of faith and of the right to believe what we wish to believe. Now a resolution not to reason is as insusceptible to logical refutation as is a resolution not to sing in public. To realize the limits of our inclinations and aptitudes in these respects is, indeed, admirable. But men who will not give the reason for their faith are not always modest. They try to save their own prestige by condemning the whole enterprise of reason, and this they do by professing scepticism as to the value of reason. But in a world of conflicting faith scepticism lends no permanent support to any creed against its destructive rivals, and a faith that becomes aware of its impotence is on the decline. Hence the poet who sings that "by faith and faith alone" we grasp the truth, finally puts it, "I will dream my dream and hold it true." [5]

This right to believe has been defended by William James on the ground that since we must take risks in life, we have a right to believe those things that we believe may help us to attain our vital interests. Assuredly, every one does risk his life on his fundamental beliefs. But this does not prove them true, or even dispose of the doubts raised by those who actually challenge the beliefs. We cannot by the will to believe in a personal God make Him come into existence. We cannot by believing it even add a cubit unto our own stature. Reason in the form of logical science is an effort to determine the weight of evidence. To tip the scales by the will to believe is childish foolishness, since the real weight of things is not thereby changed.

In defence of the view that by the exercise of the will to believe we may influence the facts believed it has been urged that pessimism may shorten our life, and optimism may lengthen it. I can see no evidence for this claim, and it is inherently doubtful whether a resolution to see

[5] If I know it is a willed dream, how can I believe it to be true?

the sunny side of life is more effective than other resolutions. Life in a fool's paradise does not last long. In the end the will to believe is as inimical to a healthy, aggressive faith which seeks to convince our fellows, as it is corrupting to our own intellectual integrity. The force of this is seen in the fact that the Catholic Church, whose worldly wisdom few question, has persistently condemned Fideism.

Chapter Two

THE RIVALS AND SUBSTITUTES FOR REASON [1]

Our rapid survey of the usual arguments against rationalism has shown not only their limitations, but also something of the partly reasonable motives which have led to the revolt against the older rationalism. We may make this clearer by examining more closely the claims of the various rivals of reason.

§ I. AUTHORITY

The prestige of authority, it is generally recognized, rests most firmly in custom. By increasing the means of travel and communication, and thus making it possible for us to visualize ways other than those under which we have grown up, the Industrial Revolution has been one of the most potent forces in undermining the prestige of the customary. By showing the people of Europe that they could take things into their own hands and change the traditional form of government, laws, and even the system of weights and measures, the French Revolution made the path of the questioning and revolutionary spirit more easy. But it also frightened those impressed with the fact that the basis of civilized society rests on habitual obedience and deference to organized authority. From the latter point of view, modern history is a fall from a social order in which every one knows his place and its duties to a bewildering chaos of conflicting claims without any authoritative guidance. The claims of authority have, therefore, been usually pressed or repelled with more poignancy than philosophic detachment.

In discussing the principle of authority we should distinguish between the necessities of conduct and those of purely theoretic decisions. This distinction is frowned upon in an age which so glorifies practical conduct

[1] The first two sections of this chapter were printed in the *Journal of Philosophy*, Vol. XXII (1925), pp. 141, 180.

as to regard purely theoretic contemplation either as impossible or as a sinful waste of human energy. Nevertheless, the distinction is quite clear and important. In matters of conduct, we are frequently compelled to decide at once between exclusive alternatives. We must, for instance, either get married or not, go to church or stay out, accept a given position or else refuse it. In theoretic issues, however, we may avoid either alternative by suspending judgment, e.g. when we realize the inadequacy of our information or evidence. It is true, of course, that most people, after having made a decision, do not like to entertain any doubts as to the adequacy of its theoretic justification. But by no canon of intellectual integrity can dislike of doubt constitute a proof of the truths assumed.

In practice, then, it is often much more important to come to a decision one way or another than to wait for adequate reasons on which to base a right decision. Often, indeed, such waiting is a sheer impossibility. Hence mankind frequently finds it necessary to settle doubts by means that have nothing to do with reason. Among such means are the throwing of a coin or of dice (of which the Urim and Thumim may have been an example), the flight of birds, the character of the entrails in the sacrifice, or the ravings of the smoke-intoxicated priestess of Dodona. Note that not only minor questions, but important ones like war, have been decided that way. The desire to find justifying reasons for adhering to decisions once made may promote the belief that these non-rational ways of terminating issues are controlled by supernatural powers on whom it is safer to rely. But in most cases, the given practice is much older than the various explanations offered for it, and its function in eliminating doubt and bringing about decision is undoubtedly the primary fact. Modern anthropology is making us realize the superficial character of the old rationalism which regarded the claims of all magicians or prophets to supernatural power or inspiration as premeditated fraud. In primitive communities magical power and the sovereignty which goes with it, are often literally thrust on certain individuals who generally share the prevalent ideas and illusions. We thrust sovereignty on others because most of us are unhappy under the great burden of having to make decisions. We can see this in our own day in the way in which the sovereignty of final or authoritative decision has been imposed in many fields upon our newspapers. Worried by doubts as to the correct dress for her boy of ten at an afternoon party, the anxious mother writes to the newspaper. Wishing to be certain as to what is the proper judgment to be passed on a play or concert, we hasten to consult the next morning's newspaper. The latter thus becomes an authority on dress, on the pronunciation and use of words, on the proper conduct for young women engaged or in love, etc.

Political authority or sovereignty has its basis in just this need to have practical controversies settled. When we are parties to a suit, we are anxious that the issue be settled justly, i.e. in our favour. But there is a general interest on the part of all members of the community in having controversies settled one way or another. Otherwise we fall into a state of perpetual war or anarchy. So important is this, especially to people depending on routine work like agriculture or industry, that mankind has borne the most outrageous tyranny on the part of semi-insane despots, rather than by revolt break the habit of obedience and face the dangers of anarchy before the rebel leader effectively asserts his own tyranny. It was not an Oriental, but the most influential of Occidental philosophers, Aristotle, who argued that even an admittedly bad law ought not to be replaced by a reasonably better one, because changing the law diminishes a prestige which is most effectively based on habitual obedience.

English and American communities are apt to flatter themselves on having eliminated tyranny or despotism by the rational devices of parliamentary or constitutional government. The men of the eighteenth century believed it possible to have a government by laws and not by men. (Some lawyers believe this today, perhaps because they do most of the governing.) It is, doubtless, possible by political devices to minimize certain of the grosser forms of tyranny; and a certain amount of discussion of a more or less rational character may profitably be introduced into the shaping of our laws. But so long as men fall short of perfect knowledge and good will, they will have to obey laws which they find oppressive and unjust—laws made, administered, and interpreted, not by an abstract reason in heaven, nor by a mythical will of all the people, but by ordinary human beings with all their human limitations upon them. If every individual refused to obey any law that seemed to him immoral, the advantages of a state of law over anarchy would be lost. This is not to deny that tyranny may go to such excesses as to make the temporary anarchy of revolution preferable. But in the ordinary course of human affairs such occasions must be regarded as exceptional or relatively infrequent. Actually, therefore, though some forms of government may in the long run work more reasonably or more agreeably to the will of its citizens, the principle of authority means that the good citizen will submit to what is in fact the arbitrary will or unwise opinion of some boss, legislator, administrator, or judge. Lawyers and sentimentalists may try to hide this unpleasant fact by such fictions as "the law is nothing but reason, the will of the people, etc." But wilfully to confuse such fictions with the actual facts is to corrupt reason at its source.

The practical necessity for authority does not mean that only an absolute monarchy or an hereditary nobility can guarantee a régime of law and order. Experience has amply shown that a titular absolute king may in fact be helpless in the hands of irresponsible courtiers, and that hereditary nobilities may be unruly as well as selfish—just as democracies may in fact be ruled by natural leaders or well-organized cliques. In practice authoritarians are people who are so afraid of the perils of change that they blind themselves to the absurdities and iniquities of the established order, while reformers and revolutionists are so impressed with the existing evils that they give little heed to the even greater evils which their proposals may generate. The true rationality or wisdom of any course of conduct obviously depends upon a true estimate of all its consequences, and such estimate is avoided both by those who will not hear of any change and by those who think that *any* change is necessarily good (because they identify change with life).

In thus recognizing the unavoidable character of authority in communal life, must our reason also abdicate and declare that whatever we must submit to is also right? That is exactly the position of those who, like Mr. Balfour, argue that since our individual reason is highly fallible, the need of order and morality demands the submission of reason itself to authority—defined as a "group of non-rational causes, moral, social, and educational—which produces its results by psychic processes other than reasoning." [2] Authorities, however, differ and in the end they cannot support themselves without reason.

In the history of the reaction against the rationalism of the Enlighten-

[2] Balfour, *Foundations of Belief*, p. 219.

ment, we find three main sources of authority to which individual reason is asked to submit. These are: (a) the church, (b) tradition, and (c) the opinions of our superiors or "betters."

(A) THE CHURCH

It would obviously take us far afield to examine all the arguments of those who have urged that our fallible individual reason must submit to the infallible authority of the church. But there is a serious difficulty common to all of them, to wit, the great multiplicity of churches, each claiming to be the one instituted by divine authority for the whole of mankind. In ancient days it was possible for men like Dante to view the Roman Catholic Church as the church of all mankind, and to regard all those outside of it as misled by "schismatics" like Mohammed. When, however, Christianity is the religion of a minor part of the human race, divided into so many sects and shades of opinion that it is difficult to say what it is that is common to all of them, and when, moreover, the authority of all supernaturalism is challenged by an increasing number of educated people, surely the old argument from catholicity, that the teachings of the church have been recognized always and everywhere, has lost its force. De Maistre, the most clear-headed of modern apologists, argues that the Pope must be infallible because practical affairs demand some one supreme arbiter. But he dashes his head in vain against the existence of Protestant countries like England, Prussia, and the United States. Any attempt to prove that the claims of the Roman Church are superior to those, for instance, of the Anglican Church must involve reference to the facts of history. These are at best matters of probability; and it is hardly possible to support a claim to infallibility on the basis of historic probabilities.

There is a popular tendency nowadays for those who think militaristic imperialism to be the logical outcome of the Sermon on the Mount, to justify Christianity on the ground that it has produced the most powerful civilization. But apart from the question how far modern western civilization is due to Christianity, the historic fact quite clearly indicates that it is only in the last three centuries that some of the Christian nations have outstripped all the non-Christian ones in material power. In the course of the last thirteen centuries whole peoples in Asia, Europe, and

Africa have been converted from Christianity to Islam, while very few have followed the opposite path. This, of course, is not to the rationalist an argument for Mohammedanism. But it certainly disarms the old argument that the spread of Christianity is itself a miracle testifying to the truth of Christian teachings.

The old rationalistic idea of finding the core of truth in that which is common to all the different religions is now generally abandoned. It is inconsistent with the authoritarian claims of every church to be in exclusive possession of the supreme wisdom. But the view of Santayana, that the diverse conflicting religions differ only as do different languages, involves the even more thoroughgoing abandonment by every church of all claim to the possession of distinctive truth. Such thoroughgoing scepticism may fit in with the complacent orthodoxy and extreme worldliness of the Lord Chancellor who told a delegation of dissenters, "Get your damned church established and I will believe in it." It may even fit in with the popular prejudice that every one ought to adhere to the religion of his fathers. But in the end no church can hope to attract thinking people or keep them unless it makes an effort to substantiate some claims to truth in the court of reason.

(B) TRADITION

The second form of the appeal to authority against individual reason is the appeal to traditional belief. This may already be seen in Burke's *Reflections on the French Revolution*. In the political life of England and America, Burke has remained the patron saint of all those who would like to see people act on settled beliefs rather than waste time reasoning as to what *is* the right. What is the use of reasoning at all if it leads not to fixed conclusions? Despite the dubious character of his historical contentions and predictions, Burke's great appeal from the reason of individual philosophers to the cumulative wisdom of the ages has become one of the persistent notes of our intellectual life. An extreme and therefore instructive form of it is seen when pious American lawyers like Judge McClain argue against *any* constitutional change on the ground that the constitution embodies the cumulative wisdom of two thousand years of Anglo-Saxon experience—and who will dare to put his private individual reason against that?

It is, of course, easy enough to meet this with the reply that our ancestors were only human and hence subject to error; and that a good deal of what they have left us, e.g. in medicine, is cumulative foolishness. Moreover, the wisdom in our heritage is largely the result of the ideas of individuals who were innovators in their day, so that the elimination of individual reason would mean the death rather than the growth of vital tradition.

But though the extreme form of the argument for tradition can be shown to be untenable, its essence is not thereby eliminated. The essence of the argument for tradition and authority is the actual inability of any single individual thoroughly to apply the process of reasoning and verification to all the propositions that solicit his attention. To doubt *all* things in the Cartesian fashion until they can be demonstrated is impossible practically and theoretically. It is impossible practically, as Descartes himself admitted, because we cannot postpone the business of living until we have reasoned out everything; and it is impossible theoretically because there cannot be any significant doubt except on the basis of some knowledge. To doubt any proposition, to question whether it is true, involves not only a knowledge of its meaning, but also some knowledge of what conditions are necessary to remove our doubt. Actually all of us do and must begin with a body of traditional or generally accepted beliefs. For it does not and cannot occur to us to doubt any one proposition unless we see some conflict between it and some other of our accepted beliefs. But when such conflict is perceived within the body of tradition, the appeal to reason has already manifested itself. It may well be contended that many errors are eliminated by the attrition of time, so that any belief long held by a large group of people has a fair presumption in its favour. Unfortunately, however, errors also strike deep roots. Legends grow and abuses become so well established by tradition that it becomes hopeless to try to eradicate them. The history of long-persistent human error certainly looms large in any fair survey of our past. We may argue that what has stood the test of ages of experience cannot be altogether wrong. But it is also true that what has found favour with large multitudes, though sound in the main, can hardly contain a very high accuracy or discrimination between truth and error.

(c) EXPERT OPINION

Unless we are to fly in the face of all human experience we must admit that some people are, by aptitude, education, or experience, wiser or better informed than others.

It thus seems unpardonable stupidity to rely on our own frail reason, when we can avail ourselves of the judgment of those better qualified. Unfortunately, however, it is not a simple matter to find out who is actually best qualified to decide a given issue. Since the practical disappearance of the doctrine that kings, because of their divine appointment, can do no wrong and are always entitled to unquestioning obedience, there remains only one class of divinely appointed ex-officio superiors, viz. parents. That the conduct of children should conform to the wishes of their parents in matters affecting the life of the family is highly desirable and unavoidable. Parents not only have the power of enforcing obedience, but, all other things being equal, have more experience and therefore generally sounder judgment. Yet nothing but mischief results when respect for parents is conceived as incompatible with questioning the infallibility of their judgment. The necessity of practical obedience does not justify closing the minds of children to free intellectual inquiry when the latter is in the least possible. An emphatically anti-rational phase of parental authority appears when parents assume the right to dictate to colleges what religious, political, or economic doctrines are too dangerous to put before their sons and daughters supposed to be engaged in finding the truth about these subjects. Parents have no right to prevent children from learning more than they know themselves, or to shut the gates of reason.

The difficulty of finding who in any realm *are* our superiors is brought out most clearly by examining that Utopia of shallow "scientific" reformers, viz. government by experts. A priori, Plato's arguments for government by the competent (as against election of officers by lot or ballot) seem unanswerable. But ecclesiastical, as well as political, history shows that government by experts or bureaucracy, from China to Germany and the mediaeval church, is no more safe against error and abuse than any other human arrangement. Rigorous training and *esprit de corps* may prevent certain abuses. But they breed a narrow class-pride and a subordination of the general interest to the routine of administration, if not to the material interests of the governing group. Against Plato's arguments it is well to remember that the method of electing officials by lot worked so satisfactorily among the Greeks that they regarded it as essential to free democratic government and never gave it up except when compelled by external force such as that of the Macedonians. It was, indeed, great practical wisdom to adapt the duties of office to the competence of possible officials rather than plan for offices that require unattainable ideal governors. The old doctrine that though people cannot govern they can choose the proper governors, finds serious difficulty when we reflect how little opportunity there is to examine with real care the precise qualifications of the candidates or their actual achievements in office.

Nor is the choice of experts by their own associates free from the limitations of human ignorance. The old homely adage: "Get a reputation as an early riser and you can sleep all day," is true in law, medicine, and other professions. Few experts have extensive opportunity of checking up the work of every other expert; and a great deal of professional prestige is based on meretricious grounds. Reputations may be based on previous achievements which happen to have hit a shining mark by an unusually favourable turn of the wind. Even in science the best work is sometimes done by unknown young men who, by the time their work becomes known and appreciated, have passed the zenith of their natural abilities.

In passing judgment on the work of scientific experts we must discriminate between data, or matters of fact, and logic, or methods of reasoning. Confronted by what seems an error of reasoning on the part of a great master such as Laplace or Maxwell, an ordinary man may well doubt whether it is not his own judgment that is at fault. But the masters *have* committed errors which lesser men have been able to discover. Certainly no scientist can openly abandon his reason and assert that a demonstrable error ceases to be one when uttered by a great master.

When we come to matters of fact, the reasonable deference to those in a better position to know is, of course, greater. This is seen best in the case of history. Here absolute proof is unattainable, and in the weighing of probabilities there is involved an element of trust in certain witnesses. If any one refuses to trust Herodotus and doubts the occurrence of the Battle of Thermopylae, we can only bring in certain corroborative witnesses. But the credibility of these witnesses also involves an element of trust. When the amount of corroborative testimony is as great as it is for the existence of George Washington or Napoleon, one who persists in his doubts and attributes the consensus of our witnesses to conspiracy or common delusion is just as unreasonable as one who should refuse to plant potatoes for fear that they might be transformed into tigers and devour his whole family. Yet, though the refusal of all trust in the testimony of others lands us in a state perilously near the insane, the careful or scientific historian is precisely the one who critically scrutinizes the witnesses on whom he must rely. He must closely question what qualified them to report the given facts and what possible motives may have led them to emphasize one phase of what happened rather than another. In history as in a court of law, therefore, the element of trust in the testimony of others has to submit to a process of weighing credibility by reference to the probabilities of human experience. These probabilities are ultimately subject to the laws of mathematics applied to such experience as every individual may in a greater or smaller measure verify for himself. For this reason large experience is a necessary qualification for the historian. But pure reason, in the form of logic, is indispensable not only in determining the weight of the various probabilities, but also in opening the historian's mind to the various possibilities to which habit blinds us.

Similar considerations hold in respect of the experimental scientist. The laboratory worker cannot go very far if he discards all the observations of others. Not only must he practically rely on the authority of the general conclusions and the observation of others, but he cannot begin his work without differentiating and attaching greater authority to some part of the tradition of science than to some other part. Suppose, for instance, that he wishes to verify the fact of the pressure of light, which is generally accepted on the authority of the mathematics of Maxwell, and the experiments of Lebedev, Nichols, and Hull. If he wishes to test this, he will have to rely on the general laws of optics and mechanics assumed in the very use of his instruments of observation. Yet no single proposition of science is authoritative in the sense that we have no right to question it on the basis of our individual reason. Nineteenth-century mathematics and physics have progressed by leaps and bounds through questioning long-established results glorified with the names of Euclid, Newton, and others.

To be sure, the vast majority of people who are untrained can accept the results of science only on authority. But there is obviously an important difference between an establishment that is open and invites every one to come, study its methods, and suggest improvement, and one that regards the questioning of its credentials as due to wickedness of heart, such as Newman attributed to those who questioned the infallibility of the Bible.

These elementary considerations show how shallow is the reasoning of those who think we can ever dispense with *all* authority and tradition. Yet rationalism obviously remains justified. Reason must determine the proper use of authority. The fact that we cannot possibly doubt *all* things at once does not privilege *any one* proposition to put itself above all question. The position of reason is analogous to that of the executive head of a great enterprise. He cannot possibly examine the work of all his subordinates, yet he can hold every one accountable. The mere fact that every one is likely to be called to account produces a situation markedly different from what would result if some were put above accountability. An even more apt analogy has been drawn between the principle of authority and the credit which makes modern currency systems possible. Rational science treats its credit notes as always redeemable on demand, while non-rational authoritarianism regards the demand for the redemption of its paper as a disloyal lack of faith.

§ II. PURE EXPERIENCE

AUTHORITARIANISM attacks reason because it is individualistic. Empiricism attacks it on the opposite ground that it is not the experience of the individual at any one moment. These two arguments (*a*) easily pass into each other, for (*b*) despite genuine differences, (*c*) they have much in common.

(*a*) The ease and frequency of the transition from emphasis on authority to emphasis on individual experience, can be readily understood, when we remember that an effective appeal to authority is never to something abstract—authority in general—but to something particular, e.g. the Church (specifically Council or Pope), a specific interpretation of the Bible, actual decisions of the United States Supreme Court, the infallible dicta of dictionary writers, etc. As these authorities themselves have to be defended, their devotees must appeal to the facts of experience or history. This is the case in Christian apologetics from St. Augustine to De Maistre, or in the Marxian and other interpretations of history. In ecclesiastical as in other social controversies all argumentation as such necessarily involves an appeal to individual conscience or sense of truth.

Nor is the converse transition of argument from experience to authority infrequent. For those who emphasize individual experience are seldom willing to admit the experience of others that is directly contrary to their own. Carrying open-mindedness to the extent of believing that everything is possible would land us in a chaotic world of superstition unlike anything we call sanity. In practice some experiences always have more authority than others and we can seldom grant full freedom of conscience to those who threaten to destroy us. The true interpretation of the Bible means the interpretation of our own particular sect or group. In general, arguments for freedom, for reliance on individual experience, must, if they are to be effective, appeal to the authority of some commonly accepted belief.

(*b*) This easy transition between authoritarianism and individualistic empiricism does not, of course, deny their opposition; and it is well, before considering the empiricist attack on rationalism, to note carefully that the philosophy of experience can be, and in its early stages certainly was, much more an aversion for arbitrary authority than a preference for anti-rationalism. This clearly seen in Locke's attack on innate ideas. Locke, as a liberal and thorough believer in toleration, is opposed to political and ecclesiastical dogmas of the type of divine rights of kings, etc. His scepticism is the scepticism of science, which is a disinclination to have the road of critical rational inquiry into genuine problems closed by plausible-sounding dogmas.

(*c*) Yet we do certainly find empiricism and authoritarianism closely united today by their common aversion for the fluid state of doubt and uncertainty which reasoning develops. The empiricist love of fact certainly often hides a disinclination to enter into the disturbing inquiry as to how much of what we commonly regard as fact is the result of traditional interpretations based on authoritatively accepted dogmas. Practical men of affairs who profess to distrust all theory and loudly call for the facts are only too frequently the last to recognize how much of the so-called facts is just old hardened theory. So empirical philosophers, believing themselves bent upon eliminating all rational or metaphysical constructions from pure experience, may really be engaged in a pitiless effort to adhere consistently to certain fashionable dogmas accepted on the authority of the supposed results of science. This is another illustration of crypto-rationalism.

Philosophy, like science, must necessarily aim at consistency, though it may recognize that its absolute attainment is denied to finite, growing mortals. Physical science has found that the pursuit of consistency is made easier by giving up the use of terms that have an honorific rather than a descriptive value. But as philosophy moves in an obscurer and emotionally thicker atmosphere its path is filled with words that attract or repel us regardless of what they denote or whether they denote anything at all. The word "experience" has certainly become one of these terms. Its bare use provides a feeling of safety and elation, of belonging to the philosophically elect, and of being able to pass all the sentinels and watchdogs of philosophical criticism. But it is very difficult to find any meaning of the term common to absolutists like Bradley and humanists like Schiller, empirio-criticists like Avenarius and supernaturalists like James. It is even difficult to find any one school always using it in some one identifiable sense. For the tendency to make a term cover everything necessarily prevents it from denoting anything definite. Originally, of course, experience stood and it still stands for something in our personal history, something that we have tried, felt, or passed through. But as no individual can in this sense experience anything but an infinitesimal part of the world, and as even the collected experience of all humanity is confined to a very small part of the vaster and, in the main, unexplored world of time and space, it is obvious that no philosophy can identify the whole world with experience except by unconscionably stretching the latter term. This process is facilitated by applying the term "experience" to objects which have not been experienced at all, but which are held to have the possibility of becoming under certain conditions objects of experience. But observe that if the mere consideration of an object in discourse or thought thereby makes it an object of experience, we must apply the latter term to the battles that Napoleon would have fought if he had won at Waterloo, centaurs, and even round squares. Obviously, in this sense of the term nothing specific could be determined by experience alone. Few adherents of the philosophy of experience would, however, accept this result. They obviously wish to regard experience as something "thicker," something more intimately or vitally connected with sensations or personal feelings. But apart from the classical difficulties of subjectivism, or of identifying objects of sensation with the sensations of objects, it is quite clear that the world of philosophy and of any science that is not merely autobiographic extends beyond what we have experienced.

Once, however, we recognize that the world is constituted by objects of possible as well as of actual experience, we are bound to recognize that rationality is a part of the objective world. For clearly, purely logical considerations as well as past experiences are involved in determining what is possible. The rationalist, who believes that logical and mathematical reasoning deals with relations that inhere in the natural world, need not assert that rationality alone, without any matter or content, determines physical existence or constitutes experience. But when radical empiricists draw sharp antitheses between experience and the objects of reason, they are certainly bent on denying the reality of the latter and therewith they commit themselves to an irrational ontology. Of course, if *reality* is used to denote psychological vividness, there can be no doubt that for most people sensational elements have a superior reality, being more stable, vivid, and massive, as contrasted with logical relations which can be grasped only after intellectual effort. But not even the empiricist can eat his cake and have it too. He cannot *select* or *abstract* the vivid elements and then deny the existence of everything else. Nor can he justly assert that the possibility of a sensation is itself nothing but a sensation. The fact that experience may be analyzed into sensational elements is no more a denial of rational relations, than the analysis of sentences into words is a denial of integrating relations of order between the words. The analysis of matters into atoms does not deny the necessity of relations between the atoms.

This brings us to the great paradox of modern empiricism. This philosophy begins with a great show of respect for "fact" as the rock of intellectual salvation. On it we are to escape from the winds of dialectic illusion. But as science critically analyzes the "facts" more and more of them are seen to be the products of old prejudices or survivals of obsolete metaphysics. Those that stand the purifying fire of critical scientific analysis come out of it in a form that common sense can hardly recognize. In any case the "facts" of science are admittedly checked and controlled by theoretic considerations—for they are characterized by rational or mathematical relations. Hence the empiricism which has an anti-intellectual animus consistently turns from the rational scientific elaboration of specific facts to a mystical pure experience in which all clear distinctions are eliminated as the conceptual fictions of the mind. Thus does the worship of fact become the apotheosis of an abstraction devoid of all the concrete characteristics of fact.

These considerations may become clearer if we apply them to some typical empiricists.

(A) AVENARIUS

The most thoroughgoing and, despite his most unfortunate linguistic apparatus, the most influential apostle of pure experience is Avenarius. Entertaining a wholesome respect for empirical biology and a truth-lover's repugnance for the fantastic effort to enthrone all sorts of supernatural or anthropomorphic entities on the basis of an idealistic analysis of knowledge, Avenarius wishes to purify our world-view by returning to the natural view of experience as it existed before it was vitiated by the sophistications of thought (in the form of introjection). But the empiricist's uncritical use of the category of the *given*, and the nominalistic dogma that logical relations are created rather than discovered by thought, lead Avenarius to banish not only animism and other myths, but also all the categories, substance, causality, etc., as inventions of the mind. In doing this he runs afoul of the great insight of Kant that without concepts or categories percepts are blind. For what can we truly say of pure experience or of anything else if none of the categories, such as unity or plurality, permanence or change, apply to it?

Moreover, as Avenarius' whole enterprise begins in the effort to explain thought in terms of biological function, his argument in fact presupposes the use of categories in the science of physiology. Some recognition of this leads him in his later work to deny that thought is a

37

biologic function and to assert that thought distinctions are as objective for parts of experience as sense data. But in the main, though he holds fast to pure or immediate experience as the all-in-all and draws the distinction between the real and the ideal within experience itself, he still concludes that sensation alone is the real, and intellectual or rational elements are in the realm of fiction.

Avenarius, even more than Mill and other positivists, represents a noble type of high and rare devotion to truth as the supreme consideration in philosophy. Yet the insistence by him and by Mach on the biologic and economic function of thought, has provided a basis for those who belittle the truth-value of physical science and dismiss it as only a practical device to organize physical experience in the most convenient way. This illustrates how even a critical empiricist can transform the "results" of science into vicious dogmas by applying them to alien fields where they have not been proved. It was Spencer who first used the argument that as thought has arisen in the biologic struggle for existence, it must serve a useful biologic purpose. Despite the fact that this argument has been accepted by Mach, Avenarius, Simmel, James, Dewey, and others, we ought to have no hesitation in denying it any force or even relevance in any discussion as to the logical nature of truth. It is but a thinly veiled form of the old theologic argument that whatever Providence or Nature brings forth must have a useful purpose. The actual existence of biologic disharmonies in the human and other organisms is beyond dispute; and in any case it is not seemly for positivists to deny that all sorts of unhygienic superstitions have actually arisen and persisted in the process of human history or evolution. We have no certain knowledge as to what races will ultimately be eliminated on account of their false philosophic views, and even if it were true that mathematicians and other intense and intellectual workers leave a smaller progeny, it would in no way bear on the truth or falsity of any mathematician's theorem.

Behind the hard technical vocabulary and the long, painful, intellectual labor, the reader of Avenarius can readily note the single-minded zeal of the reformer. It is instructive to observe the parallel between his appeal for a return to a natural world-view, uncorrupted by the artificialities of conceptual thought, and the romantic appeal for a return to a state of nature uncorrupted by the artificialities of civilization.

There is an obvious element of Utopianism in both appeals. Primitive man of past or present is in fact not a paragon of virtue, and the plain, unsophisticated man has no coherent philosophy. If the primitive state were one of virtue and satisfaction, the positivist believer in progress could never give us a naturalistic explanation of why man ever abandoned this state; nor if the plain man ever had a satisfactory coherent philosophy, how any other could have arisen in its place. Actually the thought of the plain or primitive man is, as Avenarius himself recognizes, full of all sorts of animistic and other gross absurdities.

It is easy for unsympathetic opponents to assert that this sort of Utopian primitivism is itself the product of artificiality and sophistication; and that, in the language of Santayana, nothing is so foreign to the plain man as the corrupt desire for simplicity. But a sober regard for the perennial resurgence of Utopianism will recognize its permanent elements of value—despite the absurdities of the ladies of Versailles dressing like shepherdesses, or Rousseau trying to educate Émile "according to nature" by a most elaborate set of artifices invented by an inhuman tutor to keep his charge separated from natural society. The fact is that so long as man falls short of omniscience his thought is bound to develop errors and his civilization is bound to develop absurd artificialities, even as disease is a natural product of evolution. In this state of affairs a certain imaginative retracing of our steps in the forest of error is often necessary. We must visualize a more desirable state of affairs to liberate us from brute bondage to the actual. The vice of romanticism is not its ideal, but its failure to submit its ideal construction to rational criticism, its lazy identification of the ideal with some exotic or historically remote actuality. It may well be contended that it is not of the essence of Avenarius' position that the natural worldview should be put in the past, that he himself and disciples like Petzoldt also speak of it as the view of the future. This is in a sense true, yet it is of the essence of the anti-rationalism of Avenarius, as of other romantic positivists since Bacon, to seek a return to some *tabula rasa,* to cleanse the mind by some absolute purgative before the truth can stream in on us. The true method of science is to cure speculative excesses, not by a return to pure experience devoid of all assumptions, but by multiplying through pure logic the number of these assumptions, mathematically deducing their various consequences, and then confronting each one with its rivals and such experimental facts as can be generally established. The history of science certainly shows that effective scientific search depends above all on the volume of previous knowledge. The ignorant, no matter how well cleansed of metaphysical bias, never make any discoveries in science. It is doubtless necessary for science frequently to re-examine established opinions and re-open issues supposed to have been closed. But in the main, progress is effected not by going back to a *tabula rasa* of pure experience, but by further effective thought. Mental as well as bodily health may sometimes call for purgatives. But western civilization has discovered that the main road to purification is through proper nourishment and exercise, not through trying to stop the processes of life, or trying to recover our childhood by magic elixirs.[3]

(b) MACH

The anti-rationalism which results from stressing the sensational elements of experience is illustrated even more clearly by Mach than by Avenarius.

No one who appreciates the splendid logical as well as historical achievements of Mach's *Mechanics*—especially his elimination of the ghost of absolute space from experimental physics, which prepared the way for Einstein's great work—can lightly venture to associate Mach with anti-rationalism. Yet, those who disdainfully characterize the mathematical elements of exact science as mere "fictions," "convenient but devoid of real truth," undoubtedly draw nourishment from him. How does this come about? The reading of his *Analysis of Sensations* makes this clear. It is the excessive fear of metaphysics which makes him emphasize sensations or experiential elements to the utter neglect of the organizing relations which science actually finds to be involved in the very existence of these elements. It cannot be too often insisted that elements or sensations are not really "given," but are the result of analysis of things or of the flow of mental life.

---

[3] Avenarius' views are most completely elaborated in the *Kritik der reinen Erfahrung* (1888-1890), but more clearly stated in *Das menschliche Weltbegriff* (1891). See Petzoldt, *Das Weltproblem* (1912); Bush, *Avenarius and the Standpoint of Pure Experience* (1904). Cf. Höffding, *Philosophes Contemporains* (1907), pp. 113 ff. For a bibliography of the literature concerning Avenarius see the 12th edition of Ueberweg's *Geschichte der Philosophie*, Vol. IV, p. 710.

If logical or mathematical relations are to be called unreal because they are abstracted from things or experience, elements or sensations are surely just as abstract and have no more existence apart from the physical or mental world where they are found. All scientific investigation assumes a world capable of being analyzed into elements-in-relation. Any attempt to ignore the elements or the relations can result only in confusion. Mach himself certainly fails to give a consistent or coherent account of the physical or mental world. Surely elements without causal or other relations cannot of themselves constitute the moving or throbbing world of experience. His brave effort to reduce all bodies to thought-symbols, and all mind or selves to "makeshifts designed [by whom?] for provisional orientation,"[4] cannot but break down. For these thought-symbols and provisional unities somehow engage in a real struggle for existence—a struggle in which there seem to be real purposes and causal effects of economic procedures. Indeed, it is most wondrous that the mind, a mere symbol for sensations, after all, invents or creates non-sensational elements. Nor is it consistent to put the elements in the physical world and their mathematical relations in the mind only. For a mathematical relation, like the number of the revolutions that the earth makes around its axis before it completes a single revolution around the sun, is a fact of physics not of psychology. It is certainly true that the study of mathematical relations vastly increases our power of manipulating the things of experience. But the argument that because they are so useful they correspond to nothing in the real world, is a *non sequitur*. "What," Mach naïvely asks, "have vibrating strings to do with circular functions?" "But," we may ask in turn, "how could these functions guide us to the discovery of so many physical properties of vibrating strings and other phenomena, if they had nothing to do with the physical facts?" Mach and others are misled by the fact that mathematical functions are not copies of sensational elements. This, however, need not prevent these functions from significantly representing groups of relations which do characterize physical processes. Mathematical functions are the more useful the less the relations studied are restricted to particular realms and the more widespread the processes which they characterize. Mach's resolution to treat material things as mere symbols for sensations led him to deny the

[4] *Analysis of Sensations* (3d ed.), p. 28.

41

reality of atoms—a denial which would have been fatal to a great deal of fruitful research in physics if physicists had taken this view of Mach as seriously as have idealistic philosophers and theologians.

A quaint bit of naïveté is Mach's argument that his denial of the reality of self will make for unselfishness. How can there be any selfishness to overcome if there is no self?

(c) JAMES

The difficulties of reducing actual psychic life to sensational elements alone, were realized by no one more keenly than by James as a psychologist. His radical empiricism is in harmony with the efforts of Schuppe (and at times of Avenarius himself) to show that logical relations are just as much a part of the natural world as the sensations related. How, then, does James come in the end to draw so strong an opposition between experience and logical or conceptual thought? How can experience include everything important and yet exclude conceptual thought? The point is worth examining, not only to clear up the specific confusion which the term "experience" covers, but also because it illustrates how generally fertile in confusion is every monistic effort to reduce the world exclusively to any one of two polar categories. The term "experience" usually denotes something that happens to an individual in a world where many, or possibly most, things do not happen to him at all. If the term is stretched to apply to everything, so that there is nothing which is not in experience, it fails to differentiate anything and ceases to have any definite denotation. In practice, however, it is humanly impossible to be faithful to the resolution to use an old term in a new meaning (or absence of meaning), and we, in fact, fall back on its habitual connotation. Thus in denying any reality to the objects of logical or conceptual thought (which he does when he asserts that they have no theoretic or cognitive value) James in fact falls back on the view which conceives of real experience as exclusively immediate feeling or sensation. Being temperamentally concerned with this immediate and vivid aspect of the world, he falls into the "intellectualistic" fallacy of denying the existence of any other phase or aspect. To deny the reality of anything because it cannot be deduced from a formal or explicit definition is rightly condemned by James as vicious intellectualism. But his own procedure amounts to just that sort of a denial of all that is not included in his unavowed definition of experience as immediate and vivid sensations. But though in this denial of the objectivity of conceptual relations James falls back on the traditional sensationalism or nominalism of British psychology, his effort to enlarge the concept of experience is significant of real dissatisfaction with the traditional empiricism. As an honest observer of actualities he is compelled to admit that not only the world of past and future which our ideas mean or intend, but even life itself, is more than immediate. Conscious life itself is not exhausted by what merely is. It always means something that is not yet, "an ideal presence which is absent in fact." [5] The "ever not quite" true of the best attempts at all-inclusiveness must, according to James, be true of experience itself. Pure immediate experience may be the possession of a baby or of one who has been stunned by a blow. In the conscious life of those who can be said to have any mentality and to know anything, things are already related and conceptualized. The formation of abstract concepts does admittedly make things intelligible and necessarily complements direct acquaintance. Nay, both theoretically and practically "this power of framing abstract concepts is one of the sublimest of our human prerogatives." Against the usual nominalistic view that concepts are mind-made, we have James' own admission that they are "extracts from the temporal flux." In his arguments against the reality of concepts James seems to think exclusively of class-concepts. But even here he is compelled to make the admission that "things seem once for all to have been created in kinds." [6] The great hosts of scientific concepts, however, are not of things but of processes; and it is hard to imagine that if James had in mind the procedure of actual physics or biology with such concepts as motion, velocity, energy, life, growth, etc., he could have persisted in the absurd contention that concepts are static and can have no theoretical value.

It seems to me of some importance to realize that James' hostility to logic was not necessitated by any of his psychological observations. To see the motives which led him to espouse Bergson's extreme anti-intellectualism, we must reflect on the peculiar fascination and repulsion which the neo-Hegelian, and especially the Bradleyan argument

---

[5] *A Pluralistic Universe*, p. 284.
[6] Ibid., pp. 217-218. Cf. pp. 234, 251.

for the absolute, had for him. There is a very little difference in content between Bradley's absolute as experience and Bergson's absolute intuition, which James identifies with pure experience. They all postulate the fusion of unity and plurality in an experience that passes logical understanding or discursive reason. But James, having no taste or aptitude for rigorous logic, wished to defend his theism or vitalistic view of the world by the empirical evidence of abnormal psychology. Against such efforts the pitiless scepticism of Bradley's logic is radically destructive. We are all worried more by those who wish to attain our own end in ways that are different than by those who reject our entire aim. Thus Jesus could ignore the worldly Sadducees, but was spurred to anger by the Pharisees who aimed to attain spirituality by ways radically different from his own. So James is driven to extreme anti-rationalism, not by rank outsiders who reject the idealistic or vitalistic universe, but by those who would establish it by cold and seemingly lifeless logic.

Though, under the influence of Bergson, James sometimes expresses himself as an extreme mystic, having faith only in the utterly ineffable experience from which discursive logic must be cut out in entirety, he at other times recognizes that logic has its utility. His opposition is only, except in moments of extreme wrath, to the Bradleyan logic of identity. James' essential point here is that in the world of experience things change and yet retain their identity, while Bradley's logic seems to deny that this is possible. If we start with noting the diversity and activity of ordinary life, this criticism on Bradley's logic seems unanswerable. But if we are to avoid the vicious intellectualism which James condemns, we must recognize here that both sides are right in what they affirm and wrong only in what they deny. Bradley is certainly right in insisting that the identical remains identical. If you deny that things are what they are, you annihilate all reason, sanity, and discourse. Thus James himself has to be certain that it is the "same" nucleus, that the "same" point can be on diverse lines, etc.[7] It is no error, but a useful caution, to remember that a horseman cannot be a pedestrian in so far as he is a horseman. James, on the other hand, is perfectly right in insisting that things are not only what they are, but become different without destroying their identity. To argue that a man who is at one time a horseman cannot at another time be a pedestrian, is not a necessity

[7] *A Pluralistic Universe*, p. 258.

of logic, but a gratuitous denial that the unity of time can include diversity of moments. James overshoots his mark, therefore, when he condemns all logic. Indeed, he might even have accepted Bradley's own interpretation of his logic and have argued that since unity and diversity are united in absolute reality they must be conjoined in all that in any sense truly exists. In any case all of James' objections to logical thought (because it is not pure immediate sentiency) have been stated more fully and forcibly by Bradley himself. The serious difficulty of the latter is not in clinging to logic, but in gratuitously identifying two different poles of reality, viz. the absolute totality and immediate feeling, without providing the proper locus for the mediating relations. But this is precisely the difficulty which empiricism can never avoid so long as it seeks to construct the universe out of mere immediate experience. If there is any universe at all, it must be more than immediacy. This lesson will be reinforced by an examination of intuitionism.

Before turning to this it is well to reflect on a practical aspect of the empiricist's paradox (to which we previously referred).

One of the noblest manifestations of the empiricist temper in the practical realm is James' picturesque individualism. The preference for the "thick" particular fact, as against "thin" abstractions, shows itself in an admirable insistence on the reality and supreme worth of the actual human being before us, as against all abstract labels and conventional rules. (How the latter often blind us as to the concrete fulness of life is strikingly brought out in the essay "On a Certain Blindness in Human Nature.") The supreme importance of preserving the peculiarities of the individual which distinguish him from the colourless "average" is beautifully stated in the essay on Thomas Davidson.[8] But if we ask how this precious individuality is to be preserved from the crushing effects of natural and social circumstances (and more especially from the aping proclivities within each human organism), James offers us no definite answer. For a social philosophy interested in what secures and develops individual freedom must discover the appropriate rules which will best harmonize generally conflicting interests, and this involves a rational weighing of these interests. But to anti-rationalistic individualism such weighing of interests according to some universal standard is

[8] *Memories and Studies*, pp. 73 ff.

45

impossible. It is thus left with no vision as to how its heart's desire may be attained.

Bergson, seeing how thoroughly shot through with rationalistic considerations are our daily practical affairs, keeps away from social philosophy by insisting that the interests of practical life and those of philosophic insight are entirely foreign to each other.

James, as a pragmatist, might have been expected to recoil from such a complete separation. But being bent on finding all salvation in the immediate, the demands of practical guidance as well as of rational science are eluded by a leap into the mystical sea of pure experience.

Nor ought we entirely to ignore the fact that this leap of desperation prevents James from making any contribution to a theory of art. It may well be contended that the primary function of art is to restore to us the immediacy of things that is lost in reflection and in daily hustle. But it is precisely because the artist sees more than the bare immediacy of nature, because he sees the inherently rich imaginative suggestiveness of things that he is impelled to labour arduously to add ornament and beauty to the world of actuality. Moreover, how to attain and express the effects and values of immediacy is itself a rational problem of adapting means to ends in a given medium, and is not to be solved by turning one's back on logical thought and conduct.

§ III. INTUITION

LET us begin by noting that though there has always been some conflict between pious or mystic theology and rationalistic philosophy, there had not been, previous to the eighteenth century, any marked conflict between intuitionism and rationalism. The reason for this is obvious if we remember that in the classical or Aristotelian-scholastic doctrine intuitive reason served all the functions now claimed for anti-rational intuition. Just as discursive reason established the necessary truth of propositions by deriving them from axioms according to the rules of logic, so intuitive reason enabled us to see these axioms as true or self-evident in themselves. But the spread of modern science and the development of radical changes in modern life brought to light a critical difference between these two kinds of reason. For while the question of the correctness of mathematical proof occasions relatively little difference of opinion, the rational self-evidence of various axiomatic principles is a matter of most divergent and seemingly irreconcilable opinions. This certainly seems to be the case in subjects like rational theology, ethics, aesthetics, and other branches of philosophy. Canvass the opinion of people as to what they consider inherently reasonable, axiomatic, or self-evident, and you will find that in an overwhelming proportion of cases, this quality is attributed to the familiar or to what happens in fact to have been unquestioned. The questioning of that which we have been accustomed to accept and on which we have habitually relied is profoundly disturbing. Hence we naturally resist the questioner's challenge and we hold to our primary beliefs with increased vehemence. This is plainly seen when naïve people are confronted with the demand to show the evidence for the views that they regard as certain. They answer, generally, with some emphatic: It is so; I know it is so; I am sure it is so; or: How could it be otherwise? In a homogeneous community, the challenge of the doubter or sceptic may thus be crushed by the common feeling of certainty on the part of all the respectable. But in a period of rapidly developing science in which all sorts of preconceived opinions turn out to be false, doubt cannot be so readily eliminated. Moreover, in a heterogeneous community or in a time of struggle for power between different groups, questioning the first principles of our opponents is greatly admired and encouraged. In any case modern mathematics and physics have found the systematic questioning of self-evident axioms a fruitful source of new insight.

All this makes it now difficult to hold the old doctrine of an intuitive reason as an immediate revelation of the material propositions that are infallibly true in themselves. But to be certain of our cherished views—whatever they happen to be—and to hold them as inherently superior to the views of others, is a primitive vital desideratum; and if this primitive certainty is not called intuitive reason or the faculty of innate or a priori ideas, it is called by some other name such as common sense, intuition, instinct, faith, or something else.

REASON AND THE NATURE OF THINGS

(A) INTUITION AND COMMON SENSE

These terms were adopted by Reid and his followers when Hume ventured to challenge the older rationalistic doctrine of causality and the theologic structure which rested on it. Reid, undoubtedly, made some telling points in regard to realism and the impossibility of doubting the existence of all objective order. But despite his tremendous influence in France and Italy as well as in America, his method proved utterly barren. His followers simply elevated every challenged traditional belief into a fundamental intuition of the mind.

In exact science this method has been discredited, since all sorts of propositions formerly held as indubitably self-evident are now known to be false. But in the less rigorous realms of philosophy where we lack direct experimental checks on most of our statements, this elevation of the plausible into the axiomatic still prevails. Particularly true is this in the realm of legal philosophy and ethics where the doctrine of conscience still represents the old intuitionism.

It used to be a prevailing belief that a code of all the laws that should govern human conduct could be readily deduced from a few simple principles similar to Euclid's axioms. These principles were supposed to have been inscribed by Nature in the heart or conscience of every human individual. American courts still sometimes argue: *"We need no elaborate arguments. We need only consult conscience or the eternal principles of right and wrong, etc., to see that. . . ."* But the history of laws and the fate of codes have disillusioned us somewhat in this respect. With increased study of history, even American lawyers are beginning to question whether the maxims of our bills of rights are such eternal principles.

Conscience is still generally viewed as a private oracle which gives immediate answers to all our moral questions. Note, however, that though its dictates are often regarded as final and superior to all other authority, few ever draw the anarchic implications of this. We seldom excuse an atrocious act on the ground that the conscience of the perpetrator may be different from ours. If told that the Thug or Assassin acts according to his conscience we are likely to remark: So much the worse for his conscience. Indeed, a study of actual moral judgments in any community leaves little doubt that the dictates of conscience coin-

48

cide, in the main, with the canons of generally accepted respectability. Direct teaching in childhood, natural imitation, and the general pressure of prevailing attitudes make us grow up with the habit of judging every one by the prevailing standard of respectability. As this habit is naturally carried over and applied to our own acts, we have the phenomenon of conscience.

This view of conscience as having its roots in tradition and habit does not deny individual variations of insight into moral problems. Nor does it deny the importance of maintaining our convictions where they happen to differ from those that at a given time dominate our community. The point is rather that so long as conscience is viewed as immediate and infallible, neither the social-authoritarian nor the individualistic-anarchic view of it will help us to deal with the actual mass of taboo, superstition, and moral confusion which characterize the prevailing moral judgments of any human community at any time. In the light of the actual variability of moral judgment in different groups and the ease with which the sense of right is moulded by custom, training, and our own interests, we cannot regard the appeal to prevailing intuitions of right and wrong as absolutely decisive. If it is urged that there can be no moral judgment at all unless there is at its basis an intuitive sense or apprehension of right and wrong, we can only reply that such a sense, like the sense of physical reality, is necessary but insufficient for theoretical or practical purposes. It must submit to rational criticism in which the claim of every intuition is balanced by those of conflicting ones. The duty to follow our conscience is conditioned upon making it as enlightened as possible. In practice, limited time and limited energy may compel us to stop ethical inquiry at some arbitrary point. It then becomes necessary for us to abide by the best principles that we can reach in our limited deliberation. But to stop the questioning of conscience by maintaining that its dictates are absolute and self-evident can only result in hideous moral confusion.

(B) THE ILLATIVE SENSE AND INTUITIVE ASSENT

A somewhat neglected variant of intuitionism is the doctrine of the illative sense as developed by Cardinal Newman in his *Grammar of Assent*. It would be difficult to find anywhere else such skilful or felicitous illustrations of the difference between reasoning as described in formal treatises on logic and the ways in which active minds actually jump at fruitful inferences. Napoleon perceiving at once the proper moment to make a charge, the skilful diagnostician seeing in a flash what is behind a complicated disease, the quick perception of the proper thing to say and to do which we call tact, and a host of similar examples, all seem to illustrate the intuitive rather than the formal-logical character of effective or creative thinking. But all this is futile as an attack on logic or as a substitute for it. It would be relevant if logic were a description of the way we actually think, instead of a method of determining the correctness or incorrectness of inferences. A Napoleon, an expert diagnostician, and a person of remarkable tact, can and do make mistakes in their rapid inferences; and when any one of these is challenged, it is surely no adequate reply to say: I have an intuition, or my illative sense tells me.

It has been contended by De Maistre, James and other apologists for religion that no one ever changes his belief in God or immortality as a result of arguments. This may sound plausible but it is hard to see evidence for its truth. The effect of arguments is seldom appreciated immediately. We are frequently influenced by arguments which we stoutly resisted on previous occasions when others urged them upon us. It is to fly in the face of history to assert that the arguments of philosophers and theologians like Kant or Schleiermacher have had no effect whatsoever on the course of religious opinions. In any case the claim that certain convictions rest on the assent of the whole personality and not on intellectual argument is not an argument in their favour but often a euphemistic way of admitting that they rest on no evidence. In a world where we often have to stake our life on uncertainties, one may legitimately prefer to take his chance on an attractive prejudice, on an agreeable "hunch" or intuition, rather than on a process of reasoning from evidence which may not be free of error. But such an attitude certainly does not further knowledge; and by no stretch of the imagination can it guarantee superior attainment of truth.

(C) INTUITIONISM AND ONTOLOGISM

Can the difficulties of intuitionism be averted by assuming one rather than many self-evident intuitions? This was attempted by the Italian philosopher Rosmini, who made the intuition of *being* the basis of his system.[9] Assuredly nothing can be more certain than that there is being, and that all logical thought operates only by assuming it as the matrix wherein all rational relations or distinctions subsist. But the outer fate of the Rosminian philosophy provides a clue to its inner deficiencies. Its critics like Mamiani and Gioberti had no difficulty in showing that from the intuition of mere being—actual or possible—nothing of any specific importance could be inferred. On the other hand, to the extent that the primary intuition included the being of God as its object, it inevitably led to elements of pantheism which the Catholic Church was quick to detect and to condemn—and, I venture to add, rightly so. For no genuine religion is possible for the great mass of humanity except as it stimulates and is in turn fed by the sense of transcendence, of something greater than our petty selves and beyond that which we can grasp immediately. To be sure, the Church has kept great mystics like St. Bernard. But it regards the mystic vision as a supernatural gift rather than a primitive possession. The practical import of this distinction is enormous since on it depends the continued necessity of Church discipline as a perpetual check on the vagaries of mystic visions.

(D) INTUITION, LIFE, AND INSTINCT

Few habits are so characteristically modern as the vague use of the term *life* as honorific rather than descriptive. That the nature of life (like that of other realities) has not yet been completely grasped by human intelligence is an old and tragic truth. But that by merely living we have a source of knowledge different from and higher than that attained by rational science, is a rather new doctrine. It seems to be intimately connected with the current use of the word *life* as a glorification of our impulses no matter how pathologic or pathogenic. But the very difficulties of the latter cause every vigorous people to develop

[9] See Rosmini, *Origin of Ideas*; T. Davidson, *Rosmini's Philosophical System*; L. Ferri, *La Philosophie Italienne*.

rational restraints. Otherwise life is degraded and brutally impoverished if not utterly destroyed. So in theoretic matters the first deliverances of the experience we call living are so full of illusion and contradiction that vigorous mentality seeks rational science as a deliverance (in part at least) from mortal error.

The assertion that by intelligence we know only phenomena while in mere living we know the absolute reality of things receives a certain plausibility from the widespread popular faith in introspection as the supreme and even primary form of knowledge. I take it, however, that the development of modern empirical psychology has brought ample evidence for the view that introspection is at least as full of difficulty and subject to error as any other form of observation. It certainly requires scientific training and much reflection to make its results reliable.

Life, as we know it introspectively, seems to present a greater continuity and interpenetration of parts than does the physical world. This gives a prima facie plausibility to Bergson's polemic against mechanistic explanations of life such as are offered by dogmatic positivists. Despite, however, the multitude of references to biologic literature, Bergson is hardly fortunate in his empirical illustrations.

Nor need we take seriously Bergson's central argument that because intelligence is developed in the process of evolution for practical purposes, it cannot know life. The premise of this argument—in which Bergson follows positivists like Spencer and Mach—is dubious and his conclusion is a *non sequitur*, if not in contradiction with his own premises.

It is vain to argue as to what purpose nature or evolution had in producing intelligence, or whether it had any purpose. We can tell the power of intelligence only by its actual operation before us. From the latter point of view there is little reason for doubting that intelligence plays a minor rôle as a biologic function—unintelligent species multiply and fill the earth—but in the field of natural science intelligence is the most powerful instrument for the discovery of *truth* that man has as yet developed. The reliability of any other method of reaching truth pales into insignificance compared to it. The very *utility* of science indeed depends on its truth. Bergson's argument from evolution assumes in fact the truth of scientific biology with its categories of organism, variation in time, environmental factors, etc. His premise thus assumes that by rational biologic science we do to some extent know the processes of life. From such an assumption no reputable logic can conclude that therefore the intellect cannot know the process of life.

Those who prefer to base philosophy on suggestive metaphors distrust and profess to be unconvinced by logical arguments; but you cannot both distrust logic and claim logical cogency for your own (fallacious) arguments.

In general, we may say that intelligence is the rational organization or distillation of the experience of living. Mere life apart from intelligent thought is dumb and blind. Unless intelligence illumines the meaning of our vital activity we can make no significant assertion about it nor draw any conclusion from it. That is why intuitionism has proved sterile not only in physics and ethics but also in the philosophy of art. For the essence of art is in articulate and coherent expression; and no philosophy which stresses formless feeling can throw light on the problem of artistic creation or its intelligent appreciation.

It ought in fairness to be added that by the intuition of life Bergson frequently means something similar to Spinoza's *intuitio*, to wit: a completion rather than a rejection of reason. But there can be no doubt that Bergson also draws an absolute separation and opposition between intuition and intelligence. The former is identical with instinct and extolled over the latter as the organic over the mechanical. His followers have certainly stressed this phase of his thought.

Like intuitive self-evidence or the illative sense, the expression "instinctive knowledge" is a popular device to protect favourite convictions from the necessity of rationally justifying themselves. Indeed "instinctive knowledge" seems to be but our old friend Innate Ideas in a rather thin biologic disguise. The biologic affiliation of the term *instinct* facilitates the suggestion of intimacy with life, and makes more plausible the denial that rational intelligence can know life. In the end, however, scientific biology while dealing with a subject-matter that is more complicated than inorganic physics pursues methods which are but the extension of the rational methods of physics.

Unlike most current writers Bergson uses the word instinct in a fairly definite sense. It denotes biologic acts that are congenital, attached to definite organic structures, and unmodifiable, at least during the life of the individual. With this conception of instinct we cannot speak of instinctive knowledge without stretching the term *knowledge* in a way to debase our intellectual currency. Thus Bergson speaks of the knowledge that the new-born babe has of its mother's breast. If this be knowledge, why not also call the sleeper's reflex withdrawal of a tickled limb an instance of knowledge? But is digestion or the growing of teeth and hair knowledge? There is no reason for asserting that every biologic act originates in a conscious purpose. Detailed empirical evidence is necessary to justify such a sweeping generalization; and in any case, such a generalization should not be smuggled in by simply stretching the term *knowledge.*

(E) INTUITIVE OR ROMANTIC REASON

In philosophy as in other human enterprises, not all who appeal to the same word mean the same thing. We need not therefore be surprised to observe that the most influential form of intuitionism makes its appeal in the name of reason. I mean by this the romantic doctrine which glorifies reason (as the intuition of absolute totality) at the expense of the understanding (or the discursive reason of demonstrative science). Nor need we be surprised to find this doctrine receive new impetus in the work of the sober and almost pedantic Kant, who began his career as a mathematical physicist. If we recall that among the chief influences which moulded Kant's thought were the pietism of his youth, the anti-intellectualism of Hume and Rousseau, and the large elements of irrationalism in the aesthetics of the Wolfian school (Sulzer, Baumgarten, et al.), we may better understand why Kant's immediate successors, like Fichte, Schelling, and Schlegel, were romantic philosophers. At any rate, there can be no doubt about the actual rôle of intuitionism in Kant's philosophy. We may see it best in his ethics and physics.

Ethical judgments according to Kant are absolutely and intuitively certain. Philosophic reflection cannot modify them in any way. It can only provide a formula or explanation for them. If we question the absoluteness of any individual moral rule, e.g. that against lying, and ask why should not one tell a lie to prevent a wretch from committing murder or rape, Kant has no answer except the brutally dogmatic reiteration of his position in a way that offends the finer and more flexible ethical discrimination of mankind. Kant's intuitionism is even more clear if we ask for the basis of the doctrine which is the centre of his whole positive religious philosophy, the doctrine of the *summum bonum*. The faith in God, freedom, and immortality is made to rest on the assumption that happiness must always follow moral virtue. But why assume that the social device of external rewards and punishments used in schools and penitentiaries is an eternal necessity of reason, which must govern every rational universe? The answer is clear if we refer to Kant's moral catechism in the *Tugendlehre*. Despite his insistence that moral acts cannot be directed to the attainment of happiness, Kant elevates the common feeling, which hates to see the wicked enjoying the good things of this world, into an a priori intuition. Here Kant follows not the more reflective ethical insight of the great sages, from Buddha and Jesus to Spinoza, that virtue is its own reward and sin its own sufficient punishment, but rather the sentimental intuition of the Faith of a Savoyard Vicar (in the *Émile*).

It would of course be absurd to regard these considerations as a denial of Kant's claims as a rational moralist. For few can read his writings without being profoundly stirred by his genuine respect for reason. He shows it magnificently in insisting that the mind recognize its own limitations and the necessity for suspended judgment where evidence is lacking. Yet Kant's ethical system fails precisely because, not satisfied with discursive reason as the form according to which the substance of morality must be connected if it is to form a consistent whole, he elevates the prevailing moral sentiments of his time into eternal dictates of the intuitive reason.

The appeal to intuition is also present in the Kantian theoretic philosophy in the doctrine of space and geometry. In his youth Kant caught a glimpse of the possibility of a non-Euclidean geometry. But in his maturity the specific characters of Euclidean geometry like the specific theorems of the Newtonian mechanics and Prussian ethics seemed to him absolutely certain, and this supposed certainty he could support only by intuition. How the infinite whole of space can be given in intuition, and how that which is given in sensory intuition can be necessary or apodictic, Kant never really explains. In the metaphysical deduction Kant naïvely follows the current Lockian assumption that we can be absolutely certain of the nature of things if we can derive this nature from the constitution of the mind. At other times, however, Kant realizes that our knowledge of the constitution or function of the mind is by no means indubitable. He then shifts to the transcendental question: what must be the constitution of a mind which can produce the absolutely certain synthetic a priori judgments of geometry? In the end, however, neither Kant nor any else can prove, a priori, either that actual space is Euclidean or that the subjectivity of time and space is the only possible basis or proof of the truth of geometry.

It is of some importance to dwell on this because we can here see how the intuitionism of Kant's theory of geometry led him to become the founder of the romantic *Naturphilosophie.*

If, as Kant assumed, geometry is a priori knowledge of the space of physical objects, why may there not be a priori knowledge of the laws of mechanics and gravitation? Kant himself does in fact take that step in the *Metaphysische Anfangsgründe der Naturwissenschaft.* But if the general laws of mechanics and gravitation can be deduced a priori independent of experience, why not the laws of electricity or biology? In taking just this step, Schelling, Hegel, and their followers only carried forward a method developed by Kant. If the method of the romantic philosophy of nature is now generally recognized as unsound, it is because we can no longer rely on a priori intuitions as to space, time, and the constitution of nature. It may justly be contended that the correction of these romantic claims is to be found in Kant himself, viz. in his distinction between the formal and the material, or between regulative and constitutive principles. It is certainly within the essence of Kantianism to insist that no existential proposition of physics and no material precept of ethics can possibly be deduced from purely formal principles of reason, and yet to insist at the same time that formal rational relations are genuine characteristics of the real world that is the object of science. Kant, however, undoubtedly fails to be consistent on this point; and it is instructive for our purpose to trace the cause of this failure in part at least to his absolute separation of sense from understanding and his restriction of intuition to sense alone. If, contrary to Kant, we grant that it is an experience that we think, and if understanding is seen to differ from sense not as pure activity from pure passivity, but as a more from a less developed phase of our knowledge of the world, there is no longer any difficulty in granting that we can intuit or apprehend immediately the rational connections which are the objects of exact science. In this way we can have an intellectual intuition without the vagaries of romanticism.

§ IV. CREATIVE IMAGINATION

IF GLORIFICATION of the principle of fixed authority represents one extreme of the reaction against rationalism, the apotheosis of the free creative imagination represents the other extreme. These extremes met when the leaders of German romanticism embraced the Roman Catholic Church. In general, imagination is by far the most popular of the rivals of reason. It denotes something more active than passive experience and more articulate than intuition.

Consider what a poor insignificant fragment of our world we can actually experience at any one time. Not only is it impossible for past and future events to be directly present to us, but only an infinitesimal part of the contemporary world spread in space can be directly reached by optical or other sensory contact. We can of course speak of past, future, and distant events as ideally present to the mind on the occasion when we think of them. But assuredly this is not what we ordinarily mean by experiencing things. Having an idea about typhoid is fortunately not the same as experiencing it.

If imagination, then, denotes the power to see beyond what is actually or materially present, it is the fundamental basis of our whole mental outlook. For even of things physically present, what we mentally see is not the same as what is sensibly experienced. Thus when I see a friend crossing the street, or when I read about him in a letter, all that is impressed on me through my sense organs is some patch of diverse colors. On the other hand, what I actually see is also often much less than what is sensibly present. Of the innumerable objects about us, only a few attract our attention. When those in whom we are vitally concerned change their features, how often do we continue to see them according to their past image! Imagination thus seems to build up our conscious world by adding to and subtracting from the world of actual sensible experience.

Yet as in the case of other rivals of reason, only a certain duplicity in the use of the term *imagination* makes it possible to oppose it sharply to reason. The term *imagination* does often denote something fanciful as contrasted with what is real or natural. Thus to Bacon, imagination "being unrestricted by laws may make whatever unnatural mixtures and separation it pleases." [10] To be imaginative in this sense is to be preoccupied with daydreams. In this sense we often attribute to a lively imagination the habitual lying of children who are not sufficiently developed to distinguish clearly between what has and what has not happened. On the other hand, the term *creative* or *constructive imagination* also denotes the process whereby a great scientist or historian grasps new possibilities involved in old principles, or reconstructs a comprehensive picture of past life on the basis of fragmentary remains. In these cases there is certainly no inherent opposition between imagination and reason.

Obviously the process of imaginatively building up a new view or picture of the hitherto unseen may be not only controlled but helped by a regard for the rules of logical consistency and probability. Thus it seems rather obvious that not only a great scientist like Cuvier or Faraday but a great historian of insight like Polybius, Gibbon, Mommsen, or Maitland exercises an imagination exceeding in magnitude and depth that of the inventer of impossible plots in melodramatic novels. Surely it requires greater power of imagination to discover the past causes of actual things before us than to concoct in fancy impossible or unnatural situations.

If, then, rational regard for truth were inimical to great imagination, we should expect to find the most fruitfully imaginative minds amongst the most ignorant; and the weird combinations which characterize dreams would represent the highest reaches of the creative imagination. However, despite this reductio ad absurdum, disparagement of rationality in the name of the creative imagination continues to be a distinctive trait of modernistic thought.

As the word imagination usually denotes a mental activity that is (1) constructive, (2) free, and (3) resulting in images, three arguments have been found for the romantic exaltation of the imagination over reason.

[10] *Advancement of Learning*, Book II, Ch. 12.

(A) IMAGINATION AS CONSTRUCTIVE

This argument belittles reason on the ground that it is merely analytic or takes things apart, while the imagination is constructive.

Few words have so much prestige amongst us today as the words *creative, constructive,* and *synthetic,* so that to call any intellectual work merely critical, negative, or analytical is to belittle it. The principle of polarity, however, warns us against the too ready assumption that the negative can be dispensed with or even entirely separated from the positive. May not the popular indiscriminate hankering after anything that bears the title *creative* be due to our waning faith in political and religious creeds, and our fear that doubts and criticism may shatter what is left of them? It is certainly not a very vigorous faith which says: Let each hold on to his own beliefs and let no one criticize his neighbour's. Robust faith is more active and fearless. Certainly the faith of science in its own methods makes it eagerly welcome doubts and possible negations. It knows that the most fruitful periods of modern geometry and physics have resulted from doubting of Euclid's axioms and Newton's laws of motion. Is not the cleansing of an Augean stable a task worthy of Hercules? Is not the cutting out of a cancer, "without putting anything positive in its place," a most beneficent act? Finally, are not many creations or constructions positively hideous and vicious?

If these reflections do not throw any light on the creative imagination, they should at least liberate us from the honorific bias of the word, and prepare us for a more objective consideration of the problem.

That reason is merely analytic and only cuts up what the imagination has previously built up seems to be an accepted dogma. Indeed, do not rationalistic logicians themselves admit that in logical reasoning there is nothing in the conclusion which is not already in the premises? Let us not, however, be misled by metaphoric language. The conclusion is not contained in the premises in any sense in which things are contained in a box, or as a tree or horse is often, in popular evolutionism, erroneously supposed to be contained in the seed from which it is said to unfold, develop, or evolve. Continuous epigenesis rather than preformation is the proper analogy that biology offers us here. In any case, all that is properly meant by the classical formula of logic is that the content of our conclusion must be connected with that of our premises in accordance with certain logical rules of possibility. But these rules of logic also bind the work of the imagination if the latter aims at any order or significance.

In truth the sharp contrast between imagination and reason, as if they were separated faculties or compartments, is a misleading use of language. It does not correspond to anything actual. All mental life that is concerned with what is intelligible or significant involves reasoning in the sense of passing from propositions to the contents or consequences which they imply. In a later chapter I shall try to show that even in pure mathematics, the most abstract of the sciences, reason is synthetic or creative in the sense that its results may appear novel. So long as reasoning is viewed psychologically (i.e. as a process taking place in time) its results must in some way be different from its data or starting-point. On the other hand it is generally recognized that the novelty brought forth in the imagination is also a novelty of arrangement and not of sensory elements. The proper contrast is, therefore, not between active or fruitful imagination and passive or sterile reason, but rather between, on the one hand, rationally disciplined mental operations in which there is regard for continuous transitions and necessary connection and, on the other hand, vague, irrational, undisciplined imagination or fancy. Perfect rationality and complete absence of coherency are, of course, only ideal limits. In actuality we deal with the more or less, and it is somewhat arbitrary whether we call any actual mental operation reasoning or imagination.

To appreciate more fully the confusion involved in the usual contrasts between analytic reason and creative imagination, we must note that the current conceptions of mental creation or "noetic synthesis" arise in a crude metaphor. They all involve as their starting-point a naïve picture of "the mind" and the physical world as two independent things, side by side, and external to each other as if they were both in space. According to this picture, the mind can be impressed by the things external to it only by contact or by the bombardment of minute corpuscles shot at the mind by the bodies external to it. The mind thus conceived is, like the more primitive soul, really a spatial body rather thin, like smoke or breath (*spiritus, anima,* etc.), but definitely located in an individual space-occupying body, generally the brain. However, once we recognize that the mind is not an additional body at all, the whole notion that the mind gathers its impressions into a picture of the world is seen to be only a mythologic development of a metaphor.

The notion of the mind (apart from the animated body) as an active *thing* is so ingrained in our linguistic habits—all our usual philosophic expressions have been so moulded by it—that those who explicitly reject it cannot avoid expressions which seem to presuppose it. But the fact that we cannot get rid of the habit of speaking of the sun as rising or setting need not prevent us from recognizing the truth of the Copernican astronomy. Metaphoric expressions are harmless if we fully recognize that they are not literal truths. All that is necessary in our present difficulty is to refrain from confusing the mind that by definition is not a body, with the mind which is identified with a particular animated body in which we are vitally interested. External things literally make impressions on our bodily sense-organs. Only in a vague metaphoric sense do they make impressions on our minds.

The whole issue of mental construction is still more hopelessly complicated by the introduction of the notion of an unconscious mind. If the term *mind* or *mental* has a definite and verifiable meaning as distinct from the physical, it is restricted to the conscious. An unconscious mind is neither a body nor a recognizable mind. At best it is a metaphor, denoting that certain events happen as if there were conscious direction behind them. However, unless we believe that all organic adaptations, e.g. the circulation of the blood, are conscious, there is no necessity for always postulating a mind wherever there is life. If mind is not a thing, it need have no literally continuous existence in time. There is then no necessity for supposing that a mind is literally active when we are in fact not conscious—though all sorts of processes may go on in our bodies which later do become conscious.

From this point of view knowledge need not be a construction or mechanical synthesis of elements by an external mind. (The mind apart from all knowledge is, indeed, an empty concept.) The development of consciousness and knowledge from infancy may be viewed simply as a growth of illumination, such as the increased distinctness which things assume as more light breaks upon them. Certainly the knowledge which comes with familiarity is thus a growth of distinctness where formerly things were undiscriminated. A thinking body is, of course, alive and thus active; and the growth in knowledge probably depends upon previous knowledge more than on anything else. Yet we certainly cannot consistently speak of our mind or consciousness creating itself or even its next stage. Ultimately our conscious life has its basis in a body which, though alive and sensitive, does not begin as a conscious being.

This critical attitude to the notion of creation or construction by the mind will enable us to see the fundamental weakness in the Kantian doctrine of the primacy of the creative or productive imagination—a doctrine from the womb of which have come forth Fichte's Absolute Ego, Schelling's Absolute, Hegel's Absolute Idea, and the Unconscious of Von Hartmann, from which Bergson's Elan Vital, Freud's Unconscious, and other transcendental creating deities are descended.

Kant's doctrine that the mind as productive imagination must bring together the sense elements before the conscious understanding mind can recognize an object is a logical outcome of Descartes' notion of two substances and Locke's metaphor of the mind as a *tabula rasa* dependent on impressions. You assume that the mind begins with only a supply of sensations and you introduce an unconscious synthesis to explain why our actual consciousness never takes such a form. The difference between Post-Kantian idealists and the empiricists in this respect is one of evaluation only. Idealists like Green and Gentile find the constructive work of the mind good and call it reality, while others like Bradley and Bergson, just as positivists like Mach and Vaihinger, find the work of the mind merely useful but devoid of that savour of reality which comes only with strong sensory experiences. The realistic rationalist, however, may say, "A plague on both your houses!" The creation of the order of nature out of sensations by the mind is a pure myth. There is no actual mind apart from the recognition of related things. Doubtless things as known do have the form of time and space and are related according to the categories of quality, quantity, etc. But the assertion that the mind imposes these forms and categories on the material of sense is a statement that literally passes all understanding. We know nothing and can know nothing of such a mythical material of sense, nor of any mind that creates time, space, and the categories. What indeed can we say either of the material of sense or of the creating mind if it is not already in time, is neither one nor many, nor bears any other categories? What can we say about either to distinguish it from anything else or from nothingness, without assuming that it is already subject to the categories? The mystery of creation is not removed when the mind is substituted for the demiurge or other creating deity.

Kant's fundamental assertion is that there can be no understanding or empirical recognition of an object without a previous creation or synthesis of the object by the mind (referred to as the transcendental unity of apperception). But this finds no support in any facts of consciousness. When, about to cross a street, I suddenly see an automobile in motion, not only do I not find in my mind any trace of previously having constructed the object before me, but my safety and my continued sanity depend upon recognizing that the automobile and its driver are realities having a history and purpose independent of my own.

The whole tragedy of life and human effort is meaningless if there is no order of things independent of our limited mental activity, or if there are no such things as hidden physical forces which determine human fate even when we have no idea of their existence. Who seriously believes, except when arguing philosophy, that the procession of the stars, the history of our globe, the changes of the seasons, and all the fortunes of our ancestors through the various lands and ages are the creations of the knowing mind?

You may justly protest: The foregoing is based on a caricature of the Kantian position. Kant never thought of any such absurd proposition as that our empirical mind created the world—indeed he explicitly characterized as a paralogism the proofs of a substantial soul and insisted on the dependence of the empirical self on the temporal flux of inner sense.

There is undoubtedly truth in this protest. Even Kant's most romantic successor, Schelling, insists that it is not the empirical self that creates the world but an antecedent transcendental activity continuous with it. But this, alas, does not remove the difficulty. So far as the transcendental mind is spoken of as if it were continuous with and like the empirical, it cannot be free from the limitations of the latter, and it becomes as difficult to conceive how the mind can create its own constitution as how the Squidgicum-squees can swallow themselves. On the other hand if the transcendental mind has none of the limitations of empirical consciousness which it calls into being, it becomes a confusing travesty on words to call it mind. I have already alluded to the difficulty of forming

63

any intelligent idea of a mind or self devoid of all the characters of empirical consciousness and to the difficulty of distinguishing such a transcendental entity from anything or nothing. We may press this point here by asking what there is in Kant's argument to distinguish his creative imagination, transcendental understanding, or non-phenomenal mind from the God of orthodox theology, from the logos of Stoic or Neo-Platonic philosophy, from the universal creative reason of Averroës or the unconsciousness of Leibniz, Von Hartmann, and Freud?

References to Kant's *Anthropologie* and to Baumgarten's *Aesthetica* will amply confirm an implication of the foregoing argument, viz. the close connection between the doctrine of the creative power of the transcendental imagination and the doctrine of unconscious perceptions as developed by Leibniz. The motive for the latter is obviously the desire to introduce continuity between the conscious and the unconscious. As the doctrine of continuity in development is one of the most powerful motives in modern thought, it reinforces the reifying influence of language in making us speak of the mind as a substantive, as if it had a continuous existence in time. As individual consciousness has no such continuous existence, being interrupted by sleep and other events, we fall into the habit of speaking of the existence of the mind when it is not actual, and even of the existence of thoughts when we do not think them—forgetting that thoughts are events having no more existence when they do not happen than has the killing of Caesar when it is not enacted. Orthodox empirical psychology likewise speaks of the mind of the child before it arrives at consciousness, or of the synthesis which the mind makes and which results in conscious perception. This usage is perfectly clear if we use the word *mind* in the sense in which Aristotle defines the *psyche*, viz. the activity of an organic body. But unconscionable confusion results if we attribute to the psyche in this sense conscious prevision of the results of all its activities. To interpret all organic adaptations as involving prevision is to envisage the whole after the analogy of a small part. But such analogy appeals only to what is intimately familiar and is supported by that disease of language which we have called reification, the tendency to speak of purely logical relations as if they were individuals of flesh and blood doing this or that to other things.

Calling the transcendental ego, unconscious self, or imagination the creator of our world serves the supreme romantic exaltation of making us feel like creative gods. At the same time, when pressed by the absurdity of such pretension, we have the intellectual refuge of the distinction between the transcendental and the empirical self, mind, or imagination. But if we take this distinction between the transcendental and the empirical seriously, if we realize that all our vital human interests that centre about our bodily existence can in no way serve to individualize the transcendental ego, that the transcendental imagination which creates our world is not the poor imagination which we have to exercise more or less laboriously in trying to understand our world and to anticipate its action upon us—if we realize all this, any identification of our empirical with our transcendental selves is a patent confusion. Such confusion provides us with a rather thin solace but does not cure the mortal limitations of finitude.

The way of rational understanding as well as the way of genuine piety is to begin with the more humble recognition that each one of us is but a small part of the world and not its entire creator. We are truly creative in the human arts only to the extent that we recognize our limitations as well as the limitations of our material.

The emphasis on the unconscious synthesis of the productive imagination Kant himself regarded as one of his most original contributions to philosophy. His departure from the Leibnizian tradition centres, however, about his introduction of an absolute distinction between sense and understanding, viz. passivity and activity, to replace the older one of greater or less clarity. It is a question well worth serious consideration, whether the older view is not more just to the facts of psychology. But I must restrict myself here to noting that Kant's psychology marks the passing of the high regard which the older rationalists had for clear ideas. It is certainly a characteristic of romanticism.

A Platonic realist or rationalist may fully grant that the simplest perception involves mental activity. But such activity only makes the nature or constitution of things intelligible to us. It does not create that which we apprehend in truth.

The romantic glorification of pure and even unconscious activity to the disparagement of rational deliberation naturally shows itself in belittling the rôle of criticism in art. But criticism as an effort to analyze the problem which the artist faces and the ways in which he can possibly fulfil his task, is something which every artist himself actually goes through. No one can engage in any of the arts or familiarize himself with any artistic creation of sustained power without becoming aware that artistic work is a constant exercise of judgment as to what is fitting and what must be rejected—which is precisely the occupation of the rational critic. It is true, of course, that one cannot by reflection alone, and even by the learning of rules, become a great artist. Also, many of the rules for poetry and other arts laid down in the classical renaissance were certainly narrow and illiberal. In the fine arts as in the rest of life a good many rules can be learned only after preceding trial and error. But there can be no intelligent trial or experimentation in any art without rational ideas as to what we may expect to achieve. This does not mean that the artist carries on his analysis in conscious syllogisms. But it does mean that he must reflect on the consequences of doing this or that. It is this necessity which explains the historic fact that periods of great creative artistry, e.g. the Periclean, the Florentine, the Elizabethan, and the Goethean, all show previous critical preparation. Artists grow in maturity as they reflect on the rules or necessary conditions of their work. If in addition they also aim truly to represent some phase of human life critical reflection is certainly indispensable.

(B) THE FREEDOM OF THE IMAGINATION AND THE DESIRE FOR ILLUSION

Almost all accounts of the imagination assume that it is distinguished by a certain spontaneity and freedom. But if we ask, "Freedom from what?" the answer is not very clear. The imagination, as free fancy, it seems, is free from restraining rules, and is spontaneous in the sense that it involves less effort than rational inquiry. Unsympathetic critics may call this the passive, if not the lazy, imagination. But the protagonists of the free imagination take refuge in the ancient doctrine of inspiration. By not encumbering their spirits with the dead weight of commonplace reflection, poets and seers (according to this view) keep their imagination open to the vision of a higher reality than the cold, prosaic world of hard and uninspired rationality.

Sober history shows that, in the field of scientific discovery, authenticated cases of inspiration—of flashes of truth coming suddenly upon us—are all preceded by a period of rational, systematic preparation and searching efforts. Perhaps something similar is true of artistic inspiration. But the proponents of the free imagination generally emphasize freedom from rational reality more than freedom from rational effort. In answer to the contention that imagination must be controlled by reason in order to keep us sanely in touch with reality, romantic irrationalists contend that the high function of the imagination is to liberate us from this blind, humdrum, routine reality, and that ordinary sanity is a spiritual death from which we can be saved only by inspiration and divine frenzy, madness if you please.

It is well in dealing with this attitude to realize the amazing extent to which mankind at all times has associated insanity or demoniac possession with higher wisdom. The association between prophecy and madness referred to in the Book of Samuel is still illustrated by the modern dervishes and mad mullahs. The priestess at Dodona or Delphi could not deliver the oracular utterance unless intoxicated to madness by the smoke. To this day the reverence of European peasants for idiotic cretins continues the old tradition which associates the fool with superior wisdom.

Not only common people but great thinkers from Plato and Aristotle to Shakespeare and Goethe glorify a kind of liberation from reason which they do not hesitate to call madness. Thus Plato not only identifies divination ($\mu\alpha\nu\tau\iota\varkappa\eta$) with mania, but deliberately asserts that "the sane man cannot rival the madman in the temple of the muses," and it seems hardly necessary to refer to the classic passage in the *Midsummer Night's Dream* about lunatics, lovers, and poets, "of imagination all compact."

Even Aristotle, the patron of common-sense thinkers, is reputed to have said that no great mind is free from an element of dementia; and he did certainly assert that poetry is truer and more serious than history.

We may naturally find many different reasons to explain and discount this awe for the ravings of the frenzied or insane. In the first place, the raving prophet is, after all, of the same human clay as ourselves, and his gestures, contortions, and expressions naturally produce in our organism an echoing wave of sympathy with him, just as every eloquent orator or actor carries us to some extent with him, no matter how inherently absurd his cause. The utterances of the frenzied, at least those in which we can find any meaning, are dissociated from familiar settings, and thus have a strange and often fascinating concentration. Possibly elemental pity for the obvious suffering of the insane predisposes us favourably to them. Perhaps also the fact that the behaviour of the mad prophet is out of the ordinary may arouse some of the vague fears which are generally used to explain our attitude to spirits, etc. Moreover a good deal of the ordinary regard for the sayings of the insane has a purely practical character—very much like our feeling for the throw of dice to break our suspense and enable us to decide something. The priestess, the insane, and the dice operate without apparent bias.

At bottom, however, the fundamental fact to remember is that the partial rationality which is all human beings ever actually attain is a painful acquisition and generally effective only to the extent that it has become a matter of habit. But the habitual is also what frequently prevents further insight. It is this which supports the hope that by some inspiration or Dionysiac frenzy we can break through the prison of prosaic routine and catch a glimpse of the greater reality beyond. The poetic imagination undoubtedly makes some such appeal to us. The vitality of it may be seen in the fact that when the play of our imagination is cut short by some appeal to the brute facts of actual existence we feel chilled to the marrow of our being.

Why is this so?

In the first place, it is obvious that the exercise of the imagination is in itself a pleasing activity, a manifestation of our vital human energy, and anything which shames it out of being is resisted. We can see this clearly in the displeasure of a child when shamed out of an enjoyable play activity. In the second place, and more important for our present discussion, there is the great metaphysical truth that nothing can be philosophically so deadly as the confusion of the significantly real with brute fact. The significance of facts, which is the object of philosophic search into nature, always carries us beyond actuality. From the point of view of strict logic, nothing can be so false as the positivistic notion that rational science is concerned only with the actually existing. The nature of things which rational science tries to formulate consists of laws or invariant relations of order of possible phenomena. These possibilities can never all be realized in any actual moment or thing. Moreover. the actual not only contains too little in itself but also too much in the form of irrelevancies. It is thus the function of rational science as of art, religion, and all human effort to liberate us from the charnel house of the actual and reveal to us the underlying order which governs the wider realm of possibility.

It is, of course, true that the possibilities which interest natural science are the possibilities of the actual world. But the paradoxical fact which positivistic theories of science miss is that the most fruitful scientific theories are based on hypotheses contrary to fact, though never contrary to logical possibility. Newtonian physics is based on the assumption as to what would happen to a material particle not acted on by any forces, though the law of gravitation makes it impossible for two such particles to exist. Similarly is our modern science of thermodynamics built on assumptions as to what would happen in a frictionless engine, though no such engine can exist. Similarly is the significance of historical fact revealed by asking what might have happened if things had been different, e.g. if Alexander or Napoleon had been killed in their first battles. How different might the history of the East or of nineteenth century Europe then have been? Some historians shy at asking such questions. But without raising such questions the historic significance of Alexander or Napoleon can no more be realized than can the climatic significance of the inclination of the earth's axis be realized unless we ask what would have happened had there been no such inclination.

The recognition of this opposition and yet inseparable union of possibility and actuality, significance and fact, enables us to see the rational or scientific rôle of the imagination, and saves us both from near-sighted positivism and the romantic capriciousness that wilfully blinds itself to the order of fact. The latter attitude is perhaps most characteristically expressed by Renan in one of the most perverse sentences ever penned: "Who ever got inspiration from an accurate knowledge of the text?" Of course many vague meanings that lazily float before us may be brought down and shown to be hollow by accurate knowledge. But the increase of knowledge gives the imagination many more points of support, so that sustained flights of imagination are always helped by vivid and accurate perception, memory, and much studied judgment as to what is and what is not fitting or coherent. The poet or artist may not reflect on how much rational criticism enters into his accepting or rejecting certain elements of his construction. But all imagination, poetic as well as scientific, is essentially selective, in the sense that it concentrates upon what is relevant to the task before us. Such studied selection in the interest of what is relevant takes place in scientific observation, in trained memory, and generally in all the processes of scientific knowledge, so that the growth of genuine scientific knowledge is itself a progress of the imagination, not in falsifying the order of fact, but in eliminating insignificant, accidental, and irrelevant elements.

Do not similar considerations hold true of what we generally call the poetic imagination? If poetry is more true and serious than history (in the Aristotelian sense of mere description as in the term, "natural history"), it is so because of its effort to grasp the significant. As poetry is not primarily interested in all the physical and practical effects of things, but only in certain aesthetic or emotional aspects of them, the poetic imagination can achieve its ends by seeing trees as "green-robed senators." Such vision carries a certain emotional atmosphere with it. It would become insane only if it took the identification as a physical and practical fact. The poetic imagination is not, then, a falsification but an intensified vision of things. The vision of the insane is disordered rather than intense. The poet or the painter may prefer to describe objects in a dim atmosphere rather than in the glaring sunlight. But the poet or artist himself must see clearly and accurately whatever it is that he chooses to present to us.

It would indeed be most strange if it were otherwise—seeing that great poetry, like all great achievement, requires persistent use of high intelligence. Indeed, many poets of the highest rank were by all canons unquestionably men of first-rate intelligence—witness Sophocles, Dante, Shakespeare, Milton, Goethe, and Shelley.

These reflections on the free or poetic imagination have certain obvious applications to the field of conduct. All our individual and social efforts are directed toward the realization of situations that as yet are not. The highest and most delicate reaches of human life depend upon an imaginative anticipation of potentialities not yet actualized. This is what gives significance to conjugal or parental love, devoted friendship, or the general love of our fellow men embodied in efforts at social reform. Here also we meet the near-sighted positivists who call themselves realists (because they do not see the full reality) and the lazy dreamers who call themselves idealists (because they do not grapple with basic ideas).

Just now the realistic temper seems to prevail in philosophy, at least in America. It is the fashion to disparage daydreaming and the sort of imaginative refuge that men and women find in romantic literature, Utopian politics, and other-worldly religion and ethics. The energy that goes into such Utopian dreams might, it is urged, produce durable improvements if intelligently directed to the illumination of the actual life about us. What these narrowly earnest realistic souls fail to see is that the possible improvements that can be brought into the lot of many men and women by this abandoning of their idle dreams is insignificant compared to the positive solace which these dreams themselves produce. Too often are human misfortunes so heavy that human reason has not yet found any remedy for them. To rob the poor sufferers of all anodynes that can make them forget for a moment their tortures is an insensate cruelty which can arise only out of purblind benevolence. The man who votes for a party that seeks to bring about the millennium may obtain more enduring comfort from that act than from actually electing a reform administration that reduces the tax-rate, introduces efficient government, etc. Men are not much enriched and their daily work not much lightened by the cutting out of holidays and vacations, and even the strongest often need imaginative havens of refuge from the suffocating cruelties of life.

Moreover it is better to dream than to be altogether dead. For many people the issue is not idle dreams versus rational enlightened imagination, but whether they shall to any extent be waked out of vegetative existence into any sort of imaginative life. Among cultivated people gentle and delicate stimuli may produce most intense results. But violent stimuli may be necessary to stir to life the inert clod of those stupefied by drudgery. Gladiatorial combats, bull fights, at one extreme, and virtuous melodrama at the other, all provoke the undeveloped imagination.

But while daydreams and cheap imaginative satisfactions make life more pleasant for many, they do not enable any one to escape the mortal dangers all about us. Important as it is to exercise our free fancy and see how things could be nearer to our hearts' desire, our safety also depends on being able to anticipate the tangible effects which things will actually produce. To ignore these actualities will not prevent calamity.

If heaven is the vision of the heart's fulfilled desire, nothing so effectively removes it from earth as the lazy identification of the whole nature of things with those qualities which they possess in undisciplined imagination. Those who gaily dance with illusion are most unhappy when they pay the piper.

We may all

. . . pray that Daddy Care
Wad let the wean alane wi' his castles in the air.

But

. . . mony sparkling stars are swallow'd up by night;
Aulder een than his are glamour'd by a glare,
Hearts are broken—heads are turn'd—wi' castles in the air.

(C) IMAGES AND MEANINGS

The previous arguments have extolled the imagination because it has the power of creating dim, mystic atmospheres in which the solid world melts and assumes fanciful shapes according to the heart's desire. Others, however, extol the imagination above reason for the very opposite motive. The imagination, they argue, gives us definite pictures and not mere references. Hence it is richer than reason, which operates with thin, abstract, and abstruse symbols of conceptual thought.

Little reflection, however, is needed to show the inadequacy of this view. No one denies that what some regard as reason is pure logomachy —a Barmecide feast in which no food enters into the motions of the meal. Certainly we need to apply this observation to many of our so-called practical discussions in economics, politics, and psychology, which often lose themselves in empty conventional symbols, from which they can be recalled to sense only by some concrete picture.

But homely wisdom amply warns us that we often enough lose sight of the woods for the trees. Losing ourselves in a mass of picturesque details is one way of losing their significance. The meaning not only of general ideas like identity, resemblance, negation, etc., but also of human policies and processes of nature is not to be discovered by examining our mental images.

The gift of being able to look within and report vivid pictures made James a great psychologist but limited his metaphysical and general philosophic insight. As he himself justly observed, images are what we see when we try to stop active thought about a problem and look within for resting points or *states* of the mind. The meaning of general ideas like personality, respect for law, acceleration, quanta of action, and the like is not illumined by any images which different individuals may have when they think of these ideas. Ribot and other experimental psychologists have demonstrated this beyond peradventure of doubt.[11]

If reality includes the world of our daily practical activities, it is quite clear that to dwell on our images of things would often hamper rather than promote vital strivings. Things must often be considered only in so far as they are tools for something else. To stop to consider our images of them would be fatal to our objectives. Action in business, in education, in politics, or in organized religion would be paralyzed if we could not think of numbers of men with their concrete individual differences dropped out of sight. For not all aspects of nature are relevant for any one of our practical purposes, and the exact determination of what is relevant in practical life, as in art and science, is an affair of reason.

The recognition of how often in practice we have to skip intermediate images led James in the end to accept Bergson's sharp contrast between practical life and metaphysical insight into reality. But what the content of a reality that excludes human conduct can be only the mystics claim to know, and they will not tell.

We are led to the same result (i.e. the recognition of how images must be subordinated to rational connections) by reflecting how dependent our power of imagining things is on our previous experience. No fact as to our mental power seems so well authenticated by history and experiment as the breakdown of non-rationalized imagination when we apply it to the unfamiliar. Without rational reflection the field of unfamiliar possibilities is closed. The reader can test this for himself

---

[11] Ribot, *Evolution of General Ideas*, pp. 114 ff. Burke had made similar observations at the end of his essay *On the Sublime and the Beautiful*. Cf. Stout, *Analytic Psychology*, Vol. I, pp. 80 ff., Vol. II, p. 179.

by asking how many people can imagine a piece of paper that does not have two distinct sides, or a vessel where there can be no line of demarcation between inside and outside. Moreover, when we actually produce the former of these (a Möbius strip or Kummer surface) and ask what will happen if it is repeatedly cut through the middle, the imagination unaided by mathematical considerations generally breaks down.

The history of science shows how repeatedly this happens. To popular imagination, the Pythagorean or Copernican astronomy was absurd. To the mathematically trained it is the simplest.

Finally and even more significant is the consideration that there are well-authenticated phases of the world of which we cannot possibly form any image at all. Such are motion, consciousness, the infinity of time and space, etc. When popular thinkers like Hamilton and Spencer assert that we cannot conceive the infinite, they generally mean that we can form no image of anything infinite, since an image always refers to something occupying a definite portion of space-time. But that which is so localized is always conditioned by certain processes having a wider locus. Any attempted *image* of the world, no matter how much it may contain, must always involve something beyond it which limits it, so that the image must fall short of absolute totality. The very recognition that there is something which we have not yet grasped involves something beyond imagery. This power of recognizing the limitation of imagination is most fittingly called reason. Without some such recognition of our formal power to recognize the limitations of our material imagination we fall into the absurd sophistry which ignores the most patent of all facts, viz. human ignorance, and asks, "How can we know that there is anything which we don't or can't know?" This sophistry is not only the basis of what is called subjective idealism but is the key to all that is characteristically weakest in modern thought,—the illusion that by ignoring our limitations we can deny the larger world that mocks our pettiness.

The fact that all concrete thoughts use images to jump from and to alight on, explains the tremendous hold that materialism and spiritism have on the intellectually untrained. For materialism and spiritism are the twin offspring of nominalism, i.e. the belief that reality consists of nothing but particular things of which we can form definite images.

74

Spiritism is a crypto-materialism, in that it views spiritual relations spatially, as if they were material bodies. It is thus the basis of the worship of symbols in religion, law, and social life. This makes for superstition, and the fanaticism that cannot tolerate other than the usual symbols. Reason, however, shows us that a symbol can always be replaced by another one; and this provides the basis for the liberal toleration of differences. Even in the field of art the imageless beauty of Dante's Paradiso may be ranked higher both by classicists and by romanticists like Shelley than the more definite imagery of the Inferno.

In concluding this somewhat lengthy chapter it may be well to remind the reader that we have not dismissed these rivals and substitutes for reason as mere usurpers. Neither authority nor experience, neither intuition nor imagination, can ever be completely ruled out in favour of pure reason (if the latter is identified with logical inference). All of these play significant rôles in our effort to apprehend the nature of things; but their fruitfulness depends, in brief, upon the extent to which they submit to the rule of reason. Neither brute authority nor the immediacy of experience, neither mystic intuition nor unreasoned imagery, forms a sufficient basis for an adequate human philosophy. Always we need a rational apprehension of the significance of things in their relational or intelligible contexts.

Chapter Three

REASON AND SCIENTIFIC METHOD

§ I. REASON AND THE GATHERING OF FACTS

ACCORDING to the classical conception of science as rational knowledge, the perfection of reason can best be seen in the most developed sciences. But modernistic anti-rationalism is bent on minimizing the rôle of reason in science,—even as it seeks to minimize the significance of science itself by regarding it as a mere fiction or convenient mnemonic device for arranging the substance of "experience." According to the currently fashionable view, it is of the very essence of scientific method to distrust all reason and to rely on the facts only. The motto, "Don't think; find out," often embodies this attitude. Scientific method is supposed to begin by banishing all preconceptions or anticipations of nature. In the first positive stage it simply collects facts; in the second, it classifies them; then it lets the facts themselves suggest a working hypothesis to explain them. It is only in the last stage, in the testing or verifying of hypotheses (so as to transform them into established laws) that the rational deduction of consequences plays any part. Such deduction, it is maintained, brings us no new information. It only makes explicit what experience has already put into our premises.

All this, like other conventional accounts of development through "stages," rests on a priori plausibilities rather than on actual history. Begin with collecting the facts? Ay, but what facts? Obviously only with those that have some bearing on our inquiry. Attention to irrelevant circumstances will obviously not help us at all, but will rather distract us from our problem. Now, the relevant facts of nature do not of their own accord separate themselves from all the others, nor do they come with all their significant characteristics duly labelled for us. Which of the infinite variety of nature's circumstances we should turn to as relevant to or bearing upon any specific problem depends upon our general ideas as to how that which is sought for can possibly be related to what we already know. Without such guiding ideas or hypotheses as to possible connection we have nothing to look for. For countless ages men saw things balance each other and sink or float in liquids, but not before Archimedes did men see in these phenomena the principle of the lever and the law or fact that a body replaces exactly its own weight of water. It was the ideas and reasoning of Archimedes that made it possible to see the specific gravity of substances and to use it as a test for determining the amount of gold in an alloy. Surely Newton was not the first to see that the moon revolves about the earth, and that apples and other objects fall to the earth. But no one before Newton saw embodied in all these phenomena the common mathematical relation which we call the law of gravitation. To look for and see the latter, one had to have the following in mind: (1) Galileo's law of falling bodies and Kepler's laws of planetary motion, (2) the analysis of circular motion into centrifugal and centripetal components—according to the principle of the parallelogram, and (3) the daring and unorthodox speculative idea (which Newton derived from Boehme and Kepler) of a parallelism between the celestial and the terrestrial realm. Similarly we know that it was the Pythagorean conception of the book of nature as written in simple mathematical terms that led Galileo to look for and ultimately see the simple law connecting the increased velocity of a falling body with the time of the fall. Tycho Brahe's astronomic tables did not in themselves show Kepler's laws; indeed, they suggested quite different laws to Brahe himself. Kepler could see these laws only after he brought to his vision certain speculative ideas of Apollonius (on conic sections) and of Plotinus. To be sure all these cases (as well as Darwin's discovery of natural selection) show a most painstaking checking up of preconceived ideas by accurately determined or measured facts, and a readiness to discard hypotheses that do not square with such facts. But without the well-reasoned ideas, the inquiries could not have been initiated, for there would have been nothing to verify.

It is easy for those who have not reflected on actual scientific procedure to say: Begin with the facts. But an even more fundamental difficulty faces us. What *are* the facts? To determine them is the very object of the scientist's investigations, and if that were but the beginning or first stage of science, the other stages might be dispensed with. To determine the facts scientifically, however, is a long and baffling enterprise, not only because the facts are so often inaccessible, but because what we ordinarily take for fact is so often full of illusion. Our expectations and prepossessions make us see things which do not in fact happen, and without the proper previous reflection we fail to notice many obvious things which do happen. The problem of how to get rid of illusion and see what truly goes on in nature requires that persistent and arduous use of reason which we call scientific method.

Popular empiricism speaks as if we can readily eliminate all error and attain absolute and indubitable truth by purifying the facts given in sense perception from all taint of inference or interpretation. Error, it is said, comes in judgment, not in perception. But if we did eliminate all inference and interpretation would sense perception give us any facts? Certainly not enough to constitute any science, social or physical. For assertions of fact involve all sorts of assumptions. It is doubtless true that various theoretic issues can and should be settled by reference to "hard facts" or elements of sense perception. Whether two objects of the same size and different weights do or do not fall together, whether the earth's shadow on the moon is or is not round, whether the level of still water in a canal does or does not show a curvature, are questions that can be definitely settled by properly arranged sense perception. But this does not deny that the meaning of such facts depends upon all sorts of theoretic assumptions that go beyond immediate sense perception, e.g. the assumption that physical bodies continue to exist when we do not see them and to have such properties as weight when we do not weigh them. Some of these assumptions seem imposed upon us by the conditions of our existence and it is hard to conceive of any facts independent of them. Other assumptions are found, as science progresses, to be false, and the supposed facts which involve them turn out to be illusions.

Certainly, if the term *scientific method* is used in any significant sense it cannot be said to begin with a *tabula rasa* and pure sense-impressions on it, such as the new born babe is supposed to have. Sensations, like relations, are elements in a logical analysis of what we know, and not actual starting points in scientific investigations. Science properly begins rather when there is wonder or active curiosity and an effort

78

to answer questions or problems that arise out of intellectual difficulties which reflection finds in common knowledge.

It is curious but significant that an age that has pushed the claims of the "social" in all possible and impossible realms should remain so absolutely individualistic in its theory of knowledge and regard a return to individual sensations as the road to intellectual salvation. But a social theory of knowledge will not save empiricism. Obviously the great body of what we regard as common knowledge (as distinct from rational science) of any age is funded out of traditional teachings, superstitions, and ancient metaphysics, as well as the personal impressions and opinions developed in us by our fragmentary partial experience. The tragic inadequacy of the result, the fact that it is full of error and illusion, is the primal source of the failure of human effort and aspiration. The history of magic, witchcraft, and other superstitions throughout the ages indicates that there is hardly an offspring of human fancy too absurd to be regarded as a fact by immense multitudes of men. Have not men for ages believed and do not millions even now believe it to be a fact that children get sick when the evil eye is cast on them or that they get cured when the proper words (or prayers) are pronounced? Scientific method is a systematic effort to eliminate the poison of error from our common knowledge. If common knowledge were entirely wrong in substance and method, science could not have any base from which to start or any certain direction in which to proceed. It would be impossible for science to arise out of common experience and reflection if the latter did not contain the seeds of truth as well as the noxious weeds of error and illusion. But it is precisely because common sense is such a mixture of sense and illusion, of enduring truth and superstition, and because even its truth is so vaguely and inaccurately expressed, that under certain conditions it arouses dissatisfaction in sensitive intellects and compels them to go beyond common knowledge and endure the rigours of scientific research to attain purer and wider truth. If science thus begins with the facts of common sense it is only to organize or transform them radically.

To say that science is organized or classified common sense is to be satisfied with very vague words. The knowledge in a telephone directory or in a railroad time table may be well organized so as to be readily accessible, but no one will take it as the ideal of science. Nor is classification of facts the essence of scientific method, though it may be a useful auxiliary process. A systematic catalogue of a department store in which all the goods to be sold are properly classified and described does not contain that which characteristically distinguishes a scientific treatment of a subject.

Karl Pearson, who by constant repetition has made popular the notion that science is but classified facts, adds in a footnote [1] that "classification is not identical with collection. It denotes the systematic association of kindred facts, the collection, not of all, but of relevant and crucial facts." But what facts are kindred, relevant, and crucial is precisely the question that empiricists blandly ignore. For in their dogmatic haste to explain the origin of knowledge from sensation, they will not stop to admit the obvious fact that the successful progress of any scientific investigation depends largely upon the initial or anticipatory ideas according to which it is instituted and according to which it proceeds.

When the empiricist does recognize the influence of previous assumptions in determining the course of a scientific research, he dismisses them lightly as working hypotheses. Now it is doubtless true that when we hit upon an appropriate hypothesis it will work, in the sense that it will help us to group facts together significantly. It is also true that our initial hypotheses generally have to be more or less radically modified in the course of investigation, and that serviceable anticipatory ideas or hypotheses occur to us more often as we become more familiar with our subject matter. But it simply is not true that the facts themselves suggest the appropriate hypothesis. The same facts do not always suggest the same hypothesis to every one who looks at them. The history of science indicates rather that fruitful hypotheses have generally come to certain gifted minds as musical themes or great poetic expressions have come to others. You may call them the gift of the gods to their favourites. But it is certain that it requires a plenitude of previous knowledge to enable one gifted with fortunate insights or guesses to develop them into successful scientific hypothesis.

Actually every scientific investigation begins as a question or problem. We ask whether there is any law or formula for the seemingly irregular series of lines in the spectrum of hydrogen, what is the cause of the

[1] *Grammar of Science*, 3d ed., p. 77.

heart-beat, or of the disturbances in the glandular activities. When we find that our actual body of knowledge either has no answer or offers diverse unsatisfactory ones, we institute an inquiry. To make it possible for us to find the true formula or cause, we must by logical reflection widen our view as to the various possibilities,—for unless we first consider something as a possibility we shall not look for it among actualities.

Mathematical science simply develops the consequences of rival possibilities. But natural science is interested in choosing that possibility which fits best with the existing body of factual knowledge and can make it grow. As our actual knowledge is always limited or fragmentary this desire for coherency and order in our ideas leads to the extension of our knowledge by way of observation and experiment in order to provide more adequate material for the testing of our ideas. In this process there is a constant give-and-take between what we regard as ascertained fact and possible hypotheses. We not only eliminate hypotheses found inconsistent with the facts, but we also employ theoretic arguments to correct the readings of observation or experimental results. When facts do not fit in with our idea or hypothesis, we re-examine the process by which the facts are obtained, and try to correct them by other observations and to discover hypothetical causes of the irreducible divergences or perturbations. It is only as a last resort that we modify (as little as possible) the old ideas. If we did not hold on to our old ideas tenaciously, if we threw them away the moment they encountered difficulties we could never develop any strong ideas and our science would have no continuity of development.

Reason thus plays a more active part in scientific method than is accorded to it by the usual positivistic, anti-rationalistic account. Our safeguard against fantastic speculations and hardened prejudice is not to try to clear the mind of all prejudgments or anticipations. That is neither possible nor desirable. We cannot by mere resolution get rid of all preconceptions, since most of them seem to us obvious or unquestionable truths, and it does not in fact occur to us to question one of them except when it conflicts with some other preconception. The hampering effect of narrow prejudice or prejudgment is reduced rather by logical analysis or reflection, which, by making our premises explicit, shows them to be a part of a *larger number of possible assumptions*. Reason thus enriches us with a greater number of possible hypotheses or anticipations of nature, and this makes possible a richer variety of observation. Certainly great contributions to science are not made by those who go to nature innocent of all preconceptions but rather by those have acquired the most knowledge and fruitful ideas on the subject of their inquiry.

In thus emphasizing the rôle of reason in scientific method we do not minimize the appeal to experiment and observation, but make the latter more significant. The appeal to experience is thus involved throughout: first as the matrix in which inquiry arises (as that which suggests questions), and then as that on which all theories must be tested. We start always with general assumptions and with contingent or empirical data. By no amount of reasoning can we altogether eliminate all contingency from our world. Moreover, pure speculation alone will not enable us to get a determinate picture of the existing world. We must eliminate some of the conflicting possibilities, and this can be brought about only by experiment and observation. The fact that two or more hypotheses are logically possible means that none of them involves self-contradiction. They cannot therefore be eliminated by logic or pure mathematics alone. Experiment or observation of crucial cases is needed for such elimination. When an hypothesis is first suggested we try to see whether it will explain the known facts. But we generally need new situations to determine whether its explanatory power is superior to that of other hypotheses. Hence, though no number of single experiments and observations can prove an hypothesis to be true, they are necessary to decide as to which of two hypotheses is the preferable as showing greater agreement with the order of existence. This shift from the question of whether a general proposition is absolutely true to the question of whether it is better founded than its rival is the key to the understanding of the rôle of probable and inductive reasoning.

Before taking up the question of logic or explicit reasoning in science, we need, however, to look a little more closely at the rôle that implicit reason plays in the ideal that determines the characteristic method of science.

§ II. THE IDEAL OF SCIENCE

SCIENCE may be distinguished from ordinary common-sense knowledge by the rigour with which it subordinates all other considerations to the pursuit of the ideal of certainty, exactness, universality, and system.

(A) CERTAINTY—EVIDENCE AND PROOF

Ordinary human knowledge is full of arbitrary and conflicting opinions. These opinions are broken reeds when men rely on them to escape the pitfalls of error and illusion which mark our path with destruction. Science, it is generally recognized, is an effort to eliminate baseless opinions and to establish propositions that are supported by evidence or proof. This is commonly expressed by saying that science aims at knowledge that is certain.

The word *certain* in this connection is unfortunate because of the confusion between its logical and psychological senses. Psychologically it denotes a state of feeling, as when we say we are certain that none but those baptized by our church will go to heaven, that the country will go to ruin if there is no repression of the new-fangled heresies, or that civilization will break down unless our ancient outworn institutions are forthwith abolished. Certainty in this sense is no guarantee of truth, for others feel equally certain of the direct contrary.

So often does our psychologic certainty prevent us from even entering on the pursuit of truth, that it is well to reflect that the *feeling* of certainty is often nothing but our inability to conceive the opposite of what we happen to believe. In this sense there is no certainty as great as the initial one based on complete ignorance of countervailing considerations. Thus men show greater certainty about the complicated and elusive questions of politics and religion than about simpler and more verifiable issues to which they have devoted the prolonged study of the professional expert. The feeling of certainty or conviction is also produced in us by forms of language. Pithy proverbs, the magical utterances of poets, and the sententious remarks of sages or prophets thus generally carry conviction. But reflection finds that directly opposite views can be expressed just as impressively. Consider the absolute certainty with which men have announced that nature is all life, that the good life is the one according to nature, that the years pass swiftly, that freedom is what all men crave, that the common man must be the final judge, etc. How easy to maintain that nature is for the most part inert and dead, that the good life depends upon our controlling our natural impulses, that the years drag their weary way, that freedom is precisely what men cannot endure, and that the common man always follows what some one has suggested to him. "Look before you leap," is just as convincing as, "He who hesitates is lost."

Nor is the feeling of certainty that embodies itself in a consensus of opinion through the ages a guarantee of truth. We need only think of humanity's certainty about astrology, the existence of witches, or the impossibility of men walking on the opposite side of the earth. Neither the extent nor the intensity of the feeling of certainty with which a proposition is held guarantees its truth.

The certainty which science aims to bring about is not a psychologic feeling about a given proposition but a logical ground on which its claim to truth can be founded. Certainly, if we view truth not as simply an immediate quality of an assertion in itself, but as something which has to do with what it means or implies, doubts about the truth of a given proposition can be met only by evidence involving its relation to other propositions with which it is inextricably bound up. In any case we can well say that science aims to settle doubts or debates between contending views by showing that a given proposition is the only one logically tenable or at least that it is better founded than its suggested alternative. But science does this by the paradoxical or heroic method of questioning all things that can be questioned, and in this way it seems to destroy psychologic certainty.

Man's ability to question that which he has from childhood been taught or accustomed to accept is very limited indeed unless it is socially cultivated and trained. It is a very rare individual who can perceive things for himself and trust his own experience or reason so as to question the currently prevailing views. For the typical Mohammedan child growing up in Central Arabia, there is no effective doubt possible as to whether Allah is the true God, and Mohammed his prophet. It is only when his community ceases to be homogeneous, when he comes into contact with those who do not believe in Mohammed, that doubt can take root and begin to flourish. Thus travellers, merchant adventurers, cosmopolitan cities, and the mixing of peoples having diverse traditions, play a predominant rôle as leaven in the intellectual life of mankind.

As the state of doubt is intensely disagreeable, communities try to get rid of it in diverse ways, through ridicule, forcible suppression, and the like. The method of science seeks to conquer doubt by cultivating it and encouraging it to grow until it finds its natural limits and can go no further. Sober reflection soon shows that though very few propositions are in themselves absolutely unquestionable, the possibility of systematic truth cannot be impugned.

The method of systematic logical proof, which is the essence of mathematical and exact science, was made possible when Greeks such as Thales and Pythagoras discovered the way of deducing propositions from axioms or simple principles which seemed indubitable.[2] The logical consequences of these were applied to optics, astronomy, the mechanism of the lever, hydrostatics, etc. This faith in the self-evidence of principles that required no proof was possible so long as no one did in fact suggest possible alternatives to them. (The dogmatic scepticism of Pyrrho and Sextus Empiricus, attacking as it did the whole of knowledge, was practically inapplicable and theoretically barren.) But modern science has raised effective doubts as regards Euclid's axioms by suggesting intelligible alternatives, and this procedure has been fruitful in physics as well as in geometry.

Indeed science is generally said to have begun its distinctively modern career by doubting the clear and hallowed truth (supported by the testimony of our senses) that the sun and stars move around the earth. From this it went on to doubt and to characterize as false or meaningless such traditionally self-evident propositions as that nature abhors a vacuum, that nature makes no jumps, and even that a proper part can never be equal to the whole.

How can science, if it thus distrusts self-evident principles, prove anything? How can we attain certain knowledge if nothing is inherently certain?

---

[2] Perhaps Euclid regarded axioms as conventions: *Let it be granted and the rest necessarily follows.* But even he assumes the results of geometric proof to be true in the physical world.

The answer is that effective doubt of any proposition means that we consider that some other proposition may be true and this involves the assumption of a more complex proposition containing the two alternatives. Thus the mathematician's doubt as to Euclid's parallel postulate not only leaves untouched the demonstrative certainty that, given this postulate, the Euclidean theorems must follow, but it also leads us to a wider truth in the field of mathematics, viz. that there are several alternative possibilities as to parallelism and that logical consequences can be deduced from each of them. We have lost confidence in the absolute physical truth of Euclid's parallel postulate, but the field of possibility is widened. Euclid's geometry is thus not really overthrown. As a branch of pure geometry it remains valid, and as a physical hypothesis it is seen to need certain qualifications. Similarly does modern non-Newtonian mechanics prove not that Newtonian physics is false, but only that certain qualifications must be attached to the Newtonian propositions to make them adequate descriptions of nature.

While science is thus resolved to *question any* proposition that can be significantly doubted by showing its alternative to be possible, it is, in its assumption of the existence of truth and in its reliance on logic, the very antithesis of dogmatic scepticism, which seeks to *deny* truth of *all* propositions. Absolute universal doubt cannot be consistently carried through. We cannot, of course, refute the scepticism that is just an arbitrary resolution not to believe anything. Only assertions can be refuted. If the sceptic claims to have asserted something and not its opposite, he is assuming the laws of identity and contradiction and the ideal of logical proof.

Thus by persistently asking, "Is it so?" or, "What reason is there to believe that it is so?" science pushes doubt as far as possible and so arrives at propositions which are not so readily doubted. And here we must distinguish more explicitly between proof in pure mathematics and the proof of existential propositions. A proof in pure mathematics always shows the impossibility of any significant denial of the proposition proved. This procedure is not strictly applicable to material or existential propositions. While there are certain existential propositions (e.g. that there is being) whose denial precludes the formulation of any material system, in general the existence of facts can be rendered only more or less probable. This is done by analyzing facts into their elements and showing that these elements are identical with the elements of other facts and identically arranged, so that their denial would involve the rejection of so much that we must ordinarily take for granted that effective doubt cannot be consistently or systematically maintained. Thus it is not intellectually feasible to deny that the earth existed before we came on the scene, or that Napoleon actually lived more than a century ago.

If it is urged that when no one fact is absolutely certain, we have no ποῦ στῶ or point of support on which the whole body or system of knowledge can rest, we may point to the analogous abandonment of the search for something on which the whole earth can rest. But we can account for the stability of the earth better if we view it as moving rather than resting on something that in turn needs to rest on something else. So, though individual propositions need the support of other propositions (through evidence and proof) to find a position of stability within the system of knowledge, the whole body of knowledge needs no support. Any contention that the whole body of scientific or demonstrative knowledge is false will be found to be in the long run humanly untenable, i.e. incapable of being held consistently with other propositions that claim to be true. Science can be challenged only by some other system which is factually more inclusive and, through the demand for proof, logically more coherent. But such a system would simply be science improved. Science must always be ready to abandon any one of its conclusions, but when such overthrow is based on evidence, the logical consistency of the whole system is only strengthened.

Progress in science is thus possible because no single proposition in it is so certain that it can block the search for one better founded.

We must however be on guard against the popular notion that scientific theories succeed each other, like Oriental despotisms, by killing their predecessors and their kindred. The view in books of popular science that Copernicus did in one fell swoop overthrow the whole Ptolemaic system is altogether erroneous. Not only did Copernicus and his successors keep the substance and method of Ptolemy's work, but the newer astronomy explained the apparent motion of the sun and stars around the earth by the analogous experience of the apparent motion of the whole earthly scene when we are in a carriage or boat. Science "saves appearances" by explaining them according to universal law.

It is true that, according to the theory of relativity, the statement "the sun moves around the earth" is in a sense as true as the statement "the earth moves around the sun." But this does not deny the real advance involved in the Copernican revolution. For the Copernican system not only explained everything that the Ptolemaic did, but enabled us to formulate laws of physics which the Ptolemaic system did not, and these enabled us to widen our horizon by discovering new phenomena. In fine, by always asking, "Is it so?" scientific method widens our range of vision and eliminates that logical uncertainty or inconclusiveness of common sense which leads to sectarian diversity of opinion.

(b) ACCURACY AND MEASUREMENT

Science aims at greater exactness than that which characterizes ordinary common sense.

The business of critically analyzing the exact meaning of our propositions and of determining the precise extent to which they are true is characteristic of science, and attempts to introduce it into ordinary conversation are apt to be resented as pedantry, quibbling, "finessing," etc.

Socrates made himself thoroughly disliked by showing people how vague were their ideas on *courage, justice, the good,* and other terms of popular use. Common usage gives these words vague hosts of meanings, and it requires patient work to discriminate them, as Aristotle laboured to do. It is sometimes said that people know well enough what they mean, but cannot think of an exact formulation. But the primitive vagueness of men's apprehension of the most familiar objects is seen in the drawings that children and other untrained observers make of them, e.g. a human profile with a full eye or a glass of water with a straight line or full circle to represent the base line. Obviously different and inconsistent appearances of the object are combined in the ordinary image of it, so that accurate observation generally comes only after training. The same is true in the realm of ideas. Hazy ideas and inaccurate knowledge are at the basis of the inconsistency which makes us praise an act as brave or generous at one time and condemn the same act as rash or foolish at another time. It is the lack of definiteness in our knowledge which makes us say at one time that the French are fickle, and at another that they are too logical, that the Americans are too materialistic and too idealistic, the Germans too sentimental and too hard, etc. In some sense and in some measure these opposites may be true,—but in what sense and in what measure common sense does not tell us.

Yet popular thought cannot remain satisfied with its vagueness. We may be satisfied when told that a stranger could not come to see us because he was sick. But it would be absurd for a doctor to tell us that our child is bed-ridden because of sickness. We want more definite information. It is not enough to know that a room or a boat will hold a great many people,—we want to know how many. It is not enough to know that our friend or business correspondent is a long distance away. We want to know how many miles, or how many hours by train, etc. Thus the practical demands of civilized life often reinforce the scientific craving for the accuracy that led the Greeks to the discovery of mathematics.

Workers in the exact sciences often assert that where there is no exact enumeration or measurement there is no science. As the refusal to apply the term *science* to one's work is often felt as a disparagement of it, this dictum is violently resented by those who deal with such psychologic and social problems as do not readily allow the use of exact measurement. We may sympathize with these protests. We cannot refuse the name *science* to logic or to the non-quantitative branches of mathematics such as analysis situs, projective geometry, etc. Nor is there good reason for refusing the adjective *scientific* to such works as Aristotle's *Politics* or Spinoza's *Ethics* and applying it to statistical "investigations" or "researches" that do not advance the understanding of anything. But there can be no doubt that it is of the essence of scientific method that vague terms like *large* and *small, far* and *near, hot* and *cold*, etc., shall be replaced by terms made definite by measurement.[3]

To justify their claims to be scientific, psychiatry and psychoanalysis for instance must determine exactly what it is that their patients are suffering, what is meant by a cure, in what proportion of cases certain treatments are more effective than time alone would have been, and in what proportion of cases such treatments have harmful effects subsequently. All this is, of course, difficult to find out, and the practical

[3] Perhaps it would be more accurate to say that what is really required in this respect is that all comparisons be put in terms of mathematically serial relations and that all ratios be expressed numerically,—so that in so far as any science does not deal with comparisons or ratios the ideals of enumeration and measurement are inapplicable.

demand to be "cured" must be met by more intuitional methods (in plain words, happy guesses) of those who have acquired skill by practice. But this practical necessity of acting on inadequate information does not make science out of the psychiatrist's or psychoanalyst's ideologies, no matter how technical their vocabulary.

The appropriate means for attaining definiteness vary in different fields. But two devices deserve special attention, and these are: (1) enumeration, often elaborated in the form of statistics, and (2) measurement, by which relations are numerically expressed.

(1) ENUMERATION. Things may be numbered merely to indicate their order, as when we number the pages of a book, the houses of a street, and the like. For this purpose we can use some other set of symbols of which the order is easily remembered. The letters of the alphabet are often so used. (Among the Phoenicians, Hebrews, and Greeks, the letters were indeed used as numerals.) But our Arabic numerals, being fewer than the letters of the alphabet and more readily combined, make it easier to assign a symbol whenever we interpolate a new member into our series of things ordered.

Though the assignment of numerals or other symbols to a series of things is primarily a matter of convenience of memory, it is a question of truth (and therefore of science) whether the things symbolized do have the same order as the symbols, e.g. whether the books succeed each other on the shelves according to the numbers assigned to them in our catalogue. It is one of the many great merits of Russell's *Principles of Mathematics* to have shown that both enumeration and measurement involve the ordering of a series.

Simple enumeration may reveal to us significant empirical natural relations, e.g. that 105 boys are born alive to 100 girls. But why this should be so, and what reason we have to believe that it is not merely temporary but a relatively permanent natural relation depends on rational evidence such as the physiologic causes of the greater pre-natal mortality of girls.

It would take us far afield to discuss here the logical character of statistical method. But a number of obvious points must be mentioned concerning its relation to accuracy.

(i) The exact enumeration and formulation of phenomena is frequently the only way of overcoming false popular impressions. Instead

90

of discussing how education decreases crime, or why nephews resemble their maternal uncles more than their paternal ones, it is safer first to inquire to what extent these assumed facts are so if they are so at all.

(ii) The elaborate mathematical methods of statistics may give us significant figures as to group characteristics such as distribution; but do not directly increase the accuracy of our knowledge of the individual phenomena.

(iii) Statistical computation will not remove the initial inaccuracy of our data. If incomes or the value of imports are inaccurately reported, their mathematical formulation in the form of averages, standard deviation, quartiles, etc., will not remove the initial inaccuracy, though it may help (through comparison with other and more reliable information) to make us see that there must be some error in our basic figures. We are likely, on the other hand, to be misled into thinking that changes are taking place in the phenomena studied when there has been some unnoticed change in the way we gather our statistical information.

(iv) When we deal with indefinitely large groups and formulate our correlations on the basis of a number of typical or representative instances, no amount of care to see that our samples are taken widely and at random will guarantee us against the fallacy of selection, i.e. against the possibility of our selected instances having some property which makes us pick them out but which is not characteristic of the group as a whole. To say that we have selected our instances at random is only to assert that we have not been conscious of exercising a selective influence, but the history of statistical inquiry shows how infrequently statistical workers escape the influence of unconscious selection. In studying racial "intelligence," for instance, among Chinese, Jews, or Germans in America, statistical workers seldom take into account the really determining local or social factors.

(v) High correlation between two different phenomena or factors does not always mean that they are directly or permanently connected. High rates of literacy and high rates of criminality may be closely correlated, but neither may be the cause of the other,—both being rather caused by some such factor as predominantly urban life. But sometimes a very high correlation (say of over 85 per cent.) may be entirely accidental, i.e. we can find no reason why the two factors should be correlated at all. Thus for a number of years the membership in the International Association of Machinists shows a very high correlation (86 per cent.) with the death rate in the state of Hyderabad.[4] If instances of this sort do not come to our attention more often it is because we do not look for them. We generally look for correlations where we have some reason to suppose that there is a real connection.

In general a statistical correlation even of 100 per cent. does not prove any causal relation between the phenomena or factors correlated, because (as indicated before, apropos of simple enumeration) we need evidence to support the belief that this correlation is permanent and not temporary. Statistical information obviously cannot prove anything beyond the time of observation. To prove the invariant relations implied in laws of nature we need in addition rational considerations regarding relations of identity. These relations are generally revealed by analysis.

(2) MEASUREMENT. Measurement introduces definiteness into our knowledge of phenomena by enabling us to order them in series which can be universally correlated. We can see this in a form of measurement so rudimentary that many refuse to call it measurement at all, viz. the determination of the hardness of objects. Things vary in hardness, and when any two different substances are rubbed against each other, we call that one the harder which makes a scratch on the other. Experience confirms the fact that this relation is asymmetric and transitive (i.e. if a scratches b, b will not scratch a: and if a scratches b and b scratches c, a will scratch c). If we assign some numeral or order-symbol to the softest known substance, we can then assign numerals to all other substances in order of hardness. If any substance can be assigned a definite position in this numerical scale, we say that its hardness has been measured.

In general we cannot be said to have measured anything by merely assigning an arbitrary numeral to it. We cannot, for instance, be said to have measured beauty unless we can correlate all instances of beauty by some definite tests, according to which any two instances of beauty are either equal or such that one exceeds the other, as one individual is taller or heavier than another.

The need to pay attention to these difficulties is seen in the uncritical

---

[4] From an unpublished study by Dr. George Marshall, at the Brookings Institute. Variations in the Hyderabad death rate from 1911 to 1919 are correlated with variations in the membership of the machinists' union from 1912 to 1920.

way in which many educators speak of measuring "intelligence." People are asked to answer a number of questions, and their intelligences are ranked according to the relative number of correct answers. But the nub of the question is, "What is thus measured?" Will the order of people's intelligence thus found prevail if another set of questions is asked? General experience shows variation in this respect. Some who do well in one kind of examination do poorly in another, and even those who generally score high in examinations may not show very high intelligence in other respects. The best that can be said for these tests as measures of general intelligence is that there is a high correlation coefficient between the ability to answer a certain set of questions and the ability to perform all other tasks of intelligence. But a correlation coefficient is something which applies to groups of figures, not to any individual; and this means that while intelligence tests may be socially convenient in dealing with groups they do not measure the intelligence of the individual with any high degree of accuracy, and there is too much arrogance in condemning certain individuals on the basis of such tests as permanently of low intelligence.

The case for the social value of intelligence tests is of course stronger if, instead of pretending that our "I.Q." measures general intelligence, we seek to devise tests which we have reason to believe will have a high correlation with special aptitudes required in given fields.

We can see more clearly what is involved in measurement if we distinguish between intensive and extensive magnitudes.

(i) *Intensive Magnitudes.* Intensive magnitudes, of which temperature is a good example, can be arranged in a recognizable order, but cannot be added.

The temperature of a body is called an intensive magnitude because any two bodies can be compared as to their temperatures and there is a definite test according to which the two are equal or one is greater than the other. Originally men determined the intensity of heat or cold by bodily feelings. But these are poor measures. Their range is narrow. It is practically impossible thus to determine the temperature of fire or of liquid hydrogen. Our sense organs are highly variable. That which to one hand seems to maintain a constant temperature will appear colder or warmer to the other hand if the latter's contact is interrupted by touching a warmer or colder body. Our bodily organs also fail to discriminate many different temperatures, and express this in a poor scale of objective symbols, such as *hot, tepid, cold,* modified by such vague words as *very, mildly,* etc. But above all is it difficult by the aid of our sense organs alone (aided by our poor memory of sensation) to compare as to their temperatures different bodies at different places and times. Greater discrimination and universal correlation are obtained by a thermometer based on the law that heat expands certain bodies more than others. We correlate all thermometers with each other by marking on them two points, the boiling point of water and the melting point of ice at atmospheric pressure, and then dividing the distance between these points into a convenient number of equal parts. We can then correlate the temperature of all bodies by applying to them any standardized thermometer.

Simple and reasonable as this procedure is, it is not free from theoretic difficulties, which it is instructive to note.

We assume that any two substances that mark the same height of the mercury column in the thermometer are of the same temperature. This is not at all necessary a priori, any more than that two persons each at peace with a third should be at peace with each other. It will not do to say that it is merely by definition that any two substances that produce the same reading in a mercury thermometer are called equal in temperature. This view will not explain why we are so careful to pick the melting point of distilled-water ice rather than that of ice cream or tallow candles as the lower base line of our thermometer. There is obviously some factual reason for the choice, i.e. a belief in the factual uniformity of the temperature of melting ice at atmospheric pressure, independently of the mercury thermometer which is constructed on the basis of it.

Nor do we escape difficulty if we seek for purely empirical evidence for the statement that any two substances that register the same reading on a thermometer have the same temperature. There is something clearly circular in every effort to prove that variations of temperature produce variations of volume, if we measure temperature variations by a thermometer constructed on the principle we wish to prove.

We can evade this and similar difficulties by looking at the situation historically. Before the construction of thermometers we knew by direct sense perception that with increased intensity of heat (as felt by the hand) certain substances, such as mercury or alcohol, increased in volume relatively to the glass vessels that contained them. This knowledge was extended and made more accurate and systematic by the invention of the thermometer, just as our knowledge of electricity was similarly furthered by the invention of galvanometers. Science in general does not begin with precise measurement any more than with clear and exact ideas. It begins rather with hazy ideas and inexact measurements, but greater accuracy is introduced and indeed made possible by the ideal of scientific system.

(ii) *Extensive Magnitudes.* Extensive magnitudes are generally defined as those which can be added. We may also say that they can be subtracted or multiplied. Thus the difference between 9 lb. and 5 lb. is 4 lb., which is itself an extensive magnitude, exactly like the difference between 21 lb. and 17 lb. and an infinite number of other pairs. But the difference between an intensity of heat indicated by 99° C. and that indicated by 98° is not itself an intensity of heat, and is unlike the difference between 97° and 98°. Similarly is the difference between the student who gets 100 per cent. and the one who gets 80 per cent. altogether different from that between those who get 70 per cent. and those who get 50 per cent. So also the fact that extensive magnitudes can be added, i.e. can be made up of successive units, means that one such magnitude can be a multiple of the other. This is not so with intensive magnitudes. While one temperature may be higher than another, there is no assignable meaning in saying that one temperature is three times another. So if pleasure is an intensive magnitude, and the difference between two pleasures is not itself a pleasure, there is no sense in speaking of one pleasure as three times another.

What characterizes length, weight, and other extensive magnitudes that makes them subject to addition?

In addition to some test whereby we determine that any two such magnitudes are either equal or such that one exceeds the other, the measurement of extensive magnitudes requires the existence of a standard unit, and such a process of combining such magnitudes as has the formal properties of purely mathematical addition, viz. the commutative and associative laws. Thus we add lengths if we lay off one immediately after the other along the same line. We add weights if we put them together in one scale, or at the end of a balance, and so forth. The results of such combination are subject to the commutative law if they are the same irrespective of the order in which the lengths, weights, etc., are put, and they are subject to the associative law if combining *a* and *b* with *c* gives the same result as when to *a* we add the combination of *b* and *c*. In the case of lengths and weights these laws are so familiar that we regard them as self-evident. There are, however, many processes in nature of which these laws are not true. Thus if we turn a book through an angle of 90° around each of its three edges, the order in which these three operations follow each other will make a difference in the final position of the book. The physical result when the wall of a fort is hit by two shells is not the same as when the two shells are combined into one. Chemistry and biology are full of illustrations of the difference between the effects of things taken separately and their effect when combined.

These very brief indications of the conditions that must be fulfilled in the measurement of extensive magnitudes may be sufficient to point to the necessary connection between such measurement and theoretic or systematic reason in the exact natural sciences. The important rôle of theory in the process of measurement comes clearly to light when we realize that most of our measurements are indirect,—i.e. to measure a thing we observe something else which is theoretically connected with it. Thus electric measurements depend upon galvanometers, where certain lines and angles indicate volts. So time is measured by certain lines and angles on the face of a clock or chronometer. And when we come to measurement in chemistry, biology, psychology, and the social sciences, it becomes obvious that what we observe directly in such measurements are physical arrangements and changes of which the sciences named supply the appropriate interpretation. Consider the measurement that is basic for modern physics, the determination of the velocity of light as 186,000 miles per second. Obviously we cannot determine this by direct observation. Indeed, light itself is not visible at all. It is only objects that become visible. But certain variations in the time of the appearance of the satellites of Jupiter can best be explained by supposing that it takes a second before a light signal from any object can reach an object 186,000 miles away. This assumption, to be sure, is reinforced by other considerations, and is arrived at also by other measurements, but these other considerations are also theoretical. The same is true if we measure a wave length of light or the electric charge in the electron. If the wave theory of light be false, there is no wave length, and the electron itself is not directly observable. The statement that its charge is a certain fraction of an electrostatic unit of electricity is one that follows from a number of theoretic assumptions, of which the electron theory is certainly one.

The simplest direct measurements depend upon theoretic assumptions. All measurements, for instance, assume that any two magnitudes that are equal to the standard unit are equal to each other. This, however, is not always true in experience. Thus, it frequently happens that two different lengths, a and c, that are each indistinguishable from a third, b, are visibly different from each other. We explain this by saying that the three lengths are *really* different, but that the differences between a and b, and between b and c, are each less than the minimum visible, while the sum of the two differences is greater than this minimum and therefore is visible. This explanation agrees, of course, with further experience, but other explanations might do the same. The real base of this explanation, however, is the assumption that length grows continuously and not by discrete appreciable increments.

Again, it is well known that when measurements are repeated with great care there is always a variation in the results. The height of a given building, the atomic weight of a given element, or the altitude of a star are, however, supposed to remain constant during the repeated measurements. Men of science have got into the habit of using various methods of computing what they agree to call the correct result, of which the arithmetic average of the different readings recorded by our instruments is the simplest. But this correct reading seldom if ever coincides with those observed. Why should we not, then, say that the object itself is variable, rather than that it is our experiential measurements that are in error? Such a statement would be perfectly in harmony with the philosophy of experience and the view that we know nothing but phenomena or appearances.

Such, however, has not been the method of classical science (since Galileo, Kepler, and Newton revived the Pythagorean conception of the book of nature as written in mathematical language). If what we measured were not itself constant, there would be no sense in the search for invariant laws, which is the essence of exact natural science. The assumption that the atomic weight of the element or the altitude of the star measured is a constant gives meaning to the search for the causes of the variation in the readings of our instrument. Each reading is thus a phenomenon regarded as the resultant of a number of causes, of which the object itself is only one.

The nature of our instruments, the mode of applying them, the movements of our sense-organs, are among the other causes. Thus in measuring the altitude of the star I find that a good deal of the variation is explained by the variation of the refraction of the atmosphere. In determining weight or the force of gravity all sorts of physical disturbances must be noted and explained in terms of physical laws. So in measuring length we must note the variability of our instruments according to temperature, as well as the variability of our sense perceptions and of our ways of handling our instruments.[5]

This search for laws of nature as invariant relations rather than statistical correlations is at the basis of the faith that where our readings persistently show variations which cannot be explained as within the "error" of our instruments, the variations must be due to the fact that what is measured is not homogeneous. Thus the variation in the atomic weight of nitrogen led Rayleigh to the search for and discovery of another element in the air (argon) mixed with the nitrogen. Recently the discovery of isotopes by Aston has similarly been stimulated by the presence of otherwise inexplicable variations in atomic weights. In general, mere statistical correlation of the results of measurement never satisfies the spirit of scientific research. We must analyze the variations, and such an analysis assumes some element of identity as the basis of the universal connections to which empirical results approximate.

It is true that with improved instruments and improved methods we minimize the "error" or variability and thus increase the accuracy of our results. The errors or variations in the measurement of length are minimized if, instead of relying on the old-fashioned yardstick (which generally limits our accuracy to sixteenths of an inch, and makes all sorts of errors possible in the marks between its successive positions), we rely rather on more elaborate instruments with mirrors, microscopes,

---

[5] The preference for certain kinds of readings illustrates an application of the theory of knowledge to actual scientific work, of which other examples will be given later. This recognition of the rôle of epistemology is perfectly consistent with the view that epistemology is not a prerequisite for the whole of metaphysics, of which it is rather a branch.

etc. But obviously the more elaborate our instruments, the more theoretic assumptions are involved. Indeed what way have we of telling that one set of measurements is more accurate than another, except that the results of the former are more in agreement with theoretic requirements or agree closer with our assumed laws of nature? Certainly all the technical methods of securing greater accuracy or precision ultimately depend upon the assumption that those kinds of measurements which show the least variation and the greatest approach to the constancy of scientific laws as the ideal limit are the most accurate. We can do this better in the physical sciences because we can multiply instances and manipulate our instruments more freely. But ultimately all science goes back to the classical conception of nature, according to which the variation of phenomena is to be referred to some unitary law.

(c) ABSTRACT UNIVERSALITY AND NECESSITY

Common everyday thought generally arises as incidental to our efforts at achieving our specific practical or vital tasks. Scientific thought, on the other hand, has its main roots in that wonder or disinterested curiosity which makes children annoy their parents by continually asking, "Why?" or, "What made it so?" The curiosity which is not atrophied in us by our daily routine leads to gossip, which, when cultivated, rationalized, and dignified, becomes an interest in news, in travelogues, in descriptions of how other people live, and in history generally.

If patient weighing of evidence makes any intellectual work scientific, history is often highly scientific. But there is an important distinction between interest in history or what has actually taken place and the interest in something more abstract and universal which is typical of science, or at least of the mathematical and developed natural sciences. On this point people have been confused by the fact that the history of our solar system, of our terrestrial globe, and of life on it is generally conceived as part of astronomy, geology, and biology. This, however, only means that these histories are written by those who deliberately study the laws of physics or biology which are applied in determining the facts of sidereal, terrestrial, or biologic history. There is no reason for denying the clear logical distinction between the history of events which have occurred in the past and the general laws of science which can be exemplified in an unlimited number of possible instances.

Whether this globe of ours did once support certain reptiles, whether certain rocks were once deposited at the bottom of the sea, are questions of history as much as the question whether there was an actual man Moses who led the Israelites out of Egypt in the fifteenth century B.C. This holds true of extensive or collective facts. Thus the assertion that all known examples of sulphur *have* melted at 125° C. belongs to history as much as that the Russian people have always used the Cyrillic alphabet. But the propositions of general or theoretic science cannot be confined to the past. They must be of the eternal present. Sulphur always melts at 125° C., and this means that if ever anything conforms to the category of sulphur (defined, let us say, by its atomic weight), it will also melt at 125° C. We may in fact be confident that what has always happened will continue in the future. But there can be no history of the future. A future event may show the falsity of an alleged universal law, but it cannot directly contradict any strictly historical statement as to what happened in the past.[6]

This distinction between science and history is often expressed by saying that history deals with facts, while science deals with laws. Unfortunately we also habitually use the word *fact* to denote certain general truths. That gold is yellow, that copper conducts electricity, and that mammals have lungs, are usually referred to as facts. But if *law* denotes uniform or invariable conjunction, these statements are obviously laws, stating that if anything has the properties which define gold, copper, or mammals, it will also have the other properties named in the predicates of our propositions.

Why do we not usually characterize such statements as laws? It is because we usually think of gold and the like as concrete existing things that just happen to have given properties, while the laws of developed sciences such as mechanics, optics, thermodynamics, etc., are concerned with the relations between possible or ideal things or elements such as perfectly rigid bodies, geometric rays of light travelling through perfectly continuous media, frictionless engines, etc. It may be true that

---

[6] Of course the future discovery of some historical monument may contradict our present belief as to what happened. But the monument will testify as to the time of its creation, not that of its discovery.

most of us might not be interested in these ideal relations if they did not throw light (as they do) on the actual world. But leaving aside for the moment the problem of how statements about free bodies, frictionless engines, and other actually non-existing things can give us knowledge about the existing world, there can be no doubt that the more developed a science is, the more are its laws formulated in terms of such abstract or ideal elements. Thus meteorology, despite countless millions of observations, has not yet become as developed a science as crystallography, with its ideal geometric relations, or the more recent science of radiant energy with its quanta of "action" which are formulated in terms of mathematical functions.

The obvious truth on this subject is obscured by the positivistic dogma (made popular among certain men of science by Karl Pearson) that the laws of nature are mere descriptions of the routine of our perception or habitual sequence of our sensations. Whatever may be the historic or psychologic origin of scientific laws like Newton's laws of mechanics, or Maxwell's laws of the electro-magnetic field, or Planck's quantum law, they certainly do not describe the order of our perceptions or the sequence of our sensations. For the laws are stated in terms of abstract elements like mass, time and space intervals, electric permeability, etc., which are not as such sensations. Nor do these laws assert any temporal sequence. They assert rather a mutual implication between the parts of an equation, though the elements of the equation refer to what is in time. Indeed, even what we have called factual laws are concerned with abstract elements. For obviously no fact can repeat itself in all its concrete fullness. Caesar may cross the Rubicon several times, but the circumstances are never altogether identical. The date is different each time, and that means that Caesar and others are older, and that the world is no longer the same. So it may be maintained that no amoeba repeats the identical act of cell division, and that the successive falls of a stone, dropped for purposes of physical experiment, are individually different events. It is, however, equally obvious that there is no sense in speaking of repetition unless the events repeated are in some respect identical. If the identical stone (or any other of the same volume) is repeatedly dropped, its (abstract) velocity is the same. So there are elements of identity which make us call events in different organisms the cell divisions of an amoeba. We can therefore say that in all the repetitions of an event or fact certain features or abstract elements remain the same while others change.

We may, of course, view events with different degrees of abstractness. What to the psychologist is a case of a man driven to commit suicide by jumping overboard is a case of a dying organism to a biologist, or of a falling body to a physicist. But psychologist, biologist, and physicist all view only certain elements of the total happening. To connect elements in laws according to some logical or mathematical pattern is the ultimate ideal of science.

Science, it is generally agreed, is rational or unified knowledge. It is not a mere catalogue of what has happened (which would be history), but seeks to explain why things happen in the particular way they do and not in some other way. If we notice that white cats have defective vision, and confirm this by many observations, we have made only the beginning of a scientific account of the matter. We must explain the fact, and we do so by reference to the general deficiency of the colouring matter in the retina of the eye as well as in the fur. Such an explanation puts us on the road to connect vision with the more general facts or laws of photo-chemistry, the ways in which light produces chemical changes that affect the colour of objects. So likewise we are not satisfied with merely recording that copper has certain chemical properties, is a good conductor of electricity, etc. Science is always seeking to show why it has these properties, by trying to show that they can be deduced and that they thus necessarily follow from certain more primary (generally more simple) laws. These more primary laws may themselves be just general happenings and it may be to some extent arbitrary which laws are taken as primary and which are deduced; but in any case the logical connection between these different laws shows elements of substantial identity between different facts; and this element of identity makes the connection necessary, and distinguishes the laws of science from empirical uniformities of succession.

Common sense is vitally interested in such uniform sequences of events as that a spark will explode powder, that plants will grow if watered and allowed access to sunlight, that children's teeth will decay if they eat too much sugar, or that the heart-beat will be stimulated by exercise. These may be taken as instances of the traditional wisdom by which men live.

Science, however, inquires not only to what extent such propositions are true, but also why they are true. It founds our assurance of a real connection not merely on the fact that such sequences have been observed, but on an analysis which shows elements of identity between antecedent and consequent. Thus, the energy of the explosion is the same energy which existed in the powder in an unstable equilibrium. Water forms a large part of the circulatory tissue of the growing plants; the sunlight is the energy which enters into the transformation of inorganic into organic compounds affected by the chlorophyl cells. Similarly we shall explain disintegration of the enamel of the teeth by sugar if we can show some chemical process between them, and we shall satisfactorily explain the stimulation of the heart-beat by physical exercise if we can show the threads of identity which connect the movement of the arms with the heart-beat.

It may be urged that our analysis does not eliminate all contingency from the causal relation, that the examples cited are explained either by showing that a given case is but an example of a wider factual succession or by finding intermediate steps which are also ultimately only contingent successions. This objection, we shall have occasion to note, is perfectly sound metaphysically, i.e. we do not thus remove all contingency from the total body of true propositions. But we do by scientific search into identity of substance or process introduce a relative necessity into the system of propositions which constitute our special science. This will enable us to see the elements of truth and error in the Humian analysis of the causal relation as a mere sequence of impressions.

Hume is right in holding that to the extent that the impression of the antecedent and the impression of the consequent are different there is no real connection or necessary relation (other than psychologic habit) whereby one can be deduced or analytically found in the other. In the subjective impression of powder and spark there is not found any impression of the explosion. Nor could we legitimately find it without the aid of actual experience, since the element of contingency in the factual or temporal succession could not possibly be deduced from purely logical considerations. But the fact that antecedent and consequent are different does not prevent them from also having elements of identity. When we find such elements of identity we have real connection and a basis for necessary relations.[7]

This interest of science in formulating abstract universal laws explains why as a science develops it drops the popular notion of cause and effect (which arises in a legal and anthropomorphic world) and seeks for a mathematical formulation of invariant relations, from which the numerical results of measurement can be deduced. Instead of a causal law to the effect that differences of potential produce differences of current, or differences of current produce differences of potential, physics formulates a non-temporal law of proportionality from which both statements can be deduced. Common sense knows that if a stone is hurled at a glass window it will (generally) break it. Science substitutes the more accurate, abstract, and universal formula for the strength of the glass, and for the force necessary to cause a break. Such a formula becomes connected with and a part of the laws for the strength of materials generally. For the common-sense statement that an unsupported body will fall to the earth (which is not always true), science substitutes the more universal formula of gravitation in terms of masses, distances, and time-intervals, which can always be applied to terrestrial and celestial objects at rest and in motion.

In thus deliberately sacrificing the concrete fullness of ordinary or "experiential" descriptions for the sake of abstract universality, science opens itself to the taunt that it is artificial. This is in a sense true. But only blind adherence to the empiricist dogma can draw the conclusion from this that scientific laws are not true but only useful fictions. There is no good reason or evidence for the view that the categories of physics falsify nature more than the more familiar categories of common sense or daily experience, which are demonstrably less exact.

Let us look a little more closely at the seemingly paradoxical fact that the most powerful organon for the apprehension and controlling of nature that man has as yet discovered is description in terms of ideal entities such as perfect levers, ideal gases, perfectly continuous bodies, the velocity of light in an unattainable perfect vacuum, and the like.

---

[7] Though Hume tried to bring the method of natural science into moral issues, his preoccupation with history and social affairs gave him little genuine insight into the actual methods of physics as developed by Galileo, Kepler, and Newton. His examples of causation are drawn from common practical experience, and like his followers he never caught the full significance of mathematical physics.

For explanation of this paradox we must fall back on the fact that while the so-called perceptual order of common-sense experience is the source of scientific truth it is also full of illusion and error, so that the conceptual order which science seeks to attain must depart from it to attain coherent truth. In our ordinary perception we do not always discriminate what is relevant from what is irrelevant. Things which are contiguous in time or space frequently have no direct bearing on each other's natures, and our practical interests may lead us to associate things between which there is no ontologically significant connection. Rational scientific search for real connection is thus a search for the invariant relations which constitute the nature or character of things in the variation or flux of temporal experience. Physical bodies, for instance, are never in actuality perfectly rigid; but neither are they ever completely devoid of all resistance to the mechanical pressure that would deform them. Rational analysis enables us to abstract the rigidity and to consider it separately. In this we are helped by the fact that actual substances may be arranged in a series of increasing or decreasing rigidity, so that the limiting conception of perfect rigidity is at the basis of the order in the series. Similarly do we find the ideal of frictionless motion involved in any effort to arrange the actual motions of bodies in a series of decreasing friction. So we may view the perfect vacuum, the ideal gas, and the rest as limits which condition an order in which actual substances are related in respect to certain traits, just as we may say that justice is the ideal to which some men approach more than others, and also that it is an actual trait in which existing men differ. In some fields, notably in the inorganic realm, we are helped by the fact that phenomena can be repeated indefinitely with the variation of only one factor. Thus we can analyze the factors which determine the motion of a body by independently measuring the gravitation, atmospheric resistance, the force of the initial impulse, etc. But every abstract or universal law asserts what would happen if only certain conditions prevailed and everything else remained indifferent. Prediction is possible to the extent that nature does offer us instances where the action of bodies can be accounted for by a limited number of factors, and the effects of all other influences either balance each other or are so small as to be negligible or unnoticeable. But theoretically it is true that no actual phenomenon can exclusively embody a single universal law, since in general every actual phenomenon is the meeting place or intersection of many laws. Yet every true law is actually embodied in all instances of it, and it is that which enables us to analyze phenomena and arrange them in significant order. But while abstract laws are always necessary for the understanding of phenomena, their sufficiency varies in different fields. In astronomy, Newton's laws of motion will explain pretty nearly all the actual motions. The explanation of actual terrestrial motion requires additional factors such as atmospheric resistance, etc. An even greater number of additional factors must be introduced to explain fully the motions of biologic organisms. But general interest in finding or inventing explanations also increases in this realm so that the number of proposed laws is out of all proportion to those which can in any way be verified. In verifying a law not only must we (1) deduce or explain all relevant phenomena, but (2) our explanation must have some advantage over rival explanations either in involving fewer hypothetical elements or in being characterized by what Hertz called "greater appropriateness." [8] The most effective verification of a law is the prediction of a new phenomenon that ought not to take place on the assumption of any other known law.

In any case the ideal of science is not only to find laws, but also to see that these laws are genuine universals. Science thus leads us to challenge all generalizations or abstractions and offers us a protection against hasty generalization and established superstition.

(D) SYSTEM—CONNECTEDNESS, COMPLETENESS, AND LOGICAL ORDER

Ordinary, pre-scientific, or common-sense knowledge is disconnected, fragmentary, and chaotic or illogical. Science is devoted to the ideal of system, in which these defects are to be overcome. Indeed, instead of saying that system is a characteristic of science like certainty, evidence, and proof, definiteness and accuracy, or abstract universality and necessity, we may well maintain that the one essential trait of developed science is system, and that all these other traits are incidental to it.

When we prove or give evidence for a proposition we connect it with other propositions according to some logical or rational order so that the various propositions support each other; when we make a statement

[8] Hertz, *Theoretic Mechanics*, Introduction.

definite or accurate we make it fit to enter a logical system as a premise from which precise deductions can be made; and abstract universality is necessary to give us a system which can attain a certain degree of coherent completeness.

Let us then examine what is involved in the ideal of rational system.

The first trait of a system is the connectedness of its parts. The connectedness which science seeks is more than an economic, mnemonic, short-hand, or any other convenient way of simply bringing propositions together for our attention. Scientific system seeks some intrinsic connection in the subject-matter of our science. This of course can be achieved in various degrees. Even the most external collection such as a random heap of things must have some common nature such, for instance, as occupying space together, and being (if sufficiently distinct) enumerable. We go further when we classify our objects, for all those that are of a given kind have a certain identical quality. Classification, however, is generally only incidental or auxiliary to the effort to attain greater scientific knowledge. We get nearer to the sort of connectedness that science aims at when we ask for the *significance* of a given fact or law. We decipher the meaning of a word or passage in a hitherto unknown language by its relation to its context; we determine the significance of a fossil by reference to a possible environment and an organism of which it could have been a part; and the meaning of certain irregularities in the motion of the moon is found in the slowing up of the motion of the earth around its axis. In all these cases we assume that the facts of nature are so connected that the intelligible character of anything must involve its relations to certain other (not all) facts, which are said to be relevant to it.[9] Science, then, views facts not as isolated or separate events, but as connected in essence.

It may be urged that a group of propositions which are mutually connected and consistent with each other need not be true unless they are also verified by experience. But this can only mean that our group

[9] The fact that for science the explorable nature of anything consists in its relations to other things—that all we seek to know of the electron is its interaction with other electrons—and that physics is satisfied to regard heat, sounds, and colours as all waves of different frequency, has led some to the view that modern science has got rid of substance and matter by substituting relational functions and matterless events. This view has aided "idealists" and theologians, but is inexpugnably absurd to common sense. Physical science certainly does not deny the existence of substance or matter, though it restricts itself to the study of phenomenal relations.

of propositions must be consistent with those propositions which we regard as warranted by observation and experiment. To claim truth, therefore, a group of propositions must not only be internally connected, but must include *all* the relevant propositions which the process of verification demands. Completeness is thus another essential trait of scientific system.

Scientific system is not attained by merely adding facts. Meteorology, despite the overwhelming bulk of its observation, is not a developed scientific system. We need some guiding principle to explore and take account of all possibilities and to introduce order into the chaos of unconnected facts. Thus the number of different crystals in nature seems bewilderingly large. But if in examining the way crystals grow, we take into account a few relevant physical and mathematical considerations as to the relations between axes and faces, we obtain a fairly exhaustive classification of the various physically possible forms of crystals. Analysis suggests possible elements to fill in theoretic gaps. Thus when the chemical elements were first arranged in a rational order by means of Mendeléyev's law, there were found theoretically deduced places not filled by actual elements. This led to the search and discovery of actual elements having the theoretically predicted properties. As actual knowledge is always fragmentary the demand for completeness as well as connection thus leads to the progress of science through the interpolation of hypothetical entities the actual existence of which may subsequently be verified. Thus conceptions of atoms and molecules are framed to explain the fact that the ratios according to which chemical elements combine are always integral numbers, but the development of atomic theory has led to research in diverse other fields of physics.

We may look at the question of completeness from a different point of view if we recognize with Peirce that the meaning of a proposition consists of its possible deductions or consequences. From this point of view the axioms of Euclid, the Newtonian laws of motion, Gibbs' phase rule, or the propositions of the Constitution of the United States, together with the rules of logical inference, contain a formally complete statement of the subject matter of geometry, mechanics, a certain branch of physical chemistry, or American constitutional law. By *formally complete* we mean that all the possible propositions can be derived from these axioms without the aid of further assumptions. This, however, does not exclude indefinite progress in *discovering* novel propositions (i.e. those that have never been thought of before but are recognized as involved in our working premises once they are discovered).

It may be objected that if the discovery of new truths were to depend, as we have contended, upon previous knowledge, no genuinely new truth could be discovered to correct the old. This objection, however, only illustrates the difficulties into which we get if we ignore the polar view and regard any difference between propositions as necessarily excluding all identity between them. But obviously the various truths expressed in the different propositions of any given science must be internally connected by the identity of their subject matter. Otherwise they would not deal with the same subject and we should not recognize them as belonging to the same science.

To hold fast to the element of identity amidst all variations requires the aid of logic. That is why every science aims to attain the form of a logical system, after the model of the most ancient of the developed sciences, viz. geometry. Dynamics after Newton and Lagrange, celestial mechanics after Laplace, thermo-dynamics after Fourier and Clausius, electro-magnetism after Ampère and Maxwell, genetics after Mendel, and certain parts of economics, have all more or less achieved that end, i.e. they are able to systematize their wealth of factual information by showing that from a small number of hypotheses, laws, or principles their observed results can be derived. That this is not a merely conventional form, but essential to the achievement of truth can be seen by considering the advantages of logical system.

The first advantage of logical system is that it helps to eliminate inconsistency between the propositions which it includes. Of two plausible but inconsistent propositions, either one is altogether false or both are only partially or inadequately true. Sensitiveness to contradictions is an essential characteristic of the man of science, and logical method is a way of forcing contradictions into the light. Common sense tries to achieve consistency by alleging reasons for what it wishes to believe. But if we fail to bring all these reasons together into a system we are bound to rely on inconsistent reasons at different times, as do political parties when in and out of office. The presence of unrecognized inconsistencies in our views sterilizes our knowledge and prevents substantial progress toward the truth. The popular distrust of reasoning is due to the fact that unless our reasoning is scientifically systematic or logically rigorous, we can introduce all sorts of contradictory propositions in the course of a long argument and can thus seem to prove anything while actually proving nothing. In this respect technical language and symbols are an aid. Popular language is, because of the vague diversity of meanings attached to given words, not well adapted for purposes of rigorous reasoning. It not only makes contradiction possible by the use of the same term to denote entirely different things on different occasions, but two perfectly compatible views may seem contradictory unless we have a specially developed language which enables us to make the proper distinction between them and thus reconcile them. It is essential, then, to a scientifically logical system to seek to make its hypotheses or initial assumptions explicit. Ordinarily we are not aware of the assumptions involved or presupposed in our statements; and the demand for explicit formulation of such logical presuppositions involves a progressive task. It requires careful study of the system itself to see whether the given principles, axioms, or postulates are really sufficient to deduce the propositions which constitute the body of our science. It was a long time before Greek geometry had its axioms formulated by Euclid, and modern geometers like Hilbert have greatly refined upon Euclid's own form. This effort to make explicit those assumptions from which our whole system can be deduced characterizes science from the beginning and serves not merely the aesthetic purpose of harmony (which is not to be despised for its own sake) but promotes the attainment of truth.

By making our assumptions explicit we are able to question them and to enrich our vision by thus revealing other possibilities. Our ordinary impressions as to what is true involve all sorts of unavowed assumptions which being habitual are felt with a certainty that blinds us to all other possibilities. The explicit statement of any proposition makes it relatively easy to consider its negative, or to analyze its components and to determine abstractly the various other possible combinations of these components. New forms of algebra, non-Euclidean geometry, or non-Newtonian physics are logically suggested by the explicit formulation of the axioms of ordinary algebra, Euclidean geometry, or Newtonian physics. Similarly it helps us to understand the nature of our economic system if we make its axioms explicit and then formulate possible alternatives.[10] So does ethical discussion become more enlightened if we are forced to formulate our fundamental assumptions, and so consider other axioms besides those to which custom has bound us. The progressive spirit of science which enables it to open up new fields is thus largely facilitated by the explicitness of our assumptions.

By making our assumptions explicit we not only facilitate the consideration of logically alternative hypotheses (and thus save ourselves from narrowness and fanaticism) but we make them more fruitful by reducing them to a form from which their consequences can be clearly traced. By tracing the threads of identity through the logical implications of any one factor at a time, we make possible that significant observation and experiment necessary to the process of verification. Most issues of daily life find no definite answer, because without rigorous logical reasoning we cannot isolate any one issue or formulate our question so that we can seek a definite yes-or-no answer in nature. It is only when we can put such a question to nature that we can get an answer sufficiently definite to refute one of two rival hypotheses or to show that one of these hypotheses is better supported than the other. As nature is not primarily concerned with answering our questions, such ideal crucial experiments are infrequent. But all observation and experiment (there is no hard and fast line between the two) is significant to the extent that it bears on our fundamental assumptions; and sooner or later that hypothesis prevails which best fits in with the propositions which sum up our observed results. Moreover, the significance of any one experiment in which a number of facts are involved depends upon our analysis of it into elemental factors, and upon our tracing the effects of each single factor and regarding all others as irrelevant for the determination of such effects. Thus Newton first determined the rate at which the moon falls to the earth, as if these two were the only bodies in existence, and astronomy subsequently studied the effects of other forces. The sum of all the effects can be observed in the actual motion, though there are still, and there may always remain, some residual perturbations.

Often indeed we study one aspect of nature, for instance mass, and ignore all others, such as colour. This has given certain philosophers

---

[10] See J. M. Clark on "Non-Euclidean Economics," in *The Trend of Economics* (ed. by Tugwell), pp. 86 ff.

occasion to characterize the method of science as artificial. For surely, they contend, the real universe does not consist of colourless masses. But whatever the "real" universe contains, it surely does contain masses that verifiably influence each other according to certain laws or invariant ways which are independent of colour.

While strict mathematical deduction from explicit hypotheses helps us to eliminate contradictions and facilitates the process of discovering new truths, it offers us no conclusive proof that our assumptions are free from all implicit contradictions. If the meaning of our assumptions is to be found in all their possible deductions, the full recognition of which is an interminable task, what proof have we that the future will not reveal an implicit contradiction between some of our assumptions?

But even if we admit that the bare possibility of contradiction has not been entirely eliminated in any logically developed natural science, we can definitely say that the ideal of system is necessarily valid. Even if not completely attainable, it makes scientific progress possible by enabling us to tell which of two alternative versions of the truth is the more complete and therefore more likely to be free from inconsistency.

The lurking fear that our different assumptions may involve some contradiction and the more general fact that the full meaning of these assumptions is never completely evident so that we take risks in adopting them, leads to the logical endeavour to reduce our initial assumptions to a minimum. If the first 19 propositions of Euclid can be deduced without the parallel postulate, the truth of the latter is irrelevant for our purpose. If certain psychologic phenomena can be explained by physiologic and conscious elements, then the introduction of a postulated "unconscious mind" into our system does not affect the meaning of any of its propositions and the term is, in that context, meaningless. Such is the purely logical significance of the principle known as Occam's Razor.

But it is always possible that nature is more complicated than our simplest accounts of her, adequate as the latter may appear for a while. Occam's Razor then becomes a qualified demand of prudence that we do not unnecessarily complicate the search for truth by multiplying the uncertainties and inaccuracies which follow from hypothetical entities. All terms of course involve some fluctuations of meaning, since we never at any one time grasp the full application of a universal. But the fact that every currency is subject to fluctuations of value is no reason for dealing in blank checks. In any case the relational structure of any subject-matter in which science is interested is best revealed if we have a minimum of postulates and a minimum of undefined, hypothetical, or (directly) unverified terms.

The principle of Occam's Razor ultimately implies that systems themselves should not be unnecessarily multiplied (where they are not mutually alternative). Indeed common sense may be said to be less systematic than science because it consists of a greater number of (unco-ordinated) systems. This difference is reflected in the fact that while in any advanced science we may carry through long series of inference concerned only with the question of logical validity and awaiting some crucial conclusion to test the truth of what has gone before, in popular discourse each new step in an argument reopens the question of truth. It was the amalgamation of the previously developed systems of celestial and terrestrial mechanics in Newton's *Principia Mathematica* that laid the foundations for the remarkable developments in physics. Maxwell's electro-magnetic theory of light and Einstein's general theory of relativity are similar achievements.

Though a logical system makes every proposition in it determinate, having a definite reason and certain consequences, the same body of propositions can often be arranged in different ways. In Euclidean geometry for instance, we have a certain amount of choice as to which propositions shall be taken as axioms and which as theorems. Thus instead of the usual axiom about parallels we might take as an axiom the proposition that the sum of the angles of a triangle is equal to two right angles and prove Euclid's axiom as a theorem. In law I might begin with the principle that all promises ought to be enforced except certain kinds, and get a system of propositions seemingly the same as that which would follow from assuming that no promises should be enforced except when they meet certain requirements; and in metaphysics the assertion that all is mind might lead to a system theoretically equivalent to a system based on the proposition that all is matter, for the same differences between what are ordinarily called material and mental may be recognized in both systems. This has led to the argument that logical system is merely an artifice without significance as to truth and reality. Now differences of technique are generally based on something in the subject-matter which makes one technique more suitable than the other. But if the subject-matter made no detectable difference in our two systems (as one might plausibly argue is the case between a decimal and a duodecimal system in arithmetic) it would not follow that both systems are devoid of truth. On the contrary both systems are true to the extent that they consistently explain all the actual facts, and if there is no difference between them in this respect we can call them identical in truth-value.

It is well to note that though the foregoing account of scientific system stresses the importance of logic and mathematics for all developed sciences, it does not restrict science to the numerical or quantitative aspect of nature. Logic and mathematics (the two, we shall see, are identical in essence) explore and examine the meaning or implication of any proposition, and no proposition about nature can pretend to scientific truth unless it submits to such an examination of its meaning or logical consequences.

We may see this if we take the simplest and therefore the most trivial-looking logical system, viz. the syllogism. What does the syllogism do? It calls our attention to the conditions which will make a proposition true. I may assert flatly that Jones is trustworthy. But if the issue as to its truth is raised, I must give a reason or evidence. Suppose I do so by asserting that he is a clergyman. Obviously this reason will be adequate only if it is true that Jones is a clergyman, and if there is some real connection between being a clergyman and being trustworthy. Otherwise the reason would be irrelevant.[11]

In all these ways, then, the systematization of knowledge aids in the search for truth and exhibits itself as something other than an economic device to which positivists like Mach and Pearson would reduce it. The suggested metaphysical question, "What is there in the nature of things which makes the apprehension of rational connections a source of knowledge?" will occupy us later. For the present, the fact that systematization is such a source justifies its place as an ideal of science.

[11] In law the syllogism corresponds roughly to the combination of law and fact necessary to prove a right. I claim an article as mine because I bought it in open market. This is conclusive if I prove the fact that I did so purchase it and if it is the law that such purchaser has title against every one (including previous owners from whom it may have been stolen).

§ III. INDUCTION AND DEDUCTION

SCIENTIFIC method is popularly associated with the cult of induction. But clear thinking as to what, precisely, induction is, and how it is related to discovery and proof has been blocked by three traditional confusions.

(1) The first confusion is the traditional contrast between deduction as reasoning from universals to particulars, and induction as the exact reverse. This certainly cannot be a true account of the matter.

We cannot in deduction, or in any other strictly demonstrative reasoning, always draw a particular conclusion from universals alone. To warrant a particular conclusion one of the premises must be particular. For universals may be pure hypotheses (i.e. they may assert what would happen *if* certain conditions existed, but not that these conditions do exist); and it is obvious that no hypotheses alone can give us a particular or categorically existential truth. It is a popular confusion to speak as if we can derive facts from a universal like the law of gravitation. We can derive factual conclusions from such a law only if we apply it to factual data.

The illusion that in deduction we can derive special instances from nothing but general rules is aided by the fact that our minor premises are frequently unexpressed, as when we argue that since all elements have a fixed atomic weight, radium must have one, or that the Pope must be fallible because all men are.

If we get accustomed to the idea that the logical premises of an argument need not always be expressed, it will not be hard to realize that in inductive arguments, also, our conclusion must be just as particular or universal as the combination of premises. Thus if from the statement as to the sum of the angles of a triangle I infer one as to the sum of the angles of any polygon, I seem to offer a clear case of reasoning in the direction of greater generality, and therefore an apparently perfect instance of induction. But the inference depends upon the proposition that all polygons can be made up of triangles; and if this is stated explicitly, the argument is indistinguishable from deduction.

Let us take another example. If induction is generalizing from a few instances or samples to the character of the whole class, the following is a clear case of induction. I observe that X, Y, and Z, suffering from pneumonia, have all been cured by serum A, and I infer that all pneumonia patients will be so cured. Here the conclusion will obviously be true only if X, Y, and Z are typical samples of the whole class of pneumonia patients and do not form a special class having some distinctive trait that makes the serum effective only in their case. The generalization therefore that the serum is effective in all cases of pneumonia assumes that the class of pneumonia patients is in this respect homogeneous. If this is made explicit we have the orthodox syllogism:

(a) A cure for X, Y, and Z is a cure for all pneumonia patients.
(b) Serum A is a cure for X, Y, and Z.
(c) Therefore serum A is a cure for all pneumonia patients.

The fact that the first premise is not usually made explicit is an important linguistic and psychological fact.[12] It is not, however, relevant to the question as to what premises will logically warrant the conclusion.

If an inductive inference is valid it must conform to the condition of all valid inference. If the latter is called deduction, induction is not its antithesis but a special form of it. The fact that induction and deduction are separate words does not prove that they must be antithetic.

This reduction of induction to syllogistic form may strike the reader as too easy. Surely, he may protest, there is a real difference between the reasoning of the experimentalist in medicine who generalizes from a few instances and the reasoning of the geometer who proceeds from explicit axioms or from propositions already demonstrated.

This objection can be cleared up if we get rid of:

(2) The second confusion in the popular contrast between induction and deduction, namely the confusion between reason as a logical and reason as a psychologic term.

[12] This assumption that our examples are typical of a wider class (common to all inductive arguments) is generally, without further critical analysis, dignified by the phrase "uniformity of nature." Refuge is thus found from the irksome task of taxonomy,—the determination of just what characteristics do define classes uniform in other respects. Where the taxonomic side of a science is firmly established (as for instance in Aristotelian biology, classical economics, or modern physics), there is no difficulty in determining of what class a given example is typical and the need of a special "logic of induction" is not felt. The disrepute into which taxonomy has fallen among literary scientists since the Middle Ages, however, has forced the problem of "typicalness" to re-emerge as a logical rather than a scientific question.

Psychologically, reasoning is a temporal event in an individual biography. In the logical sense, however, reason is not concerned with the manner in which ideas or propositions actually succeed each other in our consciousness but with the weight of evidence or proof. Now, in point of fact, it is very seldom indeed that in any active inquiry, even in a subject like mathematics, we can start from the right premises and go on from them in a definite order to the proper conclusion. Human life and even the growth of science would be a much simpler and more satisfactory affair if that were the case. Often we actually start with the "solution" or logical conclusion and seek for premises to support it; and more often we start somewhere in the middle and fumble backwards and forwards to discover presuppositions and implications. But as we do not know our way in the unknown, our time is for the most part spent in hesitations, false starts, and painful retracing of our steps. When we first ask a question or face a problem, we seldom have an adequate idea of what it is that we have assumed or that conditions our question. It is only after a great deal of intellectual work that we can see what are the proper premises and implications of our position. The history of geometry and physics shows how very late in their development the proper axioms or fundamental assumptions are formulated. Nor is the psychologic path from the observation of facts to inductive generalization always straight and simple. Most often it involves all sorts of analyses, reflections, and balancing of rival considerations. We seldom think in straight lines.

If, then, we distinguish between the premises which logically justify a conclusion and the psychologic starting points from which we jump to arrive at them, it becomes extremely doubtful whether there is any well-defined psychologic difference between the actual processes of reasoning in inductive sciences like experimental medicine and in deductive sciences like geometry or dynamics. Whatever difference there is must be sought elsewhere.

But if every inductive inference can be put in the form of a syllogism, what can logically differentiate it from other syllogisms? The answer for purposes of scientific method is to be found in the character of the (generally unexpressed) premise of such inductive syllogisms.

The typical form of induction is the use of a number of instances to justify a universal conclusion, as when we use a number of observed recoveries to justify the proposition that all patients of a certain class will recover. Now such evidence is, as we saw before, logically inadequate unless our instances are assumed to be representative or fair samples of the whole class. But this assumption is generally not explicitly offered as evidence, because it is generally not known to be true. It may be an unsupported guess, and only further knowledge of its consequences will verify or refute it. But when this assumption of the homogeneity of our class is well established or supported by the rest of our science, this premise becomes a link in a deductive chain (rather than a starting point). This is true in various branches of physics, where the uniformity of our classes of phenomena is attested not only by countless numbers of past observations, but also by various theoretic considerations as to why this should be so. Thus we have no doubt that every hot object will lose heat when in contact with colder objects. For not only has it always done so, but there is reason for this in the general theory of energy. Similarly, if I count the number of threads in one inch of cloth and conclude that this number prevails throughout the cloth the inference is not regarded as inductive if I know that the process of weaving the cloth makes such uniformity necessary. In mathematics the analysis of a single instance of a triangle proves the case for all triangles, because our reasoning is in fact throughout concerned with that which is by definition or previous proof common to all triangles. We approximate to this in any science if laws or uniform connections between various factors determine what is relevant and what is irrelevant to any class of objects. Hence a single observation on one specimen of a new mixture may be conclusive evidence that it is always acid.

In undeveloped sciences, however, this is not feasible. For we have no way of telling that we have not overlooked some factor which differentiates our instances (no matter how many) from the class of which they are taken as representative. Thus if I measure the intelligence, joviality, or endurance of a number of Irish, French, or Russian soldiers, there is no conclusive proof that the result will hold true of these nations generally. All sorts of selective agencies may have come into play. The expressed assumption that our classes *are* uniform in regard to these traits may put our argument into the *form* of an orthodox syllogism. But so long as that premise is itself doubtful it does not help to prove that our conclusion is true.

Induction and deduction are not, therefore, antithetic terms in the realm of purely formal logic. The difference between them is one concerning material evidence. Instead of being absolute, this generally reduces itself to a question of the *degree* of conclusiveness of the initial evidence in favour of the homogeneity of the class concerning which we wish to establish a law. Every serious attempt at establishing such laws that we are likely to meet in any science has some analogy or other suggestion in its favour. But unless the rest of our scientific system can support it through some invariant relation that simulates the invariance of logical or mathematical relations, our proposed law remains empirical or inductive. Since all the natural sciences involve unproved assumptions of homogeneity, they can all be said to be inductive. But if we remember that our distinction between induction and deduction is one of degree, we may recognize that some sciences are more deductive or less inductive than others, and their explicit use of mathematics may be a fair indication of this.

It may be objected to the foregoing view that it does violence to the common usage of the terms *induction* and *deduction*, and that it substitutes essentially vague for clear distinctions. The first objection is true, but common usage is itself hopelessly confused, as most treatises on inductive logic indicate. In answer to the second objection, it should be noted that we do not deny the absolute distinction between logic or pure mathematics on one hand and the natural sciences on the other hand; but within the latter the differences are not absolutely rigid. Our account, therefore, enables us to describe or analyze the actual historic progress of any science from a relatively inductive stage to that of greater deductive system. Thus Dalton's atomic hypothesis and Avogadro's law as to the equality of the number of molecules in equal volumes of gases transformed chemistry from a mere descriptive natural history into the beginnings of a deductive science. (The introduction of accurate weighing and volumetric measurement made this possible.) Chemistry became more deductive when Mendeléyev discovered the periodic law connecting all the elements and enabling us to predict the discovery of others. But this law itself was in turn transformed from an empirical or inductive one into a grounded or rational one when it was deduced from Moseley's law, which connects the various atomic weights with the general theory of physics.

If we thus view the difference between induction and deduction not microscopically in the individual mind but macroscopically in the history of science, we can see an illuminating example of the passage of our knowledge from an inductive to a more deductive form in the history of Balmer's law.

When the spectrum of luminous hydrogen was first observed, it was marked by three lines, and soon thereafter a fourth was discovered. These lines are always found in the same position and from certain measurements we can calculate the wave length of the light that these lines represent. By very careful measurement Angstrom, in 1851, found the wave-lengths of the first four lines to be 6,562, 4,860, 4,340, and 4,101 ten-millionths of a millimetre (known as Angstrom units and written A°). As repeated observation confirmed these measurements, they were regarded as expressing uniformities or laws of nature. But why these specific numbers,—and what connection is there between them?

The analogy between light and sound led men to seek some harmonic ratio between these numbers; but in vain. In 1885 Balmer, a Swiss school teacher, hit upon the idea that if we divide these numbers by 3,645.6A° (denoted by h) we get approximately 9/5, 4/3, 25/21, and 9/8. This is a significant series if we put it in the form

$$9/5, \ 16/12, \ 25/21, \text{ and } 36/32$$

for that means

$$\frac{3^2}{3^2 - 2^2}, \ \frac{4^2}{4^2 - 2^2}, \ \frac{5^2}{5^2 - 2^2}, \ \frac{6^2}{6^2 - 2^2}.$$

This suggests the obvious general formula $\dfrac{n^2 h}{n^2 - 2^2}$, from which it follows that a fifth line will be $\dfrac{7^2 h}{7^2 - 2^2}$, a sixth line $\dfrac{8^2 h}{8^2 - 2^2}$, etc. When Balmer developed his formula he did not know that a fifth line had been discovered in the visible part of the spectrum (by creating a vacuum) and that four more lines had been discovered by Sir W. Huggins in ultra-violet light. All these lines were found by Prof. Hagenbach to fit in very closely with the proposed formula. Since 1885 the number of known lines in the hydrogen spectrum (observed in the light of white stars) has increased to nearly fifty, and in all cases they conform to Balmer's law.

The number 2^2 in the denominator of our formula restricts its application to lines of lesser wave length than 6,562A° (if n is an integer). Balmer, therefore, suspected that $\dfrac{n^2 h}{n^2 - 2^2}$ might be a special case of $\dfrac{n^2 h}{n^2 - m^2}$ Subsequently lines in the infra-red part of the spectrum were discovered and their positions deduced from the general formula by giving m the value of 3. At this stage Balmer's law appeared as an induction from which a number of repeatable facts or laws hitherto unknown or ungrounded have been deduced; but in itself it had no other justification.

Its contingent character was emphasized by its arbitrary constant, $h = 3,645.6$A°. What is its significance? In 1890 Rydberg began a number of researches which advanced our knowledge of it by showing that if from Balmer's formula for the wave lengths of these lines (and the known velocity of light) we calculate the number of waves that pass a given point in a second of time, Balmer's constant (h) is replaced by a frequency constant which is found to be applicable to the line spectra of other elements, though the formulae for the latter are somewhat more complicated than Balmer's for hydrogen. This frequency constant (denoted by R after Rydberg), though of more general application than Balmer's constant, was still itself without any rational ground in the system of physics. But in 1913 Bohr applied the quantum mechanics of Planck and Einstein to Rutherford's theory of the atom, and deduced from it, among other things, Rydberg's constant and Balmer's law. The latter are now seen to be grounded in the most general facts or laws of physics. Some of the more complicated and residual phenomena of spectroscopy, such as the finer structures of the lines, have since been partly explained by Sommerfeld by utilizing Einstein's principle of relativity, and more satisfactorily by Schrödinger and Dirac on the basis of the newer and wider quantum mechanics. Spectroscopy is thus assuming the form of a deductive system.

Attention to such actual instances of the growth of rational or deductive system will enable us to avoid the third confusion, namely that:

(3) Induction is the method of *discovering* general truths, while deduction is merely a method of exposition.

Thus it is often claimed that the various canons of induction tell us how to discover the causes of phenomena. The reader can readily test this claim by applying these canons to discover some causes not yet known to him, e.g. the cause of cancer, or of excessive thyroid secretion, of the peculiar line in the spectrum of sodium, or of the peculiar properties of water-vapour that cannot be inferred from those of hydrogen and oxygen.

The method of agreement directs us to note the circumstances under which several instances of our phenomena take place, and it tells us that the sole common circumstance is the cause. But which of the indefinitely large number of possible circumstances under which a phenomenon can take place are we to record? The relevant ones, of course, say some modern logicians. But in the absence of established knowledge as to which circumstances are causes and which are not we have no clue as to what is relevant and what is irrelevant. Why may not a common circumstance in the case of cancer or disturbances of the thyroid be sought in the political, moral, or religious opinions of the patients, in their linguistic habits, the colour scheme of their garments, or that of the eyes that looked at them? The canon as formulated by Mill does not exclude such possibilities; and an unlimited number of most absurd causes for any phenomenon can thus be inferred by following this canon. Only if we formulate it negatively,—and if we indefinitely increase the number of instances of the phenomenon under diverse circumstances, can we hope to eliminate utterly irrelevant circumstances. We may even fail to do that much if owing to the inadequacy of our ideas our analysis of the instances is not fine enough. We may, for example, eliminate eye strain as a cause of headache because it is not a common circumstance in all cases of headache. But waiving this, we may positively say that the method of agreement amounts to the following:

The cause of the given phenomenon is A, B, C, D, or E. But B, C, D, and E are not common circumstances and cannot therefore be causes. Therefore A is the cause.

This is a valid disjunctive syllogism, but obviously we cannot by elimination alone discover the true cause unless by good fortune we have already hit upon it and explicitly noted it in the major premise.

Similarly it can be shown that the method of difference will enable us to eliminate from the suggested causes any circumstances which will be present not only when the phenomenon occurs but also when it fails to occur. But while it may eliminate falsely suggested causes it cannot of itself direct us to find the true cause. That the methods of concomitant variations and residues cannot operate unless we are already in possession of the causal circumstances is too obvious to require argument. All these methods may serve under certain conditions to increase the probability of an hypothesis suggested to us by other considerations. By themselves these canons will certainly not enable us to discover the cause of things.

In general, there is no definite method of discovering new truth, any more than there is any definite method for creating new forms of beauty or for inventing the things that will solve our practical difficulties. The discovery of what has hitherto been unknown involves a leap into the dark, and the tragic history of human failure to solve our vital problems shows how real is the darkness. In regard to the unknown, all our rules and methods involve guess-work and are certain to be inadequate.

However, as new truth like new beauty cannot be totally unlike the old—we could not recognize it as truth or beauty if it were—it follows that the proper use of old knowledge is an indispensable aid in the discovery of the new. In fact no one doubts that discoveries in any realm are usually made only by those who have active knowledge of that realm. Native genius, of course, is an element in the making of discoveries; and sometimes the solution of a problem will flash on us out of the blue without any apparent effort on our part. But for the most part previous efforts of thought precede the revelations accorded to us. Systematic deduction from previous knowledge prepares the way and may save us enormous labour by giving the proper clue in our researches. From this point of view, we can understand the fact, made evident to all by the history of science, that mathematics, or rigorous deductive reasoning, has been a most fruitful source of discovery in the physical sciences. Many phenomena have been discovered by deductive reasoning before our instruments and sense-organs were first turned to them. Conical refraction, the existence of Neptune, electric waves, the shifting of certain lines in the spectrum, and the bending of light waves in strong gravitational fields are a few of the more familiar examples.

The only clue which enabled us to anticipate these new phenomena was a proper use of deduction from previous knowledge.

We may conclude our reflection as to the rôle of induction and deduction in the discovery of new general truths with the following observations.

The usual way of putting the question, "How do we get new general truths?" involves two false assumptions, one a psychogenetic one, and the other a logical one, viz. (1) that we always know a number of individuals before we arrive by abstraction at a knowledge of the universal, and (2) that the only logical justification for a universal proposition is a previous knowledge of all the individuals included in it.

(1) That all knowledge begins with the perception of the individual and then goes on by abstraction to the universal is a widespread dogma. It probably arises from the fact that a good deal of our education consists in being taught to name and recognize certain abstract phases of existence, and as we cannot suppose that animals and children before they learn to talk have such general ideas, we conclude that they can come only after the perception of particular things. But careful attention to the actual growth of knowledge shows that it is mainly a progress not from the particular to the universal but from the vague to the definite. The distinction between the particular and the universal is generally implicit and only comes to explicit or clear consciousness in the higher stages of knowledge. Animals and infants at birth are first adjusted not to individual things so much as to certain phases of their environment, such as the warm and the cold, pressure, pleasant and unpleasant tastes, light and darkness, etc. In ordinary life we perceive trees before we perceive birches; we see Chinamen before we notice their distinctive personal features. Indeed the process of discrimination by which we learn to recognize individuals begins with a vague perception of difference. We are impressed with a stranger's beauty, agreeableness, or reliability before we can specify his features or traits. It is therefore quite in harmony with fact to urge that the perception of universals is as primary as the perception of particulars. The process of reflection is necessary to make the universal clear and distinct, but as the discriminating element in observation it aids us to recognize the individual. The progress of any science, at any rate, depends upon our ability to see things not in all their concrete fullness of individuality

but only as the embodiments of those universals which are relevant to our inquiry. A student will make little progress in geometry if his attention is solicited by the special features of his particular diagram rather than by the universal relations which the diagram imperfectly embodies. There are doubtless wide divergences between different minds in this respect. Some must depend more than others upon concrete examples. But without some perception of the abstract or universal traits which the new shares with the old, we cannot recognize or discover new truths. In the most elementary kind of inference given by Mill, namely from particular to particular, some perception of the universal or abstract element of identity is involved.

(2) With the removal of the psychogenetic error there is little that needs to be added to what we have already implied about the logical error. Suffice it to note here that the proposition, "A universal proposition can be proved only by exhaustive enumeration," cannot itself be proved in that way. Moreover to recognize that any number of instances actually before us constitute all the possible instances of the universal under consideration involves a knowledge of exhaustive possibilities, which is the essence of universality.

§ IV. PROBABILITY

ACCORDING to its classical conception science deals with a determinate universe and must therefore aim at propositions that are demonstrably true. Opposed to this there has recently grown up the view that science, unlike religion and metaphysics, deals not with absolute certainties but with what is most probable. The latter view seems to bring science into greater harmony with human life, where our practical needs and our hopes of salvation all deal with a future that is never absolutely certain.

Moreover, the introduction of statistical methods into the physical sciences and the development of what is known as statistical mechanics has strengthened the probabilistic conception of science. Ignoring pure mathematics as a science, one may maintain this new view without holding that the asserted facts or laws of natural science are less reliable than those obtained in other ways. It then means that science regards its results as always subject to correction; and this has been expressed with Gallic wit in the proposition that "science is continually making progress because it is never certain of its results."

The best-established propositions of science, such as Newton's law of motion, the law of the conservation of energy, or Maxwell's equations for the electro-magnetic field, claim to be true only because they offer the best systematic explanation of hosts of phenomena. But as there is no proof that some other hypothesis might not explain these facts just as well or better, and as it is always abstractly possible that we may find some phenomena not in conformity with these laws, the latter cannot be said to be absolutely certain.

But if the foundations of science are not absolutely secure, how can we build on them?

We may first dispose of the question how it is possible to apply certain or demonstrative reasoning to events which are merely probable.

Reasoning in mathematics, we must remember, is always hypothetical or dialectic. Pure mathematics alone can obviously never determine the probability of any actual particular event. The theory of probability as a branch of pure mathematics (i.e. of the theory of permutations and combinations) is not even competent to determine whether there are any probable events in nature. What it does is to explore the field of possible combinations of certain numbers. We give these numbers a special material interpretation when we make them represent the relative frequency of certain classes of possible events called "favourable" in relation to larger classes of events that consist of those that are "favourable" and those that are not. Thus if we assume that a penny may fall head or may fall tail in each one of three throws, arithmetic finds that the total number of ways in which heads and tails can thus appear is 2^3. Of these there are:

$$1 \text{ possibility of 0 heads and 3 tails TTT}$$
$$3 \text{ possibilities of 1 head and 2 tails HTT}$$
$$\text{THT}$$
$$\text{TTH}$$
$$3 \text{ possibilities of 2 heads and 1 tail HHT}$$
$$\text{HTH}$$
$$\text{THH}$$
$$1 \text{ possibility of 3 heads and 0 tails HHH}$$

The relative frequency of any specific combination in the total number of combinations is called its mathematical probability. But mathematics does not know or assert that a penny is just as likely to fall head as tail.

If we do assume that a given probability holds in a given case, mathematics helps us to deduce the logical consequences of our assumption. Such reasoning has two obvious uses. In the first place it prevents us from claiming a greater degree of probability for our deductions than is warranted by our premises. And in the second place it enables us to test our hypothesis by comparing its theoretic results with actual observation. Thus if in the long run of repeated throws pennies should be found to fall head less often than tails, we might reject the hypothesis that the probability of falling head is 1/2.

There are, however, certain cautions to be observed in such attempts at refutation through experience. We can always explain deviations from the theoretically expected results if we assume some disturbing factor. But the difficulty peculiar to an hypothesis of probability is that since improbable results are not impossible, their actual happening does not logically refute the hypothesis that they are extremely improbable. Thus on the hypothesis of perfect shuffling of cards and honest dealing, it is extremely improbable that in a game of whist I should get the same hand twice in succession. But since very improbable events do happen (every actual hand has a very low probability on the hypothesis that any card is as likely to turn up as any other) [13] such occurrence does not absolutely disprove the hypothesis of fair dealing. It does, however, increase the probability of the contrary hypothesis, since experience shows that such coincidences are most often the result of failure to shuffle the cards fairly, and this may be regarded as the essence of practical disproof, which is a matter of degree.

Mathematical reasoning, then, enables us (with due caution) to draw the proper conclusion from hypotheses as to probability, and while experience cannot logically refute, much less prove, a given hypothesis,

[13] It is absurd, of course, to talk of a thirteen-spade hand in bridge whist as in itself rarer than any other specified hand. It is rare only in reference to the rules of the game which make it the member of a unique class, while thousands of other hands may, in reference to those rules, be undistinguishable and hence, each, commonplace. But what is a commonplace hand in bridge whist may be a very unusual hand in some other game where the mode of dealing is exactly the same.

it can show that some hypotheses are more probable than others. This is what we call *proof* in practical life, in courts of law, in historical research, etc. Indeed, historically the words *proof* and *probability* are identical in origin (Latin, *probare*).

This brings us to the fundamental question, "What is meant by the probability of an actual event?"

According to the classical conception, nature is a determinate system, and there is no chance or probability in it. Every event is determined to happen or determined not to happen. Probability is then purely a subjective matter, viz. a state of our feeling. If I feel absolutely certain that an event is bound to happen I call its probability 1, and if I am absolutely sure that it will not happen its probability is 0. If I am entirely uncertain with no conviction either way, my feeling of probability is 1/2. The various intermediate degrees of certainty are expressed by intermediate fractions.

Now if the calculus of probability only measures our subjective feelings of expectation, it is hard to see how it can be of any help in physical science or serve in any way as a guide to natural events and such practical affairs as insurance. Certainly in any serious investigation to determine the probability, for instance, of corn surviving a certain drought or of men recovering from peritonitis, no one would think it scientific procedure to measure any one's *feeling* of certainty about it. Relative to such inquiries are entirely objective reasons, such as the various factors that maintain and destroy corn, in the first example, or the previous medical and family history of the patient, in the second. The determination of probability is too intimately connected with objective considerations to be justly regarded as only a subjective feeling; and if the probability of 1/2 is compatible with complete ignorance, how can it be of aid in scientific investigation? If I do not know anything about a die, neither how many sides it has nor what makes a given side come up rather than not, of what value would it be to say that the probability of turning up a given side is 1/2?

Difficulties of this sort have led to the rise of objective theories of probability. The best known of these is Venn's, which in substance amounts to an identification of the probability of an event with the relative frequency of its occurrence. I say, for instance, that an event A has a probability of 5/6 if I mean that the number of times it will actually occur will bear to the total series of events to which it belongs a ratio which approximates to 5/6 as the series increases.

Against this view Keynes has recently brought the serious objection that it does not really explain what is meant by the probability of a unique event. Suppose I ask what is meant by the probability of a disputed historical fact, e.g. that Louis XVI conspired with the émigrés against France. The frequency theory seems unable to give a precise meaning to the term *probability* in this case, and yet if we served on a jury to judge Louis XVI we should have to decide the probability on objective evidence.

We can avoid the difficulties and one-sidedness of both the objective and subjective views by adopting one that is both relativistic and logically objective. I shall state it in terms of Peirce's doctrine of probable inference, and suggest in passing an equivalent form in terms of classes of events.

Instead of speaking of the probability of an event, let us consider first the probability of the proposition which asserts it. What is the probability that Caesar was tall? Of course if we know that all Romans were tall it would follow necessarily that Caesar was tall. But if we know only that a majority of the Romans were tall, nothing rigorously follows as to Caesar. He may have been of the minority of the short-statured. Yet surely our knowledge as to the predominant tallness of the Romans is not utterly irrelevant or without any weight as evidence on the question of Caesar's tallness. Such knowledge renders it more probable that Caesar was tall than that he was not, or, more elliptically, renders it probable that Caesar was tall. Further factual evidence, e.g. that the wealthier Romans were the taller ones, can increase this probability, though it can never bring it to absolute or logical certainty. Suppose in addition to the foregoing we had direct testimony to the same effect by his contemporaries. Even so, the premises, "All who saw Caesar are agreed as to his tallness," and "These men have no common motive for telling us anything but the truth in this respect," do not logically prove Caesar's tallness, but only render it exceedingly probable. Indeed that Caesar in fact never existed, but was invented in the first century A.D. is not an absolutely impossible proposition. We can only say that the evidence against it makes it too remote a possibility to be worthy of a reasonable man's consideration. Despite a remaining possibility to the contrary we regard the proposition that Caesar lived as materially proved.

What do we mean when we say that certain evidence proves a proposition not with absolute or logical certainty but beyond a reasonable doubt? In legal phraseology we say that it is the kind of inference according to which a prudent man would conduct his affairs, and that if any one refused to act on it because of any remaining doubt, he would be generally regarded as unreasonable. This is a workable rule of practice. It is well to remember, however, that the same facts which to a conscientious jury prove the guilt of the accused beyond a reasonable doubt, may fail to convince the father, and the latter's faith in his son's innocence may be subsequently corroborated. What then shall we say of the inference which the jury made from the facts before it? It was a reasonable inference if it was the kind of an inference that in an overwhelming number of cases leads to the truth if the premises are true.

The probability of an inference, then, is the relative frequency with which its kind or type leads to true conclusions from true premises.

What do we mean by the kind or type of an inference?

When we say that from the proposition Sona is a man, it follows that Sona is fallible, Sona is felt to be only an illustration. The force of the argument would be the same if any other name were used. If now I assume only that most men are fallible, and derive the probable conclusion that Sona is fallible, it is still true that the individual name used is only an illustration. Any other will do. The conclusion is called probable because reasoning in such cases, it is felt, will produce true conclusions more often than not. This enables us to speak of the probability of a single event, e.g. that a given witness, who is the brother of the accused, is telling the truth. Our answer depends upon an estimate of the frequency with which brothers in a similar situation do tell the truth. If we believe that most brothers in such a situation [14] will not hesitate

[14] This formulation of the matter does not solve the question of how many or what factors of the situation considered shall be used to define the class of "similar" events which is to serve as a basis for our frequency ratio. Suppose that, for instance, in the example given above we know (1) that the witness is a Quaker, (2) that he has often quarrelled with his brother, (3) that he is forty-five years old, (4) that the evidence given is unfavourable, (5) that the crime charged is embezzlement, and (6) that the witness's name is Timothy Washington. Which of these facts, if any, shall we consider relevant? Arithmetically, there are 64 possible answers to this question (the number of combinations of acceptances and rejections of these six considerations). Each one of

to tell a falsehood we judge that the particular witness is probably not telling the truth.

More complicated individual cases, e.g. the question whether Louis XVI did or did not conspire with the émigrés, can be broken up into a number of evidential considerations, and the probability of each of these is the relative frequency with which the evidential fact is followed by the fact to be proved.

We can arrive at the same result if we define the probability of an event as the relative frequency with which the given grounds or causes will determine its type or class to occur. If we take the fall of a penny as a single separate event, there is no meaning in speaking of the probability of its falling head being $1/2$. It will fall head or it will not. But when I regard the penny as perfectly symmetrical, or if I assume that experience has demonstrated that pennies fall head as often as not, the next throw of the penny is viewed as an instance of its class or type—just as its economic value is—and to this class probability as relative frequency applies. That is why those who judge on the best ascertained probability will in the long run be more often right than those who do not.

The insistence that the probability of a proposition or event is not an absolute or intrinsic trait of it, but is relative to a given mass of evidence, enables us to do justice to the subjectivistic view of probability and guard it against the absurd consequences to which it can otherwise lead. The probability of a given event properly varies as the evidence increases. That which to the uninitiated is as likely to happen as not may become extremely improbable on further study of the subject. Only if we knew the whole of our science should we know *all* the relevant factors which determine any specific event to happen. But this does not mean that the conclusion to be drawn from partial evidence is a matter of arbitrary opinion. It is not an issue of psychology but of logic, not

these answers, presumably, will direct us to a distinct category of cases, with a distinctive answer as to the probability that our witness is telling the truth. Now as we have indicated in our discussion of induction, the question of relevance or typicalness cannot be solved on purely logical grounds. One factor of importance is the rate at which the number of instances of our situation decreases as we increase the number of defining elements or characteristics. It is likely, for instance, that in the example given above we should find no other "brothers in such a situation" if we took all the enumerated facts to be relevant. This suggests the need of introducing the concept of *degrees of relevance* as a basis for synthesizing the probability indices presented by the various classes of analogous cases.

of how we feel but of the weight of evidence as to the relative number of equal possibilities.

The most significant point, however, for scientific method in the foregoing considerations is the emphasis on the necessity of increased evidence to make our probability judgment more reliable. In a state of ignorance we have no evidence at all, our base is 0, and it does not support a probability judgment of 1/2 or any other judgment. As our evidence increases we eliminate the variations due to partiality.

The general bearing of this proposition can be seen if we connect it with our discussion of induction. Why is the reliability of (or evidence for) an induction generally greater if the number of observed instances is increased? If I once burn my finger by touching a hot stove, how can repetitions add to the force of the conclusion that hot stoves burn the fingers that touch them? This is a question which Hume asked and to which he found no answer. The answer is that if our instance is homogeneous with the class which it represents, one instance is as good as a million (and far more convenient). But where possible variations are not ruled out in advance, how shall we guard against the fallacy of selection, of assuming that the one or more instances chosen *are* representative of the whole of a given class and not rather of some special subclass? Thus if I wish to determine the time it takes for tomato seeds to ripen, a large number of seeds will strengthen the faith that the time observed is true of all tomatoes and not of those specially selected or planted under special conditions. But such general faith, though wise in the long run, is not sufficiently discriminating. Scientific method consists in seeking for possible causes of variation and planting our seeds so as to enable us to see these possible causes of variation in operation. Thus we plant them in different soils, observe them in different temperatures or seasons of the year, etc. In this way we eliminate certain differences, e.g. as to whether we plant them in one phase of the moon or in another, and show other circumstances to be determinant. We thus strengthen the probability of an induction being true if we take into our base not merely a large number of instances but a greater variety of them, such as those mentioned in the foregoing illustration. A probability is stronger if like a cable it consists of a number of independent strands.

If the reliability or probative force of an argument is thus identified with its probability, the latter has two dimensions: (1) the evidential force of any one argument, and (2) the reliability or weight possessed by an argument in virtue of the width of its base. Thus I may determine the probability of a man of 80 living ten years longer by mortality tables compiled in a given city over a short period of time, or I may use a much larger and more varied table. The results in the two cases may denote the same probability; but the latter table will be more reliable or probative.

It might be of some advantage to use different words for these two different aspects or dimensions of probability. But it is well to note that the greater probative force or reliability of an argument because it rests on a wider base comes within our definition of probability in that such arguments more frequently result in truth.

The consideration at the beginning of this chapter as to the rôle of reason in scientific method, will enable us to deal with a noted controversy as to the empirical and a priori elements in probable reasoning. Some think of probability as always discovered by empirical or statistical frequency, and others claim that certain general a priori considerations enter.

There can be no doubt that most people decide questions of probability in a more or less a priori way. When asked what is the probability of a penny falling head, or a cubical die falling on a given side, most people will say 1/2 and 1/6 without thinking it necessary to institute a statistical inquiry (in which it is uncertain how many trials are necessary). They assume that the penny or die is symmetrical and that the forces which make it fall on one side prevail as frequently as those which make it fall on another side. If by actual throwing of a penny or a die a thousand times, one side should show a marked preponderance most people would suspect some trick or disturbing cause, and would not be shaken in their faith in 1/2 and 1/6 as the *proper* ratios. A similar case of the a priori can be seen in the seeming self-evidence of the principle of a lever. If two arms of a lever are equal, we see no reason why one should weigh down the other, and we conclude that they must balance each other; and if one side does weigh down the other, we conclude that the two sides are not equal, or if they are equal in size that they are not homogeneous in density.

Now there can be no serious objection to admitting a priori probabilities in the sense of assumptions which appeal to us as rational and cannot be successfully contradicted because the contradictory instances can all be explained away. But this does not mean that our a priori principle of indifference is independent of all experience. It is only experience, as Mach has suggested, that can show that a difference in the colour of two arms of a lever is irrelevant to the question of their weight, while their length (if they are homogeneous) is relevant. Similarly it is experience which shows that, in the main, pennies do not fall head noticeably more often than tail.

On the other hand, empirical observation has a limited power to establish correct probabilities. If things have happened with a certain frequency before, it is not certain that they will continue to happen with the same frequency. Any observed frequency may have been due to a cause which has changed. It seems a fact of experience that those who rely on the study of previous distribution generally fare better than those who do not; nevertheless many go bankrupt because of their reliance on the continuance of a ratio that has prevailed for a long time.

A frequent abuse of the a priori in probability takes the form of an assumption that a number of factors whose relations are unknown to us are independent of each other. Suppose we want to measure the relative merit of scientists, poets, artists, actors, professional beauties, or monarchs. A common method today is to pick out a number of eminent persons in each field and let them make an estimate. If these estimates were in fact independent, the unanimous judgment of ten men would be overwhelmingly more likely to be correct than the judgment of one man. But are they in fact independent?

The absurdity of tacitly assuming that all traits are independent and that in the absence of knowledge the probability of any event is 1/2 leads to a classical *reductio ad absurdum*. Suppose that we know nothing at all about the possibility of life on Mars. The probability of there being no cat is 1/2 and the same for a snake, a bee, or any other animal. If then I ask the probability of there being no cat, no snake, and no bee on Mars, I get the fraction 1/8 and the more animals I add the smaller the value of the fraction, so that the probability of there being no animal at all on Mars can be made negligibly small. I can thus on the basis of complete ignorance prove beyond a reasonable doubt that there is life on Mars.

The explanation of this fallacy is obviously this: Relative to complete ignorance probability has no meaning; and where we do not know that certain events are independent of each other (the existences of different animals are not), we are not justified in assuming them to be so.

Where our knowledge is incomplete, no reasoning can give us absolute or logical certainty. But the logical theory of probable reasoning enables us to recognize the uncertain element in our inferences and helps us to evaluate the relative weights of various arguments or evidences.

§ V. THE IRRATIONAL AND THE A PRIORI

(A) THE IRRATIONAL

WHILE the foregoing account of scientific method emphasizes the rôle of reason, it does not identify knowledge with its rational element. It maintains, rather, that in the development of this rational element we have the distinctive character of science. The principle of polarity indicates that nature must be more than reason, that the actual world cannot contain form unless it contains matter that has form. We see this in the ultimate appeal to the observation of brute fact which natural science must employ in verification. Such brute fact, I have attempted to show, is not something utterly inexplicable, nor does it reveal its true or full character to unreasoned observation. We must use the critical reason and technical safeguards and instruments of science to penetrate the fogs of natural illusion and see more truly what exists; when such existents are more definitely located and examined it is always by means of abstract traits or universal connections. Nevertheless, though the existent is the locus of all abstractions which define it, it is not exhausted by any number of these universals and to this extent it constitutes an unattainable limit of analysis. Always there remains the beyond, the unexplained, the contingent.

As all proof ultimately proceeds from propositions which are themselves unproved, science must assume principles and factual data which are not themselves logically derived from anything else. We can deduce the character of certain phenomena from the observation of others if we know both sets to be subject to the same processes or transformations. But things that are alike are also different, and these differences can never be derived from their identity. If all substances are reduced to different combinations of electrons and protons, the fact of difference still remains primal and unexplained. Why do some protons take on one electron and others more? Any explanation only pushes back the contingency of our fundamental assumptions. Ultimately the universe just is what it is and contingency is uneliminable.

When the irreducible diversity of existence is viewed in its temporal aspect, it appears as the novelty which makes the future unpredictable and the historicity which makes the past irretrievable. When elements combine, some of the properties of the new result can be deduced by logical addition of the properties of the elements. But there are always also new phases—the very combination itself is new—which did not exist when the elements were separate and which, therefore, cannot be logically derived from them. To the extent that we are creatures in time and must add fact to fact, we can never logically exhaust the totality of nature. There is thus something which will always be for us beyond rational form or system, and in that sense appropriately called irrational.

We must grant, then, the contention of modern empiricism that existence in the actual world is more than rational connectedness and that it cannot be entirely grasped by mere reasoning. What is objectionable in empiricism is the notion that science cannot push its analysis beyond the point at which common perception surrenders to what it calls brute fact. The notion of the actual as the goal and limit of our conceptual analysis and as that which is contained in observation only by implication suggests that although science can never escape from fact, none of the things that we commonly believe to be facts can be immune to analysis or claim to be above question. To common perception it is an ultimate fact that the sun moves around the earth and that it is larger at the horizon than at the zenith. But science separates the element of sense perception from its familiar conceptual framework and by a new framework transforms our view as to what the facts are.

There is, then, no reasoning apart from the observation of brute facts that happen to be, but all observations contain some assumption which may be challenged upon rational grounds.

(B) THE A PRIORI

We have stressed the ineradicable contingency of existence and the necessary novelty and incompletely predictable character of the future. But this last proposition is itself a priori, in the sense that it is not inferred from the observation of the future facts to which it refers and it cannot be completely refuted by any finite number of observations (since no future fact can disprove the possibility of further future facts). The existence of a priori propositions that are not "derived" from experience presents one of the classical problems of philosophy. Here we are concerned with it only to throw light on the nature and limitation of rational or scientific proof.

(1) DISTINCTION BETWEEN THE PSYCHOLOGICALLY AND THE LOGICALLY PRIOR. The nature of the a priori has been woefully confused both by its rationalistic proponents [15] and by its empiricist opponents. For both sides in asking whether certain principles are derived from or prior to experience have failed to discriminate between psychologic derivation, which is a question of temporal priority in an individual mind, and logical "derivation," which is not concerned with a temporal priority but with logical analysis of evidence or proof. The temporal order in which I learn certain propositions about the city of Punta Arenas depends among other things on circumstances in my own biography, but these circumstances do not determine the date at which the city was founded, its present relation to Chile, etc.

The psychologic phase of the question need not long detain us. We are certainly not born with a fund of conscious general ideas and we do need experience before such ideas can develop in us. But if the babe at birth is devoid of general ideas, neither does it begin by observing the particular facts with which it comes into contact. The view that the mind at birth is a passive *tabula rasa* on which particular things write their impressions is a crude metaphor, and the view that the "mind" of the newly born babe makes a synthesis of the particular facts and by comparing them obtains general ideas is a myth for which there are no corresponding empirical facts. The truth, as we have already noted,

---

[15] Even Kant, who begins bravely enough with the explicit statement, "All knowledge begins with experience but is not derived from it," fails to distinguish between the psychologic origin of our ideas and their logical justification or validity.

is that the organism at birth is already adapted (or prepared to be adapted) to certain general phases of the physical world, to heat and cold, to smoothness and softness, to light and darkness, etc.; and the perception of particular facts is conditioned by these general dispositions of the organism which bring certain phases of nature to our attention to the neglect of others. Moreover no one human being develops all or most of his generalizations. Each one of us is born into a community and the general ideas of that community are imposed upon us through language and through other social modes which emphasize certain aspects of existence to the neglect of others.[16] There is therefore no reason to suppose that the perception of particulars always precedes the perception of the universal. Indeed if the universal were identified with undiscriminated generality one might well maintain the contrary, and support this position with the familiar fact that the child first perceives *man* (all instances of which are called *papa*) before he perceives his individual father. But it would be fairer to say that universal idea and particular fact generally develop into clearness together, the particular instance helping to give body and prehensibility to the idea, and the idea making the instance clearer and more definite. Even psychologically, however, it is true that some ideas, notably of method, must precede in time the recognition of certain particular facts which would not be otherwise sought for or observed. This is obviously true in inventions and in scientific researches which lead to the prediction of hitherto unknown phenomena, as in the case of Hamilton's prediction of conical refraction, and Einstein's prediction of the bending of light rays passing the sun. Logically, however, a proposition is a priori if it must be presupposed and cannot be proved or disproved within the system to which it is a priori.

(2) THE UNPROVABLE AND THE IRREFUTABLE. The claim that there are propositions in science which are not logically "derived" from or proved by experience, and which experience cannot possibly refute, has

[16] It is curious that the empiricist philosophy which began with Locke's rather near-sighted arguments against innate ideas ends with instincts that serve all the purposes of innate ideas, and with evolutionary race memories which often have little more factual basis than Plato's frankly mythical account of ideas as reminiscences of previous existence. A social theory of knowledge would recognize that not only are most of our ideas imposed upon us by our social environment, but that the ideas of a community may in turn be moulded by the prestige or the vogue of individual thinkers like Aristotle, St. Thomas, Locke, Kant, Hegel, or Spencer.

always seemed scandalous to empiricistic philosophers,—and not without reason. For too often have all sorts of dubious propositions sought to escape the challenging tests of verification by claiming to be a priori.

A priori propositions, it has been held since Aristotle, need no proof because they are self-evident, their opposites are inconceivable, or they are necessarily imposed by the nature of the mind.

(i) *Self-evidence.* We have repeatedly indicated how fundamental to scientific procedure is the distrust of self-evident propositions, and how often in the history of science self-evident propositions have turned out to be false or meaningless.[17] Thus despite the seeming triumph of the principle of continuity in Darwinian biology, and its marvellous fertility in mathematics, biologists are now willing to admit saltations, and mathematical physicists are willing to treat energy as well as matter as essentially discontinuous. *Nature makes no jumps* is as obsolete as *Nature abhors a vacuum.*

(ii) *The Inconceivability of the Opposite.* We need not stop long to consider the more popular test of truth denoted by this phrase. If the term *inconceivability* is used—as it has been by Hamilton and Spencer—as a synonym for *unimaginability,* we have already indicated that it varies with different persons and different climes and therefore cannot establish universal truth. But even if we use *inconceivable* as synonymous with *unthinkable,* we can well say that all sorts of propositions which formerly were regarded as unthinkable are now easily thought, e.g. that a product may be zero though neither of its factors is zero (as in quaternions and other forms of vector analysis).[18]

(iii) *The Forms of Intuition.* A variant of the argument that certain propositions are a priori because self-evident is the Kantian argument to the extent that it claims certain propositions (e.g. the axioms of Euclidean geometry) to be true because they are the forms of our own intuition. This rests on the uncritical assumption that the forms of our own intuition are fixed forever, and can be seen at once by inspection. But there is no more evidence for this than for the fixity of any form in inorganic nature, e.g. that matter has weight. If the observation of the latter is empirical, why is not the observation of the working of our

---

[17] Self-evidence cannot guarantee the truth of a universal proposition because simple inspection cannot tell us the extent of the implications which constitute its meaning.

[18] In regard to propositions whose denial involves self-contradiction see pp. 171 ff. infra.

mind also empirical? We cannot, according to Kant himself, observe the workings of our own mind except when it is working with sensuous material.

It is often said that the laws of thought are the conditions for the recognition of objects, and as nothing can be recognized which is not thinkable, the latter is the a priori condition of the former.[19] But all this is the bare truism that to be thinkable objects must be thinkable. Just exactly what is thinkable can be found only by long experience in thinking about objects, not, as Locke supposed, by preliminary thinking about the mind. We know the mind's power only when we know how it has actually been exercised in the knowledge of objects.

The foregoing considerations appear to me to give valid support to the empiricist position. We may, however, note that the empiricist is at times too absolutistic in his repudiation of self-evidence. While self-evidence is not an adequate guarantee of truth, some propositions must have more of it than others to enable us to pick them out as the most promising hypotheses or to reject others as instances of the absurd. Nor can we follow the empiricist in his ridiculous attempt to demonstrate that all propositions can be derived or proved by experience (i.e. observation and experiment). How can the universal rules of logic or pure mathematics be proved by particular instances? All such attempts at proof would presuppose these rules. Even more concrete universals like *All men are mortal* cannot be so proved, so long as the class of men continues. This is true of any universal which asserts possibilities that cannot be exhausted in a finite time.

Our arguments so far show that while the traditional proofs of the a priori are untenable, science can never dispense with unproved universal propositions. Of these unproved assumptions some can be disproved by their consequences,—while others cannot be. Ordinary hypotheses belong to the first class, while those of the second class may properly be called a priori. For an hypothesis in science is an anticipation of experience (in plain language, a guess) that may turn out to be false and refutable, while it is the essence of the a priori to be irrefutable.

Let us, however, look at the matter more closely. It cannot be too

[19] See F. Münch, *Erlebnis und Geltung*, p. 181; Cohen, *Kants Theorie der Erfahrung*, pp. 135, 214 ff.

often insisted that scientific knowledge starts not at some absolute beginning like "the first man" but like a good play, *in medias res*. The most original or revolutionary thinkers, such as Copernicus, Galileo, Newton, Faraday, or Pasteur, or the humble laboratory worker who seeks to measure the volume or velocity of the blood in the human organism, all begin with a body of past knowledge which they cannot wipe out as a whole, for out of it arise all their problems. Why, for instance, does the heart beat? Our very question as well as the discovery and recognition of an answer depends upon a fund of assumptions. But as we discover new facts, the new body of knowledge may be found no longer consistent with some of our old assumptions. Thus the hypothesis that the cause of the heart-beat is to be found in the action of the vagus nerve in the striated muscles is overthrown by the discovery that the heart can be made to beat outside of the body in a saline solution. Yet no failure or number of failures to find any explanation is adequate disproof of the assumption that the cause is physiologic or, in the larger sense, physical. We may therefore call the latter an a priori (in the sense of irrefutable) assumption.

Are a priori propositions merely conventions? The property of being neither provable nor refutable belongs to certain resolutions, such as nominal definitions, which are resolutions to use certain symbols in certain ways. Such resolutions or definitions are expressible in sentences that are not readily distinguishable from genuine propositions. Hence there is a prima facie plausibility in the view made current by Poincaré that a priori propositions are irrefutable because they are really firm resolutions to carry on the scientific game according to certain rules or stipulations. The prima facie plausibility is due to a certain real analogy and I do not wish to take issue with it except for the implication (seldom explicitly avowed) that these axioms are not genuine assertions about objective existence.

Consider such an a priori assumption as that all physical changes are conditioned by other physical changes according to some law or invariant relation. Let us agree to call this assumption a "resolution." We will then have to distinguish between two kinds of resolutions, those that are productive of consequences consistent with observable fact and those that can be maintained only by the heroic method of introducing modifying subsidiary hypotheses ad hoc. Can these two kinds of assumption be distinguished without that reference to objective reality which characterizes the notion of truth? The conventionalist may answer that the former type of assumption is preferred on the sole ground of simplicity. This again may be true, but it is certainly not true if simplicity is only an intrinsic aesthetic quality. Such simplicity could more easily be attained by the formation of purely symbolic schemes such as those used in the elementary teaching of music than by the arduous search for "essential" uniformities in nature. And if such intrinsic simplicity is the criterion of valid hypotheses, it is somewhat difficult to explain how a proposition such as *Nature makes no jumps* may be valid in one field and invalid in another field of science. The simplicity which science seeks is not something divorced from the facts explained. Scientific simplicity is the characteristic of hypotheses which seize upon factors manifesting themselves in widely diverse phenomena. The "resolution" that all physical changes are conditioned by other physical changes according to some law or invariant relation will then be "simple" (in the only sense relevant to science) only if physical phenomena do actually exhibit fundamentally and multifariously the denoted character.

We may conclude then that if the assumption that physical nature is a causal system leads us to find connections between things, it is because the connections are there to be found and not merely because we are resolved to find them. This is so obvious that it seems fatuous to insist on it. Yet it is a fact that the acceptance of the conventionalist interpretation given by Mach and Ostwald to the atomic hypothesis would have stopped the fruitful investigation which followed on the assumption that matter really is composed of atoms.

(3) THE A PRIORI AS METHODOLOGIC POSTULATE. If we maintain the distinction insisted on in the last paragraphs, we need not hesitate to agree that a priori principles are methodologic or regulative principles which enable us to organize our factual knowledge. Conventional stipulations play an important rôle in science in facilitating this work of organization. We may see this in our agreements as to the units of time and distance which are so essential in physical measurement. Even such conventions, however, we must insist, are not arbitrary; for new experimental findings may force us to change our units in order to escape inconsistencies in our system. Thus in order to maintain the accepted system of mechanics and to reconcile it with certain irregularities in the

motions of the moon, we find it necessary to suppose that the period of the earth's rotation around its axis—which used to be regarded as the constant unit of time—is being continuously lengthened; and this is supported by the considerations as to the known effects of tides.

Similar considerations hold for our unit of length. It is not without some objective ground that we pick out a certain platinum bar located in the Bureau of Standards at Washington and say that so long as it is at the same temperature and pressure it has the same length. That statement we regard as true because it enables us to anticipate actual uniformities in nature.

However, certain phenomena, such as are revealed in the Michelson-Morley experiment, indicate that we must qualify the statement of the invariance of the length of our assumed standard at constant temperature and pressure by making that constancy relative to all bodies that are at rest in respect to it, and by assuming that its length varies relative to moving bodies. Why do we do this? Because in this way we remove otherwise inexplicable anomalies and recognize all phenomena as subject to the unitary laws of physics. That this gain is not purely aesthetic is shown by the power of the new theory to predict phenomena the existence of which had not previously been suspected.

Reflection, then, shows that it is possible to view a priori principles as both expressive of the fundamental nature of things and as enabling us to organize them according to certain orders or patterns suggested by these principles.

From this point of view the rules of logic and pure mathematics may be viewed not only as principles of inference applicable to all systems but also as descriptive of certain abstract invariant relations which constitute an objective order characteristic of any subject-matter.

(4) Relative and Absolute a Priori. The laws of any special science, such as crystallography or physiology, are the invariant relations according to which the phenomena or the flux of nature may be arranged. Without such laws, phenomena are devoid of significant order. But we are not satisfied with organizing any special field in that way. We must relate it to the rest of nature by bringing its laws into relation with the laws of physics, as the most general science of nature. From this wider point of view, the invariants of the special fields are particular applications of more general laws.

A similar situation is seen in the field of mathematics. Certain axioms are necessary for Euclidean geometry or our ordinary two-dimensional algebra. For these systems these principles remain a priori. When, however, we enlarge our vision to include non-Euclidean geometry and other systems of algebra and arithmetic, the former a priori axioms become merely protases of a special field of arguments. They are therefore not invariants or a priori for the larger system, though they remain necessary and a priori for the original and more limited system.

This relativistic conception of the a priori, i.e. that a proposition is a priori for a definite field, enables us to meet a difficulty about the a priori which we have not thus far adequately disposed of, namely the problem of how one can ever be certain that future facts will not refute or overthrow what one believes to be a priori.

In the realm of logic and mathematics we prove that certain combinations cannot possibly occur, that, for example, it is impossible for the Pythagorean theorem to be false when Euclid's axioms are true, or that it is impossible for any two integers to have as their exact ratio the square root of 2. In the realm of logic and pure mathematics, then, we have absolute a priori truth.

To the extent that logic and mathematics enable us to organize any special field of science,[20] we can say that the conditions which limit or define our field necessarily preclude certain consequences inconsistent with them. Every problem or question involves certain assumptions and nothing inconsistent with these assumptions will be an answer.[21] This is all the certainty that science needs for the process of investigation—enough certainty to give direction to the growth of knowledge without foreclosing all issues and making such growth unnecessary.

For the logic of science our discussion of the a priori may stop at this point. But as the term a priori has been traditionally associated with general philosophy (metaphysics and epistemology) a few remarks on the latter may make clearer the limits of our foregoing discussion.

---

[20] Logic enables us to organize fields of experience by providing us with relations like exclusion and inclusion whose recognition constitutes sanity. A day cannot become a piece of butter and a colour cannot become a sneeze. These sound absurd because common sense has already organized our experience according to certain universal categories. Science proceeds to perfect this organization by eliminating magic and witchcraft. You cannot by the pronouncing of any word change a deer into an apple tree, etc.

[21] See F. S. Cohen, "What is a Question?" *The Monist*, July, 1929.

Philosophically the a priori has been defined absolutistically, i.e. without limiting it to any definite field. It denotes propositions which must be true absolutely or unqualifiedly and which never can be refuted by anything that has happened or will happen anywhere or at any time. Such a knowledge would presumably be possessed by a Creator of nature. Do we or can we have such knowledge? Subjectivistic philosophies answer this in the affirmative by placing mind in the place of God. If the laws of nature are really the conditions of the mind's own activity or synthesis, then if we can inspect the mind we can know in advance what is and what is not possible in nature. The difficulties involved in thus identifying our individual minds with the author of the laws of nature and in discovering the ultimate laws according to which even our own mind works have already been suggested. But it may be noted here that the question of the rôle of mind's contribution to the system of nature has historically led to the recognition of the relativity of knowledge to the human organism and, in the Kantian form, to a recognition of the regulative function of the a priori. Admitting that the more objective classical philosophies have overreached themselves in their abuse of the tests of self-evidence and self-contradiction, and in their sanctioning of false and meaningless propositions as a priori, it still remains true that in the Platonic ideas they have recognized the objectivity of the logical and mathematical forms of nature. And even as to many of their material a priori assumptions it is difficult to see how they can be avoided. We cannot seriously doubt that there are beings of some sort and propositions about them, one at least of which is true, that we exist in some sense, etc. But this indubitability is only relative to our human need to understand. They form a system of human confessions: I can't help believing . . ., or, I can't conceive any knowledge that does not assume . . ., etc. The absolute totality is not definitely determined by them.

Finally modern logically critical philosophies suggests that an a priori which is not qualified by being either formal or relative to systems of possible knowledge is a will-o'-the-wisp, because an absolute totality of all existence is not a determinate object of discourse. Any field can be defined by distinguishing it from other existences just as a figure is drawn on paper or a blackboard if it is in a colour different from that of the background. When all distinctions are wiped out between the diverse forms of being have we anything left? The existence of any particular entity is different from its non-existence. But what is the difference between being as such and non-being as such? Here we seem to reach a rarefied air where thought finds no obstacles to its flight, but, alas, no support.

Yet, though the vision into the Absolute is either into a fathomless depth in which no distinctions are visible or into a fullness of being that exceeds our human comprehension, we need the idea of it to characterize our actual knowledge at any time as incomplete and fragmentary. The wells of rational knowledge offer no magic potion to those who thirst for the absolute certainty which will solve all ultimate questions. But they do offer us the living waters which strengthen us in our arduous journey.

Chapter Four

THE METAPHYSICS OF REASON AND SCIENTIFIC METHOD

THOSE who rank the truth-value of natural science very high and wish to utilize its results in their philosophy have followed one of three ways. They have tried to build a world-view either (a) on the results, (b) on the presuppositions, or (c) on the method of science.

(A) PHILOSOPHY AS A SYNTHESIS OF THE SCIENCES

To take the generally accepted results of the various sciences and to weave them together into a picture of reality seems to many the readiest and safest way of philosophizing. The difficulties in the way of this are much more serious, however, than is commonly recognized. In the first place, it is difficult for any one but a specialist to know what *are* the results of any one special science. He who relies on popular expositions is apt to get more of the picturesque than of the true. For popular accounts necessarily simplify; and in the interest of such simplification the "results" of science are purified of the detailed qualifications essential to their truth, and separated from the technical methods by which they are obtained and which are essential to their scientific meaning.

An even more important consideration is the fact that a synthesis of the results of science is not necessarily scientific. Such synthesis may be and generally has been dominated more by practical, dramatic, and aesthetic than by scientific motives. Indeed, since it is the absence of evidence that generally compels the sciences to stop short in their synthesis and remain in the fragmentary state, it seems that attempts to carry our synthesis beyond what the various sciences have been able to effect cannot be strictly scientific. This does not deny the practical and aesthetic value of an imaginative synthesis beyond the necessarily fragmentary results that can be scientifically established. But the interests of truth demand a clear distinction between an imaginative synthesis and a strictly scientific one.

We are thus seemingly faced with a dilemma. Either scientific philosophy makes no attempt to fill the gaps in our scientific knowledge or else it must do so by methods which the sciences themselves will not use, so that our result cannot claim to be scientific even though elements of our picture may be borrowed from the sciences.

(B) PHILOSOPHY AS A CRITIQUE OF THE PRESUPPOSITIONS OF SCIENCE

One of the ways of avoiding the horns of this dilemma is to view philosophy not as a synthesis of the results of science but as a dialectic argument concerning its presuppositions. This is the road made classic by Kant, who called it the transcendental method. Kant assumes that science results in synthetic propositions a priori, and asks what must be the nature of mind (and ultimately of the world) to render such knowledge possible. Without passing judgment on the gains which have accrued to philosophy from this method, we must note two insuperable logical objections to it.

In the first place, we cannot possibly in the light of modern mathematics and physics accept Kant's assumption that in Euclidean mathematics and Newtonian physics we have a priori knowledge of nature; and in the second place, as has already been pointed out, it is a downright logical fallacy, the familiar one of affirming the consequence, to argue that any theory (like the Kantian one) that explains how knowledge is possible, is thereby demonstrated to be true. Of course the fact that a theory explains something renders it to that extent more probable, but the Kantian view will not grant any room for probabilities in metaphysics precisely because it fails to discriminate clearly between the existential propositions of physics and the dialectic or purely logical ones of pure mathematics.[1] The latter, of course, must be demonstrably true or false. Even in the computation or determination of probabilities we cannot proceed without the demonstrable truths of pure mathematics. But the latter are truths concerning the connection between

---

[1] See Book II, Ch. I, § I, infra.

possible hypotheses and their consequences. They can never prove our initial physical hypotheses or assumptions to be free from all elements of contingency or possible error.

(c) PHILOSOPHY AS AN EXTENSION OF SCIENTIFIC METHOD

A view which is both old and widely accepted today claims that philosophy can be scientific only by applying scientific methods to its own subject matter which is distinct from the subject matter of the other sciences. This subject matter may be *being* as such, *reality* as distinguished from appearance, the nature of the *mind*, or the nature of *knowledge*, etc.

No one can well deny that in this way great gains have been made for philosophy, especially in the careful analysis of specific concepts. But a philosophy that excludes the subject-matter of the special sciences, natural and social, cannot satisfy that interest in the cosmos which has at all times been the heart of philosophic endeavour. This is shown by the work of Locke, Hume, and the psychologic school of philosophy to whom scientific method means the method of natural history applied to mental and moral issues. In this way a great deal has been done for psychology, but the philosophy of nature has certainly been impoverished. Nor can the use of the older deductive method *more geometrico* be fruitful in natural philosophy unless illumined by the factual knowledge of the sciences. Without such knowledge we are apt to accept propositions as self-evident when they are in fact questionable or vague or even meaningless.

These difficulties in the paths of scientific philosophy are not readily surmountable. Yet difficulties are not vetoes to courageous effort at the "supreme good of the intellect." Philosophy, seeking the most comprehensive vision, cannot ignore the insight gained by the sciences, but must go forward to envisage their possible synthesis. Though such synthesis is necessarily speculative it may be well to note: (1) that a certain speculative element is necessary for the substantial growth of science, and that the various sciences have in fact thus been nurtured by philosophy, and (2) that a scientific philosophy corrects the dangers of speculation by a rigorously logical analysis of fundamental concepts and assumptions, so that it should be aware of how much certainty can be attached to its wider speculative reaches. In recent years the work of the mathematical or neo-Leibnizian philosophers has remarkably clarified such traditional concepts as infinity and continuity, and the logical nature of inference and proof. We know better, thanks to the labours of Peirce, Frege, Whitehead, and Russell, what is requisite for rigorous proof, and we can be more honest in estimating the degree of probability that may attach to our various answers to the questions which science is not yet in a position to attack directly. In this respect philosophy is continuous with science in method. For, contrary to popular impression, science does not eschew speculative ventures into the unknown, though it is very cautious not to confuse anticipation with verification. The nature of number, matter, and life does not cease to be a concern of philosophy because definite light has been thrown upon these problems by modern logic, physics, and biology. But as the sciences grow by constantly correcting their content, it is the inescapable task of the philosopher to use the invariant principles of scientific method to go back to ever more rigorous analysis of the elements or rudiments of our knowledge, to examine the ideals which guide scientific effort, and to anticipate where possible what science *may* conquer in the future.

§ II. THE PRINCIPLE OF SUFFICIENT REASON

SCIENTIFIC method, it is generally recognized, depends on the principle of causality. This, however, is only a special instance or application to temporal events of the wider principle of sufficient reason. The latter, as applied in mathematics, as well as in natural science, may be formulated as follows: *Everything is connected in definite ways with definite other things, so that its full nature is not revealed except by its position and relations within a system.* This is a familiar commonplace. Yet I venture to assert that its precise meaning is seldom justly appreciated in metaphysical discussion.

Let me illustrate some of the most widespread and influential of its misapprehensions.

(A) REASON AND CONTINGENCY

Contrary to the usual views of it, the principle of sufficient reason as actually relied on in scientific procedure is not only compatible with a domain of chance, contingency, or indetermination, but positively demands it as the correlative of the universality of law. We may see this in the application of any law of mechanics or physics. For the most thoroughgoing mechanism necessarily involves: (1) contingent data, e.g. the actual position of the elements, (2) abstraction from other aspects or elements which are thus regarded as irrelevant and independent, and (3) rules of connection which themselves have a contingent aspect.

(1) CONTINGENCY OF SCIENTIFIC DATA. That the data in any physical inquiry are in fact contingent, that they are discovered by observation and measurement and cannot possibly be rationally deduced except by reliance on other data, no one can possibly doubt. We cannot derive facts or existential properties from pure formulae without any data. But, you may insist, these data are undetermined only for us. In themselves they are subject to laws. Granted! But note that we do not thereby really eliminate all contingency. We merely push it back one step. Thus the present arrangement of particles in the universe may be the result of mechanical forces operating on a previous distribution. But as mechanical forces cannot be supposed to operate except on some given distribution of material particles the contingency of the latter is as ultimate as the existence of the laws.

(2) CONTINGENCY OF IRRELEVANT FACTS. Every mechanical or physical law asserts that a certain phenomenon or characteristic depends on one or a few factors and on nothing else. The acceleration of gravitation depends upon mass and distance. Everything else is indifferent to it. The freezing of water depends on temperature and pressure, and nothing else is relevant. To hold seriously to the popular dictum that everything is connected with everything else would make the scientific search for determinate connection meaningless. There would be no use seeking for *the* cause of cancer, or *the* reason why sugar disintegrates the enamel of the teeth, if all things were causal factors. We put this in the accurate language of mathematics by saying that the laws of nature must be expressed in functions containing a limited number of variables.

(3) CONTINGENCY OF SCIENTIFIC LAWS. Why the particular laws of

nature which we have observed to prevail do so, rather than others, is a question which can never be answered without assuming certain contingent characteristics of the universe. For we cannot prove all propositions without assuming certain undemonstrated premises. It is always desirable to reduce the number of physical laws to a minimum. But the assumption that in this way we can eliminate all contingency is inadmissible. Even if all the laws of the universe could be derived from one —an assumption which the study of deduction shows to be impossible— that one would still be itself contingent. This uneliminable character of contingency is but the logical expression of the metaphysical fact of individuality. There is no universe without a plurality of elements, of atoms, of moments of time, etc. It is a blind hostility to pluralism, a preference for a lazy monism wherein all distinctions and differences are swallowed up, that leads to blatant panlogism from which all contingency is banished. But the latter attempt defeats itself. In the end the universe of existence has the particular character which it has and not some other; and contingency is not removed by being funded in the conception of the whole universe or made into the essential characteristic of reason itself.

(B) REASON AND THE ABSOLUTE

This brings us to an obvious but important observation: the principle of sufficient reason as actually employed in science is incompatible with the view that regards the total universe as the cause of any of its constituent facts. For the scientifically adequate cause of anything must be something determinate and the universe as a whole is not determinate in the sense in which any given fact in it can be made determinate. The total universe is by definition never actually complete in any moment of time and the principle of causality means that something occupying a given position in time and space can be determined only by something else also occupying a definite position in space and enduring over a definite time-interval. This is not to deny the determinateness of the physical universe in its distributive sense, i.e. in the sense that each thing in it is determinate. But the absolute collective whole is—at least from the point of view of scientific method—undetermined by anything outside of it; nor can the absolutely total universe be said to have any definite character such that from it we can infer that some particular entity has one rather than another determinate trait. Attempts to characterize the universe as a whole, as one (not many), continuous (not discontinuous), conscious or purposive, and the like, all involve a stretching of the ordinary use of words to include their opposites, and from this only confusion rather than determination can result.

We may put this in a different form by saying that scientific determinism is concerned with the definite character of things rather than with their brute existence. Rational scientific investigation is not concerned with the mystery of creation whereby existence may have come into being out of the void. It is concerned rather with the transformations whereby things or events acquire a determinate character within a given system. Even if the various parts of the universe influence each other and their relations can be rationally formulated, their brute existence cannot be thus derived.

(c) REASON AND THE REALITY OF UNIVERSALS

If the foregoing observations have stressed the incompatibility between organicism and the principle of sufficient reason it is because the incompatibility between the latter and the various forms of atomism or individualism is more obvious. Thus it ought to be obvious that the application of laws to phenomena presupposes the existence of real classes, that many things and processes are really alike. If there were no real likeness, no examples of identity in different instances, the formulation of scientific laws would be without any possible application. The great convenience of classification and the fact that the same things can be classified in different ways for different purposes, do not justify the conclusion that there is nothing in the things classified corresponding to the properties which serve as *principia divisionis*. There is no evidence for the nominalistic or phenomenalistic view that the universe *really* consists of atomic sensations and that scientific laws are fictions or nothing but convenient shorthand symbols for groups of separate facts that have nothing real in common. The scientific pursuit of rational connection presupposes that things do have certain common natures and relations. The economic efficiency of scientific knowledge is based on something in the facts.

(D) REASON AND DISCONTINUITY

While the application of the principle of causality thus implies the genuine existence of constant class properties it is well to note that in order that these classes be recognizable there must be discontinuities in nature. Thus we could never recognize any biologic species if there were not gaps between certain classes of animals (or plants) and others. The same is notably true in chemistry where the laws of multiple proportion and the properties of the different elements all involve discontinuity. If there had been other elements indistinguishable from hydrogen except in atomic weight, we should have called them all hydrogen and could not have suspected the constancy of the atomic weight of what we now isolate as hydrogen.

Our insistence that the interdependence assumed in the principle of sufficient reason is limited and involves independence as well as dependence, thus warns us against vicious forms of atomism, organicism, and mysticism. Atomism is vicious if it makes every entity a complete and independent universe and thus annihilates real relations between them. Such vicious atomism may be seen in the various forms of individualistic anarchy or pleas for irresponsible self-expression of every momentary impulse. Vicious organicism, the denial of all relative independence or externality of relation between the constituent elements of a system, may be seen in the various forms of social fanaticism, of which indiscriminate worship of the social or group "mind" is a characteristic example. Mysticism is vicious or obscurantist if it denies the definite or determinate character of things in the interest of beliefs which cannot stand the light of reason.

Science in thus emphasizing interdependence exercises a function analogous to that of social sense or sanity aware of the demands of a situation, while in refusing to eliminate independence, its function is analogous to that of the sense of beauty in which the individuality of things is intensified (though the atmosphere wherein they exist is not destroyed). In its opposition to obscurantism science operates very much like a keen sense of humour which is quick to note the absurdity of claims or the delicate lines which separate the sublime from the ridiculous.

(E) REASON AND PROBABILITY

The recognition that the material truths of physical science are more or less probable both corrects and enriches our conception of a metaphysic based on the requirements of rational or scientific method. It makes us less pretentious or arrogant in our claims and leads us to recognize the necessity of supporting our inferences by a multitude of diverse considerations rather than by a single dialectic chain. But more positively it calls our attention to the capital fact that in constantly increasing the relative probabilities of its results science is essentially a self-corrective system. The apodictic certainty of science is not the absolute certainty of any specific result or material proposition, but the dialectic demonstration that any inaccuracy or false step can be corrected only by relying on principles inherent in the system of science itself. This is a position unassailable by any argument that can pretend to have any evidence in its favour. We can discuss the rational alternative to any single proposition, but we cannot claim to have drawn any valid and significant inference without recognizing rules of implication or evidence. On purely historic grounds also, we have ample reason for the view that the methods of science which make it self-corrective form a more permanent feature of its continued existence than any specific results.

A metaphysic of scientific method is, then, concerned with the nature of a world in which the result of scientific investigation is always subject to contingency and error, but also to the possibility of self-correction according to an invariable ideal. This ideal is the direction from which correction must come if more adequate truth is to be attained. The analogy of remediable organic and social ills suggests a universe of parts closely interconnected in certain ways and yet in some degree independent of each other. Neither sensations (or other forms of immediacy) nor mediating relations by themselves can exhaust the full nature of existence; but every true existent has a domain of uniqueness and a domain in which its true existence is beyond itself: to wit, the larger system of which it is a part.

Complete nature cannot reveal or exhibit itself in any moment or interval of time as far as that moment excludes other moments. But in so far as the meaning and content of each *here* and *now* necessarily involves some essence or character that is more than merely here and now, we have a point of view in which the whole of time is included. The point of view to which the whole of time and space has a meaning may be called the eternal (as distinct from the everlasting, which applies to what endures in time and space). It is true that in no moment or interval of time can we grasp or see as actually present to us the whole content of time and space which we call *the* universe. But in knowing the meaning of any fragment as a fragment we know the direction of completion. In this sense there can be no valid objection to the assertion that a knowledge of the absolute is involved in any true knowledge of phenomena.

Eternity may thus also be viewed as the limit or ordering principle of a series of expanding vistas. Such a limit may be called ideal. But the objective validity of such an ideal is not to be dismissed as merely mental. It is a genuine condition of the series of stages in the self-corrective system of natural science.

§ III. THE NATURE OF THINGS

IF WE thus take the principle of sufficient reason seriously we are justified in examining the nature of things without worrying about the ego-centric predicament of how we know that such knowledge is possible. The assumption of the critical philosophy that we can know only our own ideas is itself a dogmatism which involves an infinite regress. If the fact that I know a given entity does not determine any of its specific characteristics—and it is hard to see how it can determine any one known trait more than any other—then the fact of knowledge can be eliminated from the most general formula for the nature of things, though the existence of knowledge is itself a most important fact in our universe.

The fundamental metaphysical issue between rationalism and the various forms of anti-rationalism may be stated thus: Is the nature of things revealed most fully in developed rational science, or is it so well known in non-rational ways that we are justified in saying that science is a falsification or a merely practical device for dealing with dead things? Actually the various forms of anti-rationalism dogmatically assert the nature of things to be "really" individuality or continuous experience, spontaneity or practical experience, etc. But an attempt to justify any one of these formulae by evidence commits the anti-rationalist to the canons of scientific method.

The main metaphysical contention of anti-rationalism with its banal shibboleth about life, the organic, the dynamic, etc., is that things have no constant nature, that everything is pure change and nothing else. Historically we can understand the motive for this when we reflect how many of the old constancies have had to be abandoned in the progress of physics and biology. But though the principle of identity has undoubtedly been abused, the effort of the reaction to draw an account of the world without any element of identity in it is clearly self-defeating. Changes cannot have any definite character without repetition of identical patterns in different material. If the growth of science dissolves the eternity of the hills or the fixity of species, it is also discovering constant relations and order in changes which previously seemed chaotic and arbitrary. In daily life we find no difficulty in asserting that an individual or object maintains its character in the stream of change. Scientifically this constancy is expressed in the accurate language of mathematics by the concept of the *invariant*, not the isolated constant but that which remains identical amidst variation. We may say then that the nature of anything is the group of invariant characters.

From this point of view we are justified in making the ordinary distinction between the nature of anything and its manifestations. The fact that science seeks the invariant properties amidst the flux makes clear why science is never satisfied with empirical fact, and always seeks for explanation why things are constituted or behave in their particular way. The answer to the question *why*, is always a reason which puts the fact to be explained into a system, so that knowing the nature of the system and certain data (or given existences), we can deduce or form a rational account of the events to be explained. The fact that the abstract law makes the concrete particular intelligible does not, of course, prove that this law is less real or is more the product of our arbitrary fiat than the fact explained.

This view as to what is meant by the nature of things necessarily assigns a large and necessary rôle to the realm of possibility. If the actual is identified with the immediate, and the immediate with the sensuous, then the actual is certainly an infinitesimal portion of the wider world which it is found to presuppose. The sensuous vividness of the immediate may often be precisely what is meant by reality. But scientific reflection must and always does assume a larger world than that which is immediately before us or actual. Most of the sensuous material of the past, and of remote space, is beyond us and yet conditions the actuality before us. To be sure, the realm of possibility may be partly anticipated in actual imagination. But this is at best necessarily fragmentary. We may also denote by the word *actual* the historical order—all the things that have happened or will happen in all time and space, including men's dreams as themselves events. But shall we include in actuality the relations or implications of things which no one has perceived and which no one will, because of human limitations? Clearly we must distinguish here between knowledge by reference and knowledge by realized acquaintance. The totality of nature through all time and space is a limit which we can never attain and yet the idea of it is a necessity of scientific method. For the explanation of any part of the world always presupposes still other parts necessary to its complete determination. A completed rational system having nothing outside of it nor any possible alternative to it, is both presupposed and beyond the actual attainment of any one moment. It coincides in part with the Bradleyan Absolute, but it is an ideal limit rather than an actual experience. Unrealized possibilities are within it precisely to the extent that it contains endless time.

§ IV. RATIONALISM, NATURALISM AND SUPERNATURALISM

IT is frequently asserted that the principle of scientific method cannot rule out in advance the possibility of any fact, no matter how strange or miraculous. This is true to the extent that science as a method of extending our knowledge must not let accepted views prevent us from discovering new facts that may seem to contradict our previous views. Actually, however, certain types of explanation cannot be admitted within the body of scientific knowledge. Any attempt, for instance, to explain physical phenomenon as directly due to providence or disembodied spirits, is incompatible with the principle of rational determinism. For the nature of these entities is not sufficiently determinate to enable us to deduce definite experimental consequences from them. The Will of Providence, for instance, will explain everything whether it happens one way or another. Hence, no experiment can possibly overthrow it. An hypothesis, however, which we cannot possibly refute cannot possibly be experimentally verified.

In thus ruling out ghostly, magical, and other supernatural influences, it would seem that scientific method impoverishes our view of the world. It is well, however, to remember that a world where no possibility is excluded is a world of chaos, about which no definite assertion can be made. Any world containing some order necessarily involves the elimination of certain abstract or ungrounded possibilities such as fill the minds of the insane.

From this point of view, what may be called the postulate of scientific materialism, viz. that all natural phenomena depend on material conditions, is not merely a well-supported generalization but the requirement of an orderly world, of a cosmos that is not a chaotic phantasmagoria.

As materialism has served as a sort of "bogey-man" to scare immature metaphysicians, it is well to make more explicit its relation to rational scientific method. If materialism means the denial of emotions, imaginings, thoughts, and other mental happenings, it is clearly not something worthy of serious consideration. It is contrary to facts of experience and clearly self-refuting. But this in no way disposes of the materialism of men like Democritus, Lucretius, Hobbes, or Spinoza, or of the assumption that every natural event must have a bodily or material basis.

The truth of the latter proposition is obscured by the popular confused concept of mental efficiency. Even technical discussion of the relation of mind and body is often vitiated by an inadequate analysis of the principle of rational determinism and a consequent misapprehension of the force of principles such as the conservation of energy. If, as it was contended before, scientific causality applies only to certain abstract aspects of entities, there is no reason why entities that determine each other in one way in a certain system may not be bound together in another way in another system. The presence of causal relation does not involve the denial of teleologic relations, while the assertion of the latter presupposes the former.

The principle of conservation of energy, for instance, leaves a wide realm of indetermination that other relations can make determinate. For the assertion that amidst all the transformations of a system the total amount of energy remains constant, is an assertion that clearly does not determine the character of the transformations other than in the one trait explicitly mentioned. The second law of thermodynamics does endeavour to indicate a general direction of phenomena, viz. towards a maximum of entropy. But this is largely problematic in certain regions, and in any case it often leaves room for all sorts of indetermination. Whether energy shall remain potential or be transformed into energy of motion is not completely determined by either principle, since either principle remains fulfilled whether the transformation takes place or not. Teleologic determinations are therefore not ruled out by the laws of energy.

Now the ordinary conception of mental efficiency combines elements of teleologic with strictly causal determination. We achieve certain purposes by taking advantage of natural mechanism. But this is in no way inconsistent with the proposition that every material change is correlated with a previous material condition in accordance with certain laws. What is inconsistent with scientific procedure is the argument that the existence of a mental motive makes the coexistence of a physical cause unnecessary.

Besides the objection from the existence of mental efficiency there is another line of arguments against materialism, viz. that matter is purely passive and cannot explain the activity of the world. Now it is doubtless true that the close connection between the notion of matter and that of mass or inertia, leads to the view that matter by itself cannot explain the active processes of nature, and this leads to the introduction of forces which are the ghosts of spirits or of the volitions often connected with our own bodily movements. This argument, however, is based on a logical fallacy, taking the nature of matter to consist solely in its exclusive and passive aspect. But there is no valid reason against supposing that a purely material system without external influence can contain motion within it; and there is no conclusive argument against the view that under certain conditions material systems such as those which constitute the human body are capable of organic processes, feeling, etc.

The one serious objection to materialism from the point of view of the requisites of scientific method comes into play when materialism allies itself with sensationalistic empiricism and belittles the importance of relations and logical connection between things. The identification of empiricism with the scientific attitude is just a bit of natural complacency. The excessive worship of facts too often hides a disinclination to enter into a genuine inquiry as to whether they are so. A rationalism that is naturalistic must, of course, agree with empiricism in maintaining the factual or immediate aspect of existence. But scientific rationalism is incompatible with the complacent assumption that in sensations or in self-sufficient "facts" we have the only primary existences.

We do not have pure particulars any more than pure universals, to begin with. We begin with vague complexes which raise difficulties when we wish to give a rational or coherent account of them. It is scientific procedure itself which enables us to pass from vague impressions to definite propositions. Definite individuals are, therefore, the goals or limits rather than the data of scientific method. When we attain knowledge of particulars we see that their nature depends upon the universal connections which make them what they are.

To realize that the substance or nature of the individual consists of universals we must get rid of the Lockian confusion between matter and substance, and return to the Aristotelian distinction of ὕλη and ὀυσία. Ὕλη or matter is a relative concept. Bricks are ὕλη for a building but are formed substances for one who makes them out of clay. Absolute primary ὕλη or matter is a limiting concept, not a starting point. The intelligible substance of things, however, is not pure formlessness or empty possibility, but the actual universals which, though arrived at as a result of inquiry, are conditions of what exists. Indeed, inquiry like all other forms of human effort must begin with the partial and can attain the whole or universal—if at all—only by seeing how the parts are conditioned.

The view that identifies the genuine substance of things with those relations or structures which are the objects of rational science is so opposed to the nominalistic tendency of our time, that a host of objections to it is naturally to be expected.

The most serious objection is one that cannot be answered—it is the habit which associates substance with reality and reality with the sensuously or psychologically vivid. But, however decisive the appeal to the subjectively vivid may be in practice, it is after all no evidence as to the objective constitution of the natural world.[2] More specific objections to our identification of the order of scientific ideas with the intrinsic order of things are the following: (a) Rational scientific method is devised for practical purposes only. Its fictional devices cannot give us truth. (b) The abstractions which science employs have no correspondence with or real existence in the natural world. (c) Reasoning supplies us only ground for belief not ground for existence.

(A) FICTIONALISM

Philosophers as diverse as Bradley, Mach, and Bergson rely heavily on the so-called fictions of science, e.g. corpuscles of light, the ether, etc., which have proved useful, though not literally true. With regard to this we may observe that many of these so-called fictions, e.g. atoms, have turned out to be very much like other empirical entities. We count them, we weigh them and study their behaviour—philosophers to the contrary notwithstanding. I have elsewhere tried to show that contrary to the contention of Vaihinger, none of the so-called fictions of science involve any contradiction.[3] If they did so, they could not be useful, since no consistent inferences could be drawn from them. Even when not completely true, they are analogies which offer useful suggestions just to the extent that they are true. To the extent that they fail they are subject to the process of correction.

(B) CONCEPTUALISM

The form of reasoning to which science always seeks to attain is mathematics. But do the steps of a mathematical process correspond to anything in the objective world—even when the initial premises and final conclusions do? Mathematical physicists like Duhem and Mach categorically deny this. What have our equations, differentiations, and integrations to do with natural objects? If the result of a mathematical calculation gives us a true account of objective nature, may not the

---

[2] "If reality," argues Bradley, "consists in actual sequence of sensuous phenomena then our reasonings are all false because none of them are sensuous." To which we reply that if reality consists of sensuous material arranged in certain form or order then all the reasoning which is faithful to that form or order is true.

[3] See "The Logic of Fictions," *Journal of Philosophy*, Vol. XX (1923), p. 477.

mathematical process correspond to the sharpening of tools or to the mixing of colours, processes which surely do not correspond to the features which the artist wishes to represent? We must not however allow analogies to lead us away from the facts. Mathematical propositions do relate to the properties of all possible objects. Valid mathematical reasoning therefore deals with processes to which the objects before us are as subject as any others. It is often difficult to recognize these universal aspects in the particular, just as it may be difficult to recognize that an enemy is also a human being, yet truth requires the recognition of just such obvious general or universal aspects. Mathematical reasoning may indeed be too general for a specific situation (if we lack the proper data) but if true it always has objective meaning.

(c) IRRATIONALISM

Bradley has argued that while reasoning determines the ground of our belief, it does not even pretend to determine the ground of existence. In our reasoning, he claims, some datum suffers alteration. Why assume that reality transforms itself in unison? This objection is largely based on the false suggestion of the word *transformation*. Logical reasoning does not produce any temporal change in the object reasoned about. The latter remains the same when we make progress in the recognition of its nature. But the ground of a true belief differs after all from the ground of a false belief precisely in this—that the former *is* connected (though not directly identical) with the ground of existence. For this purpose the most favourable example, for Bradley's objection, would be the case where I conclude that my idea of a given object is false. But even here the ground of existence and the ground of true belief are not independent. I begin in fact with an hypothesis as to the nature of the object. I consider what consequences it (with other things) has if it is true. I find the consequences are impossible because in conflict with actual existence, and I conclude that my original hypothesis is false. If my reasoning is valid it is because it has come into contact with actual facts and the transformation of the entities reasoned about do correspond to reality.

We conclude, then, that if the abstract is unreal, reality is of little moment. For what that is humanly interesting is not abstract? Mr.

Bradley has gone through the whole gamut from qualities, relations, and things to our precious selves, and shown with a logic that is more readily ignored than refuted, that all these things are but abstract or detached parts of the absolute totality. But the conclusion that everything short of the absolute totality is appearance and not reality is a logical consequence of an arbitrary view of reality, which identifies it with purely immediate feeling or experience. But though the craving of the flesh for strong sensations and feelings is an important element of life, it is certainly not conclusive even as to the guidance of life. Even the hedonistic ideal cannot be realized except by organizing life on the basis of an intellectual recognition of our possibilities and the rational evaluation of the different factors which determine our happiness.

The contention that abstractions or logical relations form the very substance of things does not, according to the foregoing account, involve panlogism. Rationality does not exhaust existence. The relational form or pattern points to a non-rational or a logical element without which the former has no genuine meaning. For to deny the existence of any irrational elements is to make rationality itself a brute, contingent, alogical fact. The fact that we can rationally use terms like *irrational, alogical, inexpressible*, and the like has given rise to interesting paradoxes. These paradoxes, however, disappear if we recognize that a word may point to something which is not a word at all, and though the pointing is rationalized fact the thing pointed to may not be so. Rational distinctions and relations and all expression hold in the field of being which is thus presupposed but never fully described—just as the various lines on a blackboard may indicate the various objects represented on it but do not fully represent the blackboard itself which conditions them. If this doctrine that our universe thus contains something fundamental to which we may point but which we cannot fully describe be called mysticism, then mysticism is essential to all intellectual sanity. Language ceases to be significant if it cannot indicate something beyond language. But if we use the word *mysticism* to denote this faith in a universe that has ineffable and alogical elements, we cannot too sharply distinguish it from obscurantism. For the former denies our power to know the whole of reality, while the latter holds reality to be definitely revealed to us by non-rational processes. Rationalism does not deny that clear thoughts may begin as vague or obscure premonitions. But the essential difference between rationalism and obscurantism depends upon whether our guesses or obscure visions do or do not submit to the processes of critical examination and logical clarification. Our reason may be a pitiful candle light in the dark and boundless seas of being. But we have nothing better and woe to those who wilfully try to put it out.

§ V. THE PRINCIPLE OF POLARITY

THE foregoing considerations are all applications of a wider principle, viz. the principle of polarity. By this I mean that opposites such as immediacy and mediation, unity and plurality, the fixed and the flux, substance and function, ideal and real, actual and possible, etc., like the north (positive) and south (negative) poles of a magnet, all involve each other when applied to any significant entity. Familiar illustrations of this are: that physical action is not possible without resistance or reaction and that protoplasm, in the language of Huxley, cannot live except by continually dying. The idea is as old as philosophy. Anaximander expressed it in saying that determinate form arises out of the indefinite (το ἄπειρον) with the emergence of opposites like hot and cold, dry and moist, etc. And Heraclitus insisted that strife was the father of all things and that the balancing of opposite forces, as in the string of the bow or lyre, gave form to things. The essential Hellenic wisdom of Socrates and Plato which viewed justice and the other virtues as conduct according to measure (Aristotle's mean) involves the idea of adjustment of opposite considerations. The relativity of form and matter, according to Aristotle, is determinative of all existence (save the divine essence).[4]

[4] The reading of Plato's *Parmenides* first impressed upon me the lesson taught to the young Socrates, viz. that it is impossible to arrive at sound philosophy without experience in tracing the diverse and opposed dialectic implications of such propositions as "unity exists" and thus learning to guard oneself against their pitfalls. Propositions are not bare tautologies but significant predications because non-being has being of a sort and *the one* is inseparable from, though not identical with, *the other*.

I am indebted to Professor Felix Adler for the figure of the scissors to denote the fact that the mind never operates effectively except by using both unity and plurality like the two blades which move in opposite directions. Professor Marshall, in his *Principles of Economics*, has used the same figure to express the mutual dependence of the two factors of supply and demand. We may, if we like, also use the figure of the pestle and mortar, of our jaws in mastication, or of applying brakes when going down a hill.

This principle of polarity seems to me to represent what is sound in the Hegelian dialectic without the indecent confusion at which we arrive if we violate the principle of contradiction and try to wipe out the distinctions of the understanding. The being and non-being of anything are always opposed and never identical, though all determination involves both affirmation and negation. Far from overriding the distinctions of the understanding the principle of polarity shows their necessity and proper use. Thus physical science employs this principle when it eliminates the vagueness and indetermination of popular categories like *high* and *low, hot* and *cold, large* and *small, far* and *near,* etc. It does so by substituting a definite determination such as a determinate number of yards or degrees of temperature. The indetermination and consequent inconclusiveness of metaphysical and of a good deal of sociologic discussion results from uncritically adhering to simple alternatives instead of resorting to the laborious process of integrating opposite assertions by finding the proper distinctions and qualifications.

Under the head of polarities we may distinguish between contradictions, antinomies, and aporias or difficulties. Strictly speaking, contradictions are always dialectical, i.e. they hold only in a logical universe. Thus if I say a house is thirty years old, and some one else says it is thirty-one years old, the two statements are contradictory in the sense that both cannot possibly be true at the same time and in the same respect. Both statements, however, can certainly be true if we draw a distinction, e.g. thirty-one years since the beginning and thirty years since the completion of its building.

Thus two statements which, taken abstractly, are contradictory may both be true of concrete existence provided they can be assigned to separate domains or aspects. A plurality of aspects is an essential trait of things in existence. Determinate existence thus continues free from self-contradiction because there is a distinction between the domains in which these opposing statements are each separately true. When opposing statements are completed by reference to the domains wherein they are true, there is no logical difficulty in combining them. In the purely logical or mathematical field, however, we deal not with complexes of existence, but with abstract determinations as such. Here two contradictory assertions always produce a resultant which is zero, i.e. the entity of which they are asserted is absolutely impossible.

166

Of incompletely determined existence—as in the case of the total universe—contradictory propositions do not annihilate each other (since they refer to a complex of existences); and yet they cannot always (because of the indefiniteness of the subject) be reconciled with each other. This gives rise to the antinomies of metaphysics.

In general, the opposite statements that are true in regard to existing things give rise to difficulties when we cannot see how to draw the proper distinction which will enable us to reconcile and combine these seeming contradictions. Thus we frequently find certain facts in a scientific realm calling for one theory, e.g. the corpuscular theory of light, and other facts calling for a diametrically opposite one, viz. the wave theory. Such difficulties are solved either by discovering new facts which give one of these theories a preponderance or else by discovering a way of combining the two theories. Sometimes an intellectual dilemma is avoided by rejecting both alternatives. This is illustrated by the old difficulty as to whether language was a human invention or a special revelation. The difficulty was avoided by introducing the concept of natural growth.

Nature also presents us with seeming impossibilities in the form of practical difficulties, e.g. how to live long without getting old, how to eat our cake and yet have it too, etc. Such contingent or physical impossibilities may baffle us forever. Yet some of them may be solved by finding the proper distinction. Thus the invention of boats enabled us to eliminate a former impossibility—namely, how to cross a river without getting wet.

This analysis puts us on guard against two opposite evil intellectual habits: on the one hand to regard real difficulties as absolute impossibilities, and on the other to belittle such difficulties by calling them false alternatives. Thus it is not sufficient to say that the old controversy between the claims of the active and those of the contemplative life represents a false alternative, and that we need both. It is in fact most frequently impossible to follow both and the actual problem of how much of one we need to sacrifice to the other often requires more knowledge than is at our disposal.

If it be urged that this after all is the essence of the Hegelian logic I should not object—provided it does not include Hegel's explicit identification of the historical and the logical, the real and the rational. The heart of Hegel's philosophy is, after all, the attempt at a synthesis calculated to do justice both to the classic rationalism of the Enlightenment and to the inspiring sweep of the romanticism of Fichte, Schelling, and their associates. Such a synthesis seems to me to be the great desideratum of our age. We cannot today accept Hegel's methods and results precisely because they are not—despite all their pretensions—sufficiently rational or logically rigorous. But his tremendous influence in law, art, and religion, as in the development of all the social sciences, shows that he grappled with a vital problem. If, as I think we should all admit, he was guilty of indecent haste due to intellectual ὕβρις, it is for us to face a similar task with greater patience and honest resoluteness not to minimize the obstacles to rational inquiry.

These suggestions of a possible metaphysics may be objected to either as commonplace, as unimportant, or as unjustified. Against the first objection we must note that sound metaphysic like science itself should begin—though it should not end—with the commonplace. As against the second and third objections we may urge that the full meaning, importance, and justification of a metaphysical doctrine can be seen only in its development. Towards this development the present chapter can offer only the barest hints.

END OF BOOK I

Book II

REASON IN NATURAL SCIENCE

Chapter One

THE NATURE OF MATHEMATICS[1]

ATHEMATICS is the first of the sciences. Of course men had knowledge of stars, metals, plants, and animals long before Thales and Pythagoras discovered the art of proving theorems. But not only is this art of proof or demonstration itself one of the outstanding achievements of the human mind, but the distinction between science and common knowledge largely depends on it. Apart from proof human knowledge is always subject to the uncertainties of the future. Not only may new stars, metals, plants, and animals appear, but the old ones may reveal new properties. Such painfully acquired knowledge as that the stars are fixed relatively to each other, that the mass of any metal remains constant, that plants are rooted in the soil, and that fishes live only in the water may all turn out to be in different ways somewhat illusory. But never will a rational number turn up that will denote exactly the square root of two. The proof of this proposition was already ancient and classical in the days of Aristotle.[2] Let us consider it here as a model to keep in mind when we refer to mathematical demonstration.

The proof consisted in considering what would be the case if there were such a number. It would of course have to be a fraction, since there is no integer between 1 and 2. If we denote such a fraction by s/t we shall have the equation $s/t = \sqrt{2}$. We may assume that s and t have no common factor, since if they had one we could divide them both by that common factor without changing the value of the fraction. Now if the equation $s/t = \sqrt{2}$ be true, we can square both sides and, multiplying them both by t^2, get $s^2 = 2t^2$. It follows that s must be an even number, for if it did not contain 2 as a factor its square (s·s) could not contain it; and if s is even, t must be odd since s and t are prime to each

[1] This chapter embodies some material published in "The Present Situation in the Philosophy of Mathematics," *Journal of Philosophy*, Vol. VIII, 1911, p. 533.
[2] *Prior Analytics*, Bk. I, Ch. XXIII.

other. But if s is even, it is divisible by 2 and equal to twice some other number, which may be denoted by n, i.e. $s = 2n$, and $s^2 = 4n^2$. Hence instead of $s^2 = 2t^2$ we may write $4n^2 = 2t^2$, or, dividing by 2, $t^2 = 2n^2$, which means that t is even, and therefore s must be odd. If, then, there were a number equal to the square root of two, its numerator and denominator would each be both odd and even, i.e. divisible and not divisible by 2; and this is of course absolutely impossible.

While other sciences are not always so rigorous in their demonstrations, they differ from common sense to the extent that they utilize rigidly demonstrative reasoning.[3]

The fact that mathematics is thus both demonstrative and constantly applied to nature, leads to the following seemingly paradoxical triad:

(1) The propositions of mathematics deal with the material world.
(2) Material propositions are not necessary truths.
(3) The propositions of mathematics are necessary truths.

Each of these propositions seems to be obviously true, and in common sense they might dwell amicably side by side. But they are logically inconsistent, that is, any two logically imply the falsity of the third, so that a coherent philosophy demands that at least one of these seeming truisms be denied. Thus Kant argues that since the first and third propositions are obviously true, the second must be false, and concludes that there must be some intuitive source of a priori truth as to the material structure of the universe. Mill, accepting the first and second of these propositions, is forced to maintain that mathematics does not give us necessary truth, so that in some other universe, for instance, two times two might equal five. Others, e.g. Mach and Duhem, accepting the second and third propositions, come to the conclusion that mathematics tells us nothing about the material universe, but is merely a way of manipulating symbols. Each of these positions is revolting and unnecessary. We can avoid all paradox by distinguishing between the formal and the material aspects of mathematics, and noting that the first

---

[3] This difference is of course one of degree, since all coherent discourse involves some rational connection, indicated by such words as *because, for, hence, it follows*, etc. But one who studies Oriental literature before the advent of Hellenistic influence cannot but be struck by the absence of that rational classification and methodical or systematic order which is characteristic of the Greek spirit and only found its clearest expression in mathematics.

proposition is true only if by *mathematics* is meant *applied mathematics*, that the third refers to *pure mathematics*, and that consequently the second proposition is consistent with the truth of the other two.

This distinction raises three problems of real significance: (1) What is the difference between pure and applied mathematics? (2) How can a purely formal science exist? (i.e. How can *novel* truths be derived by deduction?) (3) How is applied mathematics possible? (i.e. How can mathematical methods increase our knowledge of the material world?)

§ I. THE DISTINCTION BETWEEN PURE AND APPLIED MATHEMATICS

THE distinction between pure and applied mathematics has been brought to light by recent developments not only in mathematics but also in physics and logic.

In the first place, the discovery of non-Euclidean geometries and of algebras and arithmetics in which the old familiar laws such as the commutative ($ab = ba$) no longer hold, has made it clear that the question of the logical consistency of the Euclidean system or of ordinary algebra and the question whether physical objects conform to these systems are distinct issues. Then, also, the rise of non-quantitative branches of mathematics (e.g. the theory of groups, projective geometry, and analysis situs) has shown the inadequacy of the traditional definition of mathematics as the science of quantity or magnitude, and has naturally suggested that the essence of pure mathematics may be its logical form rather than any particular subject-matter. The distinction between pure and applied mathematics is also forced on us by the development of physics, in which the use of mathematics is not merely formal but depends upon giving to logical indefinables an interpretation in terms of physical objects and operations, while other branches of mathematics, such as the theory of prime numbers, find no such interpretation or application in physics. Finally the movement which began with Gauss to make mathematics logically rigorous, and the movement to generalize logic begun by Leibniz and renewed in the nineteenth century (by DeMorgan, Boole, Peirce, et al.) have united to make it clear that mathematics as a pure formal science is indeed identical with logic. Since logic alone is admittedly incapable of deciding such questions as that of the nature of space, the question whether actual space is Euclidean must belong to some branch of natural or physical science that may be called *applied mathematics*.

Let us consider this distinction more fully in the field of its first historic appearance, geometry. Geometry, as the term itself indicates, began as a practical science dealing with the measurement of land. The common sense of surveyors and builders consisted of useful but often inaccurate formulae until the subject took on a more abstract form at the hands of Thales and Pythagoras. Euclid succeeded in giving it a logical coherency which has served as a model of scientific system ever since. Realizing that while diagrams are an aid to the imagination they are not essential to rigorous proof, and that the appeal to sensuous intuition has indeed frequently been a source of fallacious reasoning, modern geometry has refined upon Euclid's procedure. In the works of men like Hilbert, Veblen, Pieri, and Huntington, it assumes the form of a perfectly deductive system where every proposition is logically deduced from the fundamental axioms or postulates.

Formerly these axioms were viewed as propositions whose truth was guaranteed by their self-evidence, but the persistent and unsuccessful efforts to prove Euclid's parallel postulate showed doubts as to its self-evidence. Since Riemann, other axioms of Euclid have been doubted and various systems of non-Euclidean geometry have thus been built up. By suitable transformation formulae Cayley, Klein, and Whitehead have shown that for every proposition in the Euclidean there is a corresponding one in the Lobachevskian and the Riemannian geometries, so that if there is an inconsistency in any of the latter it is also to be found in the former.

But though these geometries are logically alike as to consistency, they differ in their assumptions. Which is true?

The pure mathematician does not need to answer this question. He does not assert either the Euclidean or the non-Euclidean axioms to be true. He asserts only that from the Euclidean axioms certain theorems necessarily follow, and that from other axioms other propositions must follow. In the physical world, however, we are not satisfied with systematic consistency. We want to know which axiom or system is true in fact. From a point outside of a straight line can only one parallel be drawn, or can more than one or none at all be thus drawn?

Obviously we can determine this physically only by going out into the field and verifying by means of measurement with proper instruments under proper conditions some consequences of these axioms, e.g. whether the sum of the angles of a triangle is equal to, greater than, or less than two right angles. Note, however, that in determining which are the proper measuring instruments, in choosing a platinum yardstick rather than an iron one, or a light ray and clock rather than either, I must assume certain truths as to the physical nature of these entities; and such truths are beyond the scope of pure mathematics. More generally still, you cannot test the physical truth of Euclidean geometry without giving its really undefined terms [4] such as *point, line,* etc., some specific interpretation by identifying them with some existing things such as the crossing of two chalk marks, a taut string, or a light ray. But this restriction on the meaning of the undefined terms that enter into the axioms of geometry is entirely irrelevant to the purpose of the pure mathematician. His assertion, as we have seen, amounts to saying that if *anything* has the properties specified in the axioms, it will also have the properties asserted in the theorems. The physicist exemplifies this *anything* in certain idealized entities, such as rigid bodies, with which certain existing bodies are qualifiedly identified. But so far as the pure mathematician is concerned an indefinite number of other entities may exemplify his propositions just as well.

Thus instead of interpreting the undefined term *point* as a position in physical space, we may, without changing the deductive order of Euclid's propositions, make the word *point* denote the degree of holiness of an angel. *Right* and *left* can denote greater and lesser holiness. A line, as a series of points, will then be a series of degrees of holiness. A plane, as a series of lines, will be a series of such series. And so we may go through the whole geometry and interpret every term and proposition as some function of angelic holiness. The logical consistency of our geometric system will not thereby be in any way changed, since no term or proposition of that system (as perfected by modern refinement) depends upon any definition or intuition of *point*. If our sug-

---

[4] Every deductive system must, of course, contain terms which it does not define, just as it must contain propositions and rules of inference which it does not prove.

gested interpretation of the Euclidean point, line, and plane seems strange, consider the more orthodox one in which *points* are interpreted as pairs of real numbers. The latter gives us the well-known double or two-dimensional algebra as the logical equivalent of plane geometry.

If this conclusion is not obvious it is only because we are generally not accustomed to think of geometry except as concerned with a definite subject-matter in the physical world. But this restriction goes beyond the formal statements and definitions of Euclid to a physical interpretation of his undefined terms. Such a restricted interpretation of geometry is important (as well as historically primary and for practical purposes perhaps most useful). But it is also important that we note the difference between geometry as pure mathematics and geometry as a branch of physics.[5] What the older rationalists like Kant have failed to note is that in passing from pure to applied geometry we leave the apodictic certainty which attaches to the former and reach the empirical problems of that geometry which Newton already characterized as the simplest branch of mechanics. The test of truth must now be applied not only to the logical assertion that something *does* follow from something else, but also to the material assertion that our spatial entities have certain definite properties. We can, as pure mathematicians, prefer to restrict our investigation to what rationally follows from certain possible situations, but we cannot as physicists assert on a priori grounds that these possible situations are actually exemplified in space.

The presence of undefined terms in our various pure geometries makes it possible that with suitably diverse interpretations of these terms the same physical facts can be described equally well in Euclidean, Lobachevskian, and Riemannian geometry. Thus in the problems of navigation it makes no difference whether we use Riemannian plane geometry or Euclidean spherical geometry—the Riemannian straight line is indistinguishable from the Euclidean arc of a great circle in so far as both are the shortest distances on the surface of the earth. Similarly on the surfaces studied by Minding,[6] a Lobachevskian straight

---

[5] Note, in passing, that physical particles which may for certain practical purposes be viewed as points approximate but do not exactly conform to the requirements of pure geometry. Thus in the latter realm we can always place a point between any two, and hence an infinite number; but no existent physical entities can exemplify this property.

[6] Lobachevsky's paper on "Imaginary Geometry" appeared in *Crelle's Journal* in 1837. Two years later in the same journal Minding published a paper on the trigonometry of

line will have all the properties of a Euclidean geodesic or minimal distance.

If, however, we give precisely the same physical denotation to all the corresponding terms in the various geometries, the latter become incompatible physical theories. And difficult as it may be to devise appropriate experiments to test their comparative merits, it must be assumed that a material difference in our hypotheses will sooner or later bring forth a difference in their consequences.

It is instructive to bear the last point in mind when we consider the position of Poincaré. No one has pressed more forcefully the meaninglessness of empiricism in pure geometry or has so cut the ground from under Kantian intuitionism. For Poincaré shows clearly that the claim for an a priori intuition that space is Euclidean rather than Riemannian is no more intelligible than a claim that we have an a priori intuition that space conforms to a decimal rather than to a duodecimal arithmetic. Yet he insists that Euclidean geometry will always be preferred because it is inherently more simple, and if any future optical, astronomical, or other physical phenomena fail to conform to it, humanity will always prefer (for reasons of simplicity) to explain departures from the Euclidean pattern as due to special physical circumstances or causes.

This shows some confusion as to the proper rôle of the principle of simplicity in physics, and a failure to realize that if a physical interpretation is given to a geometric system the latter becomes part of physics and must be integrated with other physical theories. It is true that the metric formulae for distances, areas, and volumes are inherently simpler in the Euclidean geometry than in the others (as judged by the number of elements that these formulae contain); and to the extent that the Newtonian mechanics which assumes them satisfies our physical observations there is no reason for abandoning them. But we must distinguish between the inherent simplicity of our initial propositions taken by themselves and the simplicity of the entire system based upon them. Of what scientific value is it to start with simpler propositions if to do so involves subsequent introduction of complicating features? If geometry is to be applied to the actual physical world, there can be no

certain peculiar saddle-backed surfaces (now called pseudospherical), in which all the formulae were exactly those of a Lobachevskian plane triangle. Yet this fact was not noticed until Beltrami called attention to it in 1868.

a priori certainty that the Euclidean system will always give us the simplest results.

In fact since Poincaré wrote, Einstein has shown reason for assuming a non-Euclidean geometry in his physics.

Classical mechanics regarded space, time, and mass as independent variables. Just as we study lines, then planes, then volumes, to get the geometric properties of space, so we add the time element to geometry to get kinematics (the "geometry of motion"), and when to this we add the consideration of masses, we have mechanics. But Riemann and Helmholtz had already suggested that our geometric ideas, such as superposition, are moulded by our idealized picture of the motion of rigid bodies. In his special theory of relativity Einstein established that the spatial and temporal co-ordinates of physical bodies are not in fact independent of each other and that both are dependent on the velocity of the body. According to the conditions of measurement all bodies in motion must be foreshortened in the direction of their motion, so that a circular body at rest becomes flattened or elliptical when in motion. But as the same body can be said to be at rest in respect to one center of co-ordinates, e.g. one fixed in the earth, and in motion in respect to another, e.g. one fixed in the sun, the question whether a given body on the earth is circular or elliptical in a given plane depends on whether we regard the earth or the sun as in motion. The geometric properties of bodies on the earth depend therefore on whether we adopt the Ptolemaic or the Copernican astronomy.

In his generalized theory of relativity, however, Einstein goes further and shows not only that the measurable geometric properties of bodies must depend on the presence of masses and gravitational fields, but that if we regard the earth as rotating around an axis fixed with respect to the sun and stars, the ratio between the circumference and the diameter of a circle of latitude on it can no longer be π, and our geometry cannot be regarded as Euclidean.[7] For as a body rotates, the arc perpendicular to any radius contracts in the direction of motion, but not the radius itself.

Now it is possible to hold on to Euclidean geometry and to explain this and other phenomena, such as the movement of the perihelion of Mercury, and the bending of light rays and the spectrum shift in a

[7] Einstein, *Geometrie und Erfahrung* (1921), pp. 6-7.

gravitational field, as due to special physical forces. But such special forces certainly do not simplify our physics, and they are peculiar in that they are independent of the physical structure of any of the objects affected; they cannot be screened and they affect all objects alike. Are not these properties which are common to all objects alike irrespective of their special physical structure exactly what we ordinarily call kinematic and geometric? If the geometric properties of space are to be determined by physical means,[8] say light rays, clocks, yardsticks, and moving bodies, then Einstein has shown reason for holding that actual space is not Euclidean within the highest degree of attainable precision.[9]

To sum up, geometry as a statement of the properties which *any* entity must logically have if it has certain other properties is pure mathematics; but geometry as a systematic description of the nature of space is applied mathematics, or applied logic, or physics. (If the indefinables of our systems are interpreted in other terms, we may have a chemical,

[8] Certain metaphysicians may object that Einstein is concerned with the properties of physical objects in space, not with space itself, and that the latter is the true object of geometry. But if space is more than an object of pure mathematics, if it is something really existent, its properties cannot be studied apart from all bodies in it. In this respect Einstein's metaphysics, though largely influenced by Mach, is at one with that of Plato, Philo, and St. Augustine when they insist that space and time do not exist apart from the motion of bodies.

[9] Since ancient errors die very slowly and there are still distinguished men like Aliotta, Nelson, and L. J. Russell who argue that the Euclidean system is the only one that we can represent in intuition, it is well to be reminded that the difference between the Euclidean and the other geometries concerns what happens at infinity, which is something beyond the scope of sensuous intuition. But if we limited ourselves to the intuition of finite space, the Lobachevskian geometry would be the only one that would commend itself to us. This may be seen in the following diagram.

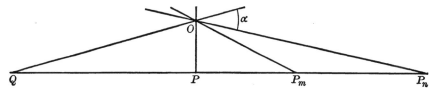

Consider the line OP, a perpendicular through the point O to a given line. If it be revolved about the point O, say counter-clockwise, the points of intersection P_m, P_n, etc., will move rapidly to the right, but after an interval will appear to the left of P, say at Q. Now so long as we are restricted to finite spaces there is always an angle, a, between OP_n and QO. This angle, to be sure, diminishes with the extension of our space. But so long as our space is finite there is such an angle and all lines in it fail to intersect our given line and are therefore parallel to it.

theologic, or legal system, which is, in an equally valid sense of the terms, applied mathematics or applied logic.[10]) Kant and his followers are right in their insistence that the material description of the universe is impossible without some immediate intuition of physical nature, and that pure mathematical truths are apodictic propositions. They err, however, in supposing that pure mathematics rests on any spatial or non-logical intuitions, or that applied mathematics requires an infallible a priori intuition (rather than one that is piecemeal and only probable). Mill, on the other hand, is correct in affirming that mathematics as a material description of the universe cannot be apodictic and that as a body of material truths it depends upon piecemeal and uncertain experience. He is mistaken in asserting that pure mathematics depends upon experience of the physical world and that its truths (e.g. that two times two is four) are merely probable. Finally those symbolists or conceptualists who deny that mathematics deals with the material world are right in holding that in so far as mathematics asserts apodictic truth it does not refer to any *particular* field of experience but is purely formal. They err, however, in failing to see that form without the possibility of matter or content is a metaphysical monstrosity, and that although our application of logical form to any existential material is empirically uncertain, it is nevertheless metaphysically possible and actually fruitful.

The distinction between arithmetic as a branch of pure mathematics and physical or applied arithmetic follows the lines indicated in regard to geometry, and enables us to discern the truth in the rival contentions of those who regard propositions of the type $7 + 5 = 12$ as a priori (Kant) and those who view them as inductive and empirical (Mill).

Pure arithmetic asserts that from certain purely logical rules aided by various definitions, it will necessarily follow that $7 + 5 = 12$. To say with Mill that on some other planet this might not be true is demonstrably nonsensical. And if any one calls our attention to the fact that on our own planet 7 pints of water and 5 pints of alcohol will not make 12 pints of the mixture, we reply that this physical truth is as irrelevant to our mathematical assertion as the statement that two boys together can do ten times as much mischief as one. Arithmetical addition is a

[10] The Greeks, with characteristic insight, recognized this and named the sciences accordingly.

logical process in which the entities added retain their identity. Physical operations which do not conform to this type are simply not within the scope of our addition theorem or equation. The latter does not assert that anything exists, but only that if any group of entities is characterized by being the *sum* of 7 and 5 (according to the accepted meanings of these three terms) it will also have the property which we call 12.

But to convict the empiricist of confusion is not to deny the soundness of his feeling that some (applied) arithmetical assertions are categorically or physically true, that, for example, 7 pounds and 5 pounds on a scale balance 12 pounds, and that such physical truths ultimately rest on observation, no matter how wide the formula (e.g. that of the conservation of mass) which describes them.

In point of historic fact, the axioms of pure arithmetic have come to light only after a long development of special number theorems which arose to answer the needs of practical life. Very early men found it useful to match various groups with some standard series of things like the fingers of our hands or the words of a song, and thus to correlate the various groups with each other. But it is only in recent times that we have clearly recognized in this matching, similarity, or one-to-one correspondence of groups the logical essence of number.

We may go further and show that the extensions of the number-concept to include fractions, negative numbers, and irrationals were first dictated by the needs of computation and that they found their logical justification or derivation later. But this is only a partial truth. We must also remember that more than to any practical need, it is to the logical principle of continuity of algebraic operations (incorrectly stated as the principle of the permanence of form [11]) that we owe the introduction of complex numbers. Practical needs have often stimulated theoretic works in mathematics, yet it is also true that the logical or theoretic demands of our science have always been wider than its practical applications, and many baffling problems of physics and engineering have been solved on the basis of previously developed theoretic arithmetic. Thus the formulation of the fundamental equations of electricity and the solution of practical problems of utilizing electric and magnetic forces were made possible by the previous elaboration in pure mathematics of

[11] Dantzig, *Number, the Language of Science*, Ch. X.

the functions of $\sqrt{-1}$.[12] We may conclude, then, that though the assumptions of ordinary arithmetic were found by us only after some of their physical applications were recognized, the meaning of these theoretic assumptions is wider than any particular exemplification of them; and whether tacit or express these assumptions are protases which logical analysis reveals as essential to the truth of the propositions of pure arithmetic.

It is well also to note that corresponding to non-Euclidean geometry we have newer or "non-Pythagorean" arithmetics in which we no longer assume the commutative law or the proposition that the product of two numbers cannot be zero unless one of them is zero. These newer arithmetics also find interpretations in physical nature. Vector analysis is one of these newer arithmetics and its fruitfulness is becoming more widely recognized every day.

If we recognize the necessary distinction between pure and applied mathematics, we can see more accurately into their common nature, and how mathematics can be applied to the material world. But before attacking this problem we must inquire further into the nature of pure mathematics and dispose of some radical misconceptions concerning the character of mathematical reasoning.

§ II. HOW CAN A PURELY FORMAL SCIENCE EXIST?

(A) THE LOGICAL CHARACTER OF PURE MATHEMATICS

OUR discussion thus far has emphasized the fact that while all mathematics is concerned with drawing necessary consequences from premises and some mathematicians have so defined their subject-matter,[13] the premises of pure mathematics are themselves logical prop-

---

[12] Some philosophers, misled by the word *imaginary*, still regard $\sqrt{-1}$ as a fiction. But it is a logical consequence of our two-dimensional algebra, it involves no contradiction at all as an arithmetic entity, and it is a most apt description of certain physical relations. See Whitehead, *Introduction to Mathematics*, pp. 87 ff.; Stolz and Gemeiner, *Theoretische Arithmetik*, II, pp. 219 ff.

[13] "Mathematics is the science which draws necessary conclusions," says Benjamin Peirce at the beginning of his famous memoir on "Linear Associative Algebra," reprinted in the *American Journal of Mathematics*, Vol. IV, p. 97. Similarly Whitehead, in his *Universal Algebra*, p. vi, declares: "Mathematics in its widest signification is the development of all types of formal necessary deductive reasoning." This is for-

ositions, i.e. rules of inference. In a number of impressively rigorous demonstrations, Veblen, Pieri, Frege, Russell, and Whitehead have actually exhibited the purely logical derivation of the fundamental theorems of arithmetic and geometry. But though no one has successfully attacked [14] the carefully wrought chains of reasoning by which this thesis has been established, it has not as yet won universal recognition. Many still *feel* that there is some distinction between pure mathematics and logic. Such feeling rests not only on the failure to distinguish clearly and consistently between pure and applied mathematics but also on the habit of mind which identifies mathematics with its characteristic symbolism, and on the fact that the words *logic* and *mathematics* have historically represented two different subjects.

(1) Is MATHEMATICS THE ART OF MANIPULATING SYMBOLS? This suggestion may at first seem trivial. For some of the older classical mathematical treatises, from Euclid's *Elements* to Galileo's *Discorsi* and Newton's *Principia*, have relatively little symbolism, while treatises on logic can now be written in symbolic form. But the view that mathe-

mally and accurately expressed by Russell: "Pure mathematics is the class of all propositions of the form 'p implies q,' where p and q are propositions containing one or more variables, the same in the two propositions, and neither p nor q contains any constants except logical constants" (*Principles of Mathematics*, p. 3).

[14] Poincaré (*Science and Method*, Bk. II, Chs. 3-5) seems to have set the fashion of denying the identity of logic and pure mathematics on the grounds of impressionistic psychology rather than on the basis of rigorously logical or mathematical demonstration. I have tried to indicate this in a review of Brunschvicg's *Les Étapes de la Philosophie Mathématique* (in the *Philosophical Review*, Vol. XXIV, 1915, p. 81).

Poincaré's one serious contention, unfortunately repeated by Hadamard and others, is to the effect that there is a petitio principi in the logical derivation of number. But this rests on the failure to distinguish between the identity of an object and what is meant by the cardinal number 1. The *Principia Mathematica* (Pt. II, § A) makes that clear. It is true that we ordinarily begin with the more familiar properties of numbers before we construct or derive them from their logical elements. But in scientific work we must not confuse the temporal or psychologic order in which we learn things with the logical order in the things learned.

It is of course to be expected that a new and radical thesis such as the identity of logic and pure mathematics will meet with really serious difficulties. But those that have so far come to light are of the kind that we may reasonably regard as problems that workers in this field must face rather than as justifications for the abandonment of the whole enterprise. This applies to the argument that the axioms of infinity, selectivity, and reducibility in the *Principia Mathematica* are not of strictly logical character. It is in a sense true that they do not have the formal, universal, and self-evident character of the other axioms of the *Principia*. But they have the essential character of logical axioms, viz. they are rules of inference which make a logically demonstrable and coherent system of mathematics possible. In regard to the axiom of infinity we may say that men like Weyl labour under a misapprehension when they interpret it as asserting matics is essentially the manipulation of symbols that may be meaningless is seriously held [15] and no one who reflects on the tremendous importance of language will lightly brush its consideration aside. There can be no question that appropriate symbolism has made certain mathematical developments possible which otherwise could not have been humanly carried through. Special symbols are free from the penumbras of associated popular meanings and are therefore capable of precise denotation. They can also be manipulated more easily than ordinary words, and by bringing them together we may make complicated relations more readily visible.

All this, however, in no way supports the symbolist school in their denial that mathematical operations are of the nature of logical inference, or in their contention that mathematics is an art of arbitrary or extra-logical manipulation of meaningless marks. Consider this doctrine as stated by Professor C. I. Lewis: "A mathematical system is any set of strings of recognizable marks in which some of the strings are taken initially and the remainder derived from these by operations performed according to rules which are independent of any meaning assigned to the marks." "The question of logical meaning, like the question of empirical denotation, may be regarded as one of possible *applications*, and not of anything internal to the system itself." [16] This is admittedly "mathematics without meaning." But the behaviour-

"the actual existence of infinitely many objects in the real world" (Rice Institute Pamphlet, Vol. XVI, 1929, p. 248), though this misapprehension is somewhat justified by the unguarded language of Cantor and Russell. Clearly the existence of infinites in pure mathematics can mean not the presence of real or actual objects in the physical or mental world, but only freedom from contradiction in the realm of logical possibility.

As to the axiom of selectivity (or multiplicative axiom) we may admit that it seems necessary only for theorems as to the transfinite, and yet there is so much difficulty in formulating a consistent system of mathematics without the latter that we may regard this axiom as necessary for mathematical reasoning as a whole. The axiom of reducibility, that "any combination or disjunction of predicates (given in intension) is equivalent to a single predicate," seems in its present form to lack the self-evidence which we generally expect of logical axioms. But self-evidence is not a reliable criterion for axioms, and despite the work of Ramsay (*Proceedings of the London Mathematical Society*, Vol. XXV, 1926; *Mathematical Gazette*, 1926, p. 185) and Waisman it seems reasonable to expect that future mathematics will (as in the case of the Euclidean parallel postulate) discover a simpler and more satisfactory form of it rather than abandon it completely.

[15] Lewis, *Survey of Symbolic Logic*, pp. 355 ff., and Hilbert, *Die Grundlage der Mathematik* (1928). This view was first proposed by De Morgan in his *Trigonometry and Double Algebra*, pp. 89 ff.

[16] Op. cit., pp. 355-356.

istic argument is made that an external observer does not need to know what the mathematician *has in mind;* he can see that all that the latter *does* is to set down certain marks and to manipulate them. Reflection, however, shows that this is not the case. An observer ignorant of the meaning of mathematics could not distinguish the mathematician's use of symbols from shorthand writing, from the notation of chess games, or from the meaningless marks that students make in their notebooks during tedious lectures. Obviously also two men may use entirely different symbols to denote the same mathematical process, e.g. the Leibnizian and the Newtonian ways of representing the operations of the calculus. Moreover if the manipulation of our symbols is to have any definite character, like the movement of the pieces in a chess game, the rules must be consistent with each other and consistently applied. Such a game, therefore, cannot be played without assuming logical consistency.

These considerations are so obvious that the significant question is, What leads the symbolist to overlook them? The answer is that the symbolist makes the common confusion of so many formal logicians between independence of any one special meaning and independence of all meaning. Assuredly pure mathematics, dealing with the realm of all possibilities, must be independent of special intuition or empirical facts. But this does not mean that its symbols and their operations are devoid of all meaning. On the contrary, precisely because mathematics is interested in the possible types of order, the arrangement of our symbols represents the order of all possible entities that can be thus symbolized. Marks that do not stand for or point to anything beyond them are just physical entities and not symbols at all.

But the fact that mathematics thus uses significant symbols does not make it, as Mach, Duhem, and Poincaré have at times maintained, merely a form of language. For if pure mathematics were nothing but a set of symbols, how could we speak of real discoveries in it or by it? Can we discover the properties of the equation of an ellipse or solve a problem as to prime numbers by merely giving things new names? No one seems to have the courage to say so. But many seem to think with Schopenhauer that reasoning, whether mathematical or logical, "cannot actually extend our knowledge, but can only give it another form," [17] which means that "the differential calculus does not really

[17] *The World as Will and Idea*, Ch. I, p. 18.

extend our knowledge of the curve," but only transforms what we already know by intuition. No one who has ever used the methods of the differential calculus in the solution of actual problems need have the absurdity of this position pointed out to him. Mathematics is full of instances in which analytic methods have enabled us to discover not only new properties of curves, but even such as on grounds of intuition seemed utterly impossible. To take only one example: it had always been supposed—and on grounds of intuition nothing seemed more certain—that any element of a continuous curve could be extended to form a tangent. But by purely analytic methods Weierstrass discovered a whole host of continuous functions which have no derivative at all. In any discussion of pure mathematics, then, we must hold on to the fact that reasoning is not an arbitrary manipulation of symbols but does help us to extend our knowledge. And this means that we are dealing with what can be symbolized.

(2) Is Mathematics Based on Intuition? This brings us to the contention that mathematical reasoning differs from logic by involving some special intuition and that it is because of this fact that mathematics is constantly growing while logic seems relatively fixed, if not sterile. No man since Gauss has been able to keep up with the progress of mathematics in all its fields, while many since Kant have asserted with some plausibility that logic has made no progress since Aristotle.

These claims, it will be found, usually rest upon the confusion between pure and applied mathematics to which we have already directed attention. It will be useful, however, to consider at this point three attacks upon the formal or analytical character of mathematics, in each of which some special type of intuition is put forward as a basis for distinguishing mathematics from logic and as a ground of explanation for the fact that pure mathematics produces genuine novelty.

(i) *Kant.* It is essential to the Kantian view to distinguish between logic and mathematics on the ground that the former proceeds analytically (according to the principles of identity and contradiction), while the latter is synthetic or constructive. Kant does not clearly distinguish between inference, reasoning, or proof, and judgment in mathematics. He seems to regard the whole of mathematical knowledge as constituted by a series of synthetic judgments, each of which is a construction based on intuitions of time and space as the form of sense. Being debarred by various metaphysical considerations from entertaining the possibility of a purely logical or intellectual intuition, and being committed to the view that mathematical judgments are apodictic, these intuitions of time and space must be a priori and yet refer to the material or sensible world.

Now as regards theoretic arithmetic we can say definitely that since Gauss its proofs have become so logically rigorous that there is today no ground for asserting any special intuition of time to distinguish them from other logical demonstrations. In the light of Russell and Whitehead's *Principia Mathematica* the proof of the proposition that $7 + 5 = 12$ is as purely logical as any other proof.[18] Similarly we can now say quite definitely that modern proofs in geometry (as a branch of pure mathematics) eschew any appeal to an intuition of space. It is true psychologically that some people cannot reason effectively on geometric or even arithmetic problems without some images or diagrams to help them, and we have been accustomed by the mathematical and physical teachings of over two thousand years to interpret these diagrams in Euclidean terms. But such diagrams, it must be insisted, are no part of the mathematical proof, and all sorts of fallacies result from appealing to the diagram rather than to the axioms and previously demonstrated propositions.

It is true in a certain sense, as I shall try to indicate later, that the most abstract demonstration of logic itself involves an element of intuition, but such intuition or apprehension is intellectual and no different in a mathematical than in any other proof or chain of reasoning. Kant's fundamental insight into the necessity of some intuition to support the process of reason, as the air is necessary to support the bird in its flight, is profoundly sound. His error or failure to see the identity between logic and pure mathematics in this respect was due to the fact that the distinction between pure and applied mathematics had not yet been discovered and that the mathematics of his day was progressive rather than rigorous. Yet something of the truth he saw when he insisted that any consideration of a subject was scientific to the extent that it contained mathematics. Leibniz, though hampered by his unsatisfactory distinction between analytic and synthetic judgment, had a firmer grasp of the nature of mathematics and saw further into its future. And in the light

---

[18] Cf. Couturat, *Les Principes des Mathématiques*, p. 255 ff.

of the modern development of mathematics even professed Kantians like Hermann Cohen, Natorp, and Cassirer have abandoned the whole doctrine of intuition as the productive agency in mathematics.

(ii) *Poincaré*. A modified form of the Kantian attempt to explain the fruitfulness of mathematical reasoning on an intuitive rather than logical basis is that of Poincaré. He rejects the Kantian a priori intuition of space as Euclidean and in his account of the nature of time (in which he partly anticipates Einstein) there is certainly no room for a priori intuitions. But he attributes the fruitfulness of mathematical reasoning to the principle of mathematical induction, which he holds to be an intuition of our mind's power to repeat its own operation ad infinitum. The principle of mathematical induction is undoubtedly of wide application, though there are many regions even in arithmetic where it is difficult to see its relevance, e.g. the science of prime numbers. But the important thing to observe is that this principle of mathematical induction is entirely different from the induction that prevails in the physical sciences.

The essence of induction as practised in the physical sciences is clearly that in the presence of the empirical observation that all examined x's are y's, we conclude that all x's are y's. Such a conclusion can never be more than merely probable. It does not exclude the possibility of some x turning up that is not a y. But mathematical reasoning deals with possibility and with inferences which are necessarily true. It is absolutely certain that if a proposition is established by mathematical induction, it will never be disproved, i.e. if a *general* proposition is true of n + 1 whenever it is true of n, and is also true of 1, then no possible finite integer can arise of which this proposition is not true; for all finite integers are by definition such as can be reached by the addition of one.

It is of course true that many theorems in mathematics are *discovered* inductively, but they are never admitted as mathematically valid unless deductively demonstrated. Thus for over two thousand years before Lindemann in 1882 definitely proved π to be transcendental, mathematicians were unable to derive a rational or algebraic value for π. Some took the pains to show that π could not be expressed in decimals of 700 places. Yet this latter fact was never held to be mathematically sufficient to establish the transcendental character of π. So, also, it was nearly 100 years before Fermat's theorem that $2^{2^x} + 1$ is always prime was shown to be false (through Euler's discovery of 4,294,967,297 as divisible by 641), and yet at no time during the interval between 1637 and 1732 was this theorem properly a part of mathematical science— simply because no deductive derivation of it had been offered.[19]

If, therefore, mathematical induction has a place in pure mathematics it is because it is a form of logically rigorous demonstration. Poincaré's effort to make a special intuition of it is motivated by a desire to prove that numbers exist without distinguishing between psychologic and mathematical existence. The failure of modern logistics to produce any satisfactory proof is, in his opinion, its great defect. If numbers, however, can be deduced from the principles of logic it is vain to ask for any other proof of their existence, for that is to ask that the principles of logic be themselves proved, which obviously involves an infinite regress. Admitting that there are difficulties in proving the self-consistency of the axioms from which numbers are derived, that difficulty is in no way obviated by transferring the problem to the mental realm and there postulating an intuition by the mind of its power to repeat certain mathematical operations infinitely. It is no easier to test the consistency of the infinite implications of this intuition than it is to determine the consistency of the purely logical assumptions of modern arithmetic.

(iii) *Recent Intuitionism.* The latest attempt to dispute the logical nature of pure mathematics and to base it on intuition is that of Brouwer. Brouwer is led to this absurdity through his failure to discriminate under the heading of *formalism* between the view of Hilbert according to which pure mathematics is concerned with the symbols themselves

---

[19] In an effort to question the deductive character of mathematics Professor Smart (*Philosophical Review*, Vol. XXXVIII, 1929, p. 240) adduces as an example of induction the fact that the equality of the sum of the angles of a triangle to two right angles was first proved for each particular kind of triangle, equilateral, isosceles, and scalene. This, if historically true, is certainly not a case of induction at all. An argument in which all the possibilities enter into our premises is a clear case of the disjunctive syllogism. In none of the other mathematical examples adduced, such as the theory of differential equations or functions, is there a genuine case of an induction such as we meet in natural science, the essence of which is that some future fact is likely to disprove it. True mathematical generalization is not subject to that contingency because while it generalizes by dropping inessentials it always proceeds from an exhaustive analysis of *all* the possibilities.

The one genuine case of inductive generalization mentioned by Professor Smart is taken from Egyptian "mathematics," and characteristically enough leads to an inaccurate result. Such inaccurate guesswork has practical value and may suggest genuine mathematical problems, but it does not itself belong to mathematics.

and the conception of pure mathematics as essentially logical. Obviously if logic is merely a symbolic calculus it must take certain mathematical truths for granted. But this is to ignore the more comprehensive meaning of logic. Brouwer does this because he naïvely assumes the current anti-realistic conceptualism according to which the exactness of mathematical truth does not apply to the natural world but only to ideas in our mind—if this exactness is not to be confined to the marks on paper. But like other conceptualists he does not ask how mathematical ideas can be true apart from the objects with which they are concerned. Never for a moment entertaining the possibility of an objective meaning of logic or of the field of possibility, he cannot see that a proof of freedom from inconsistency is sufficient to establish logical or purely mathematical existence. Brouwer further claims that logic is based upon mathematics and not vice versa.[20] But it is easy to see that logic or demonstrative reasoning in general cannot be a species of demonstration through special intuition. What is more significant about Brouwer's position is the feeling that the purely logical or formal method in mathematics is responsible for the paradoxes or seeming antinomies into which the newer developments of the *Mengenlehre* and especially Cantor's transfinite numbers have led. Many other orthodox mathematicians are scandalized that there should be any part of mathematics in which proofs are subject to dispute. And they note with a touch of irony that this occurs in the field where abstract logic is most in evidence. It has not, however, been shown that the evil of these various paradoxes is insurmountable. Patient logical analysis is solving them and there is no reason to despair of the future.[21] Disputes about the adequacies of proof and even downright errors existed in mathematics even more abundantly before Gauss, Weierstrass, and Dedekind introduced the example of greater logical rigour in demonstration.

But if the present evil is great, what remedy does the school of Brouwer propose? Shall we substitute the intuition of self-evident principles of time, of number, of the smallest infinite ordinal, and of the

[20] Brouwer, *Bulletin of the American Mathematical Society*, Vol. XX (1913), p. 85, and *Jahresberichte der deutschen Mathematiker Verein*, Vol. XXVIII (1919), pp. 203 ff.; Weyl, *Symposium*, Vol. I (1925), pp. 1-32; Bäumler and Schröter, *Handbuch der Philosophie*, Vol. II, pp. 3-50. A complete bibliography is to be found in Fraenkel, *Einleitung in die Mengenlehre* (1928), pp. 394 ff.

[21] Cf. Paul Weiss, "The Theory of Types," *Mind*, Vol. XXXVII (1928), pp. 538 ff.

linear continuum for logical proof? How will that prevent such errors as those into which eighteenth century mathematicians, including the great Euler, fell? In mathematics as in every other field of science there is need for penetrating minds with the gift of anticipating the outcome of research, and we may quote Gauss himself to the effect that he had worked out certain views and that all that was left was to supply the proofs. But this can in no way minimize the indispensability of the proof. Without rigorous logical proof there is no mathematical truth.

The intuitionist school generally put their thesis in the form of a restriction—we must not in mathematics reason or operate with any symbols unless we can on the basis of intuition construct what they denote. More specifically let us beware of existence-proofs as to infinites. This sounds like a counsel of caution. Unfortunately, however, they use the term *construction* in an essentially hazy manner, and the line between what we can and what we cannot intuit is so vague and arbitrary as to be utterly unsuitable as a basis for rigorous mathematics. Moreover in their attack on the logical principle of the excluded middle, they confuse the indetermination of our knowledge with the indetermination of its object, and their positivistic test of existence would if consistently carried out make it impossible for pure mathematics to go beyond mere description of perceived but contingent objects in the physical or mental world. Let us briefly examine these points.

Brouwer begins by recognizing that not only non-Euclidean geometry but even the earlier Cartesian reduction of geometry to algebra has made it unnecessary for the mathematician to assume any a priori intuition of space. (Incidentally, is not Brouwer's intuition of the linear continuum spatial and a priori?) But despite the criticism of Herbart and others, he clings to the equally dubious intuition of time as basic for the fundamental number series. It is easy to see that the finite integers are familiar to us, but what precisely is meant by saying that we have an intuition of them? Is it by a special intuition or by logic that we recognize that the series of integers has no last term and is therefore infinite? If we admit, as Brouwer does, an intuition of the least ordinal infinite, why may not higher orders of infinity be possible? And why is not mere possibility or freedom from contradiction sufficient for pure mathematics?

An even wider group of mathematicians (especially French) insist that we shall not admit any numbers unless we can construct them. But what is meant by constructing a number? Shall we deny the fundamental theorem of algebra which would make an equation of the fifth degree have five roots because we cannot in general find these roots? Nor is it clear what is meant by constructing a number like 9^{9^9}. Surely not to write out the hundreds of millions of places which constitute it. Indeed when we say $374 + 693 = 1067$, we do not construct these numbers in our minds. We use the symbols with confidence because we can prove that given operations on the symbols will always lead to the correct result as to the numbers symbolized.

Weyl, the most avowedly philosophic of our contemporary distinguished mathematicians, has recently substituted for intuition and construction the notion of *exhibiting the individual objects of an aggregate.* And he argues that since we cannot exhaust an infinite it is meaningless to deny that "all numbers are even" or to assert it as a real judgment that "there is an odd number." [22] Though he argues for a spiritually-minded philosophy against servile positivism, and regards Brouwer's intuitionistic mathematics as "the restoration of mind to its old and sacred rights," his test of significant assertion would if thus consistently applied lead to abject dogmatic positivism. For it would deny the significance of all universals that go beyond actually observed particulars. We could not even say that all men are mortal, but only that those individuals who have exhibited their mortality are so.

The narrow conception of existence (or the failure to distinguish between purely logical and physical or mental existence) which makes Weyl deny that "there is an even number" is a judgment, also makes him attack the law of excluded middle as existential absolutism. But as a logical law it does not tell us what specifically exists, but only the character that any existence must have if it is to be determinate. Weyl, like Kant, instead of viewing mathematics as a demonstrative chain or system, views it as a series of propositions each of which is categorically true. Take as an example Fermat's theorem that there is no number n above 2 which will satisfy the equation $x^n + y^n = z^n$ (x, y, and z not being 0). It *seems* true but has never been proved. Is it necessary to assume that it can be proved or disproved? Weyl thinks it

[22] Rice Institute Pamphlet, Vol. XVI (1929), pp. 245-247; cf. his Symposium article, pp. 9, 20, 24.

possible that there is nothing in the axioms of arithmetic to necessitate either the truth or the falsity of this theorem.[23] Certainly whether true or not, Fermat's theorem *seems* significant and categorical, not hypothetical.

It is a sound attitude to consider every proposition as asserting something before inquiring whether it can be proved or not. But what does Fermat's theorem assert? Obviously something about numbers. But what are numbers and when is a proposition about numbers true? If numbers were physical things our theorem would be a physical hypothesis to be tested by observation and experiment. But obviously mathematicians do not take it in that sense, and have not tried to "prove" it in that way. An integral number is a logical characteristic of all possible classes or multitudes that are similar or can be matched. Fermat's theorem, then, asserts that a certain type of relation indicated by the equation cannot possibly exist, not merely in the physical but in the rational or logical realm, i.e. that there would be an inconsistency or self-contradiction in the assertion that there is such a *number* n greater than 2. Agreement with the fundamental axioms about the nature of number is, therefore, the very essence of the meaning of our proposition, and if the known axioms of arithmetic are in fact such that they do not necessitate either the truth or the falsity of any theorem about number such as Fermat's, they are insufficient to define the nature of number and we should complete them. If they do define the nature of number, then there is a determinate answer to the question whether Fermat's equation is true or not, though we have not been able to find out how to prove it, i.e. how to show the connection between the axioms of arithmetic and the particular theorem. The discovery of the proof that π is transcendental did not create any logical relation but showed us what the relation always has been.

Is $2^{2^7} + 1$ prime or not? The fact that we have not as yet found the answer does not justify us in asserting that there is no determinate answer to the question. The intuitionist who says that it is meaningless to ask whether a given number is algebraic before we succeed in constructing an equation of which it is a root [24] is guilty of confusing a num-

---

[23] *Das Kontinuum*, pp. 11-12.
[24] Baldus, *Formalismus und Intuitionismus in der Mathematik*, p. 26, similarly **Wawre** in *Rev. de Metaphys. et de Morale*, Vol. XXXIII (1926), p. 426.

ber with our knowledge of it. It is the determinate nature of numbers which makes our knowledge true. Mathematical problems are not indeterminate during the period when we see no way of solving them.

(B) THE NATURE OF LOGICAL NOVELTY

The crux of all the serious objections to the identification of pure mathematics with logic is the conviction shared by empiricists and rationalists that in a logical argument there is nothing in the conclusion which is not contained in the premises, that all logic consists in asserting that A is A. If, therefore, mathematics is genuinely productive—and the enormous number of contributions published every year testifies to this—it cannot be purely logical. But to rely on this is to be misled by a traditional but unreflective use of words.

In what sense, may we ask, is the conclusion of a deductive argument contained in its premises? Surely not in the literal sense in which physical things are contained in a room.[25] The only relevant and significant meaning that we can attach to the relation of *containing* between premises and conclusion is that of logical implication. From this point of view it is an insignificant tautology to maintain that deduction cannot bring us propositions that are not already implied by our premises. But it is not a matter of definition but a fundamental fact of human experience that we are not actually aware of all the consequences or implications of our statements or assumptions, and that the discovery and study of such implications is a large part of the search for truth.

Take the classical example: A priest says, "My first penitent was a murderer," and a baron soon thereafter entering the room says, "I was the reverend father's first penitent." The dramatic conclusion is surely not contained in either premise and if we say that it is contained in the two when they are put together we only mean that it stands to the premises in that peculiar logical relation of implication which denies the possibility that the premises are true and the conclusion false.

[25] A popular view which goes back to St. Thomas (*Quaest. Disput. de Veritate*, XI, 1), holds axioms to be the germs from which the body of knowledge develops. This is a helpful analogy in suggesting a parallel between the determinate character of organic development from germs and the development of logical arguments from axioms. But the popular (preformist) impression that the whole adult organism lies hidden in the germ is paralleled by the absurd notion that all the Euclidean propositions are somehow tucked away in the Euclidean axioms.

The psychologically startling novelty of the conclusion in this case is an apt illustration of the fact that in every syllogistic or other logical inference the conclusion, though its content is in a sense identical or continuous with the content of the premises, is nevertheless in some respects different. This difference may often be psychologically or practically invisible and theoretically infinitesimal; but it is an infinitesimal which, when integrated or summed up over a long line of arguments (in which intermediate steps are generally omitted because of obviousness) makes the logical distance between the simple Euclidean axioms and a multitude of complicated theorems.

The difficulty of seeing how pure mathematics can be both productive and logically deductive is obviated even more if we view the propositions of pure mathematics in their character as rules or formulae of procedure. The number of specific rules applicable to specific situations in the different possible chess games that can be played with our present fundamental rules is practically inexhaustible. Yet every possible or legitimate move in every game is one that is determined by these original rules. Now the rules of chess are restricted to a very small number of entities, while the rules of logic are applicable to all objects of any sort, physical or mental, material or formal, real or ideal. It is therefore precisely because the propositions of pure mathematics are formal—logical rules of inference telling us that if *anything* has the property *a* it will also have the property *b*—that they have such a wide range of application and that so many new possibilities are constantly being discovered in their field.

From the fact that pure mathematics (as logic) refers to the field of possibility (and does not restrict itself to temporally existing objects) it follows that there is a type of novelty which it does exclude, and that is the novelty which upsets physical generalizations. Physical knowledge or generalization of what has been observed to happen can never exclude the possibility of something novel and different appearing in the future. But as true logical or mathematical propositions exhaust the field of possibility, their denials are logically impossible, i.e. if we can show that a certain conclusion follows from its premises, we show that it is meaningless, logically impossible, or self-contradictory for the conclusion to be false and premises to be true.

A clear understanding of the fact that pure mathematics is both deductive and productive has been rendered difficult by two paradoxes about formal logic, viz. (1) that it asserts nothing but identities, and (2) that it is empty of meaning.

It is true that when a proposition is proved (for instance, the one about $\sqrt{2}$ at the beginning of this chapter) and its denial shown to be self-contradictory, the proof may in a sense be reduced to an assertion of identity between premises and conclusion. But such an identity of form or logical function does not, as we have seen before, exclude diversity in the subject-matter. And it is precisely when we can trace the element of identity which makes things or events repeatable that we have the most fruitful approach to the nature of things. Abstract formulation isolates the elements of our inquiry and helps us to slough off the obscuring accidents of existence which hinder our vision of what is relevant. Naturally powerful minds can get at essentials without the aid of such abstract formulation of the identities in different things or events; but less masterful minds need the help of such abstract framework in the solution or exploration of new situations or problems.

The notion that formal propositions are empty of meaning is a persistent radical confusion. It is true in a sense that every form is independent of its matter. But a formal act is one that is the same for all regardless of the individual differences in the class to which it applies; and so the rules of logic or pure mathematics universally apply to all propositions irrespective of differences of their material content. But this does not mean that logical or mathematical forms can exist apart from all reference to any possible content. On the contrary the most formal propositions are those which apply to all kinds of entities, and reference to such possible application is essential to their meaning.

It is not then claimed that the strictly logical reasoning of mathematics can by itself produce facts of sense perception. But this, of course, does not mean that we should blindly swallow the dogma that sense experience is the sole source of truth. No dogma can get rid of the fact that mathematical reasoning does extend our knowledge if anything does.

The principle of polarity warns us that while the rational and the sensory elements of our intellect are inseparable they are distinct. We may grant that in every case of actual analytic reasoning some sensory element, no matter how faint, is present, and yet we must insist on the relative independence of the rational element. If you ask me whether a triangle whose sides are respectively 105, 252, and 273 units is right-angled, I can by reasoning answer that in Euclidean space it must be, and I need no physical observation, measurement, or experiment. The dogma that all reality is sensory or can be apprehended through sense organs alone is as gratuitous as the assumption that all reality is audible, or that all reality is odorous. The fundamental paralogism of empiricistic systems is the argument that because I cannot by pure reason apprehend particular sensible existents, therefore all reality is nothing but sensation. One might as well argue that since I cannot through my sense of sight apprehend the hardness or softness of things, all reality is tactile and colours do not exist. Obviously, however, we have one phase of reality in sensory experience and one phase is apprehended through reason. Human experience functions with sense and reason, and the effort to do without one of these elements is no less ridiculous than the effort to make an army march exclusively with the left or exclusively with the right foot.

To the question, then, "How can pure mathematics as logic extend our knowledge?" we may answer that it undoubtedly does so in fact and that we must not let difficulties of explanation obscure that fact. We may also say that in the assertion of any logical inference, the *therefore* expresses a fact of intellectual perception or, if you like, intellectual intuition, viz. that a given conclusion does follow from or is implied by its premises. For it is out of what is asserted by our proposition and not by the reasoner's fiat that the conclusion follows. The assertion that all propositions are mental is at best irrelevant here since any specific conclusion follows from its specific premises and not from any others.

Logically, then, the relation of implication in its elementary form is as ultimate a fact as that a particular object looks red or tastes bitter. If my neighbour cannot see the redness of an apple or taste its bitterness, there is no basis for further argument. In some special case I may suggest to him that he is mistaken and that his experience does not agree with that of others or with his own at other times. But ultimately there must be an appeal to immediate apprehension. Similarly there can be no further argument if one cannot see that the two propositions *All men are fallible* and *He is a man* imply *He is fallible*.

The fact that logic and pure mathematics deal with the possible and are not restricted to those few possibilities that are actualized enables us to see the logical consequences of propositions that are not true. It is not necessary that the propositions *All men are mortal* and *Socrates is a man* should be true in order that they should imply a conclusion. Worthy people have believed that Elijah did not die, and a certain school of popular anthropologists (De Gubernatis et al.) ought not to be surprised if some one were to prove that Socrates was a solar myth. But there can be no denying the fact that the conclusion follows from the premises.

We conclude then that pure mathematics is both logical and fruitful because the relation of implication is something as objectively valid (i.e. independent of our individual psychology) as any of the facts of physics, and demonstrative reasoning is a series of intellectual intuitions or apprehensions of these implications. This also explains why creative genius seems necessary to bring about a substantial widening of any field of mathematics. The creative mathematician is one who has a genius for dealing with our particular species of facts, the ability to grasp long chains of reasonings in their entirety, and the intellectual imagination to anticipate results which less gifted minds reach only by painfully treading every step.

§ III. HOW IS APPLIED MATHEMATICS POSSIBLE?

WITH the foregoing considerations in mind we may turn to a question that is fundamental in the analysis not only of mathematics but of all exact science, the question, namely, of how mathematical reasoning can give us knowledge of the external world,—in other words, how applied mathematics is possible. Here again we must not let difficulties of explanation blind us to the certainty of the fact. On any theory we must admit that I can by reasoning from the known density of iron (assuming the general relation between the velocity of sound and the density of the medium) predict the exact number of seconds it will take for a sound to be transmitted through an iron bar or wire of any length or, on the basis of a few observations on Polaris, tell exactly where and when the sun will rise in a particular locality.

On the prevailing view that reasoning is merely a mental process, controlled by psychologic laws, the question arises, "Why does nature conform to the results obtained by following these laws of thought?"

The three answers which have prevailed in this field are those of empiricism, transcendental idealism, and pragmatism.

(1) EMPIRICISM. The answer of empiricism, so far as men like Mach and Pearson meet this question, is that an applied mathematical science like mechanics is simply a convenient description of phenomena, and the process of reasoning, through equations and the like, only gives me back what I put into it in the shape of the results of observation and experiment. But this account, if strictly adhered to, would make mathematical reasoning useless as a method of extending our knowledge of nature or of solving any physical problems. Why, indeed, go through arduous processes of reasoning to get back only what we put into our mathematical machine? To get them back in a more convenient form or order, the empiricist often says. But is not the "more convenient form or order" a form of the object studied? Our analysis of the nature of pure mathematics prepares us to reject the dogmatic assumption that the result of mathematical reasoning cannot extend our knowledge of nature, and prepares us to believe that it can do so to the extent that nature embodies mathematical form.

What, asks Mach, have vibrating strings to do with harmonic functions? Clearly the two are not identical. Yet if they did not have anything to do with each other is it not amazing that the consideration of one should so help us to understand the other?

The empiricist admits that the results of our physical measurements and observation can be expressed in the premises of our mathematical argument and that the results of the mathematical deduction can be translated back into terms of physics. But the intermediate steps of mathematics, he maintains,[26] do not have any objective reference. And it must be admitted that on the face of it it is difficult to see anything in the physical realm exactly corresponding to the mathematical processes of differentiation, integration, and the like. Obviously, however, mathematical symbols like other symbols, e.g. paper money or tokens used as money in card games and other ways, do not have all the properties of the things they symbolize and have certain properties which

[26] Duhem, *La Théorie Physique*, pp. 337 ff.

the latter do not have. It is because of such differences, in fact, that it is more convenient to use them than the things they represent. Yet there would be no use in working with symbols where there was no correspondence between operations on the symbols and operations on the things symbolized.

A geographer wishing to make a true map of a country will be careful to see that there is an exact correspondence between the order of his symbols and the order of the things symbolized. Yet there will be certain operations on the symbols themselves, e.g. the colouring of lines representing mountains, which do not directly refer to the things symbolized.

With this analogy in mind we can distinguish in any mathematical treatment of a physical problem those logical operations which refer directly to phases of the physical transformations studied and those which have to do directly only with the symbols themselves. Let us take as an example the mathematical solution of the problem as to the gravitational effect of a spherical body. If we take Newton's elegant solution of this fundamental problem, we have no difficulty at all in seeing the objective reference of each step. Newton resolves the sphere into a number of thin shells, and by showing that each shell has the same gravitational effect as if it were all concentrated at its centre, he proves that the whole sphere has the same effect as if its whole mass were concentrated at its centre. In its modern form this proof is stated in the language of the calculus, which has the great advantage of enabling us to see elements of procedure common to this case and to others more complicated, but which involves not only statements or definitions as to the equivalence between physical ideas and the new symbols but also statements or convenient agreements as to the manipulation of the symbols themselves.

There is therefore nothing in the empiricist view that vetoes the suggestion that the reason we can apply mathematics to natural problems is because nature is mathematical.

The most acceptable definition of pure mathematics to mathematicians themselves is that which views it as concerned with the study of relations in any ordered multiplicity or "manifold." [27] And what reason is there to doubt that nature has realms of order in it? Before further de-

[27] Paperitz in *Jahresberichte der deutschen Math. Verein*, Vol. I.

veloping this point, it is well to note the attempts of transcendental idealism and pragmatism to deal with the problem.

(2) TRANSCENDENTAL IDEALISM. According to transcendental idealism my reasoning is able to anticipate nature because nature is the product of absolute mind or of the laws (or categories) of the understanding impressed on the material of sense. But a mind which is the author of the system of nature must be distinguished from the mind of the fallible empirical individuals *in* nature who are struggling to understand some fragment of it. We need not, therefore, here examine the metaphysics of transcendental idealism as to the nature of the absolute reason, and its relation to the understanding of our empirical selves. We must insist in any case upon what all scientific work implies, namely that there is an order in nature which we can understand by the use of our reason, but which is there whether we understand it or not. The mathematician at any rate must insist that on any metaphysics the laws of mathematics can be no more subject to our will than are physical laws, that the laws of convergent series are no more constituted by our minds than are the laws of moving bodies.

(3) PRAGMATISM. In speaking of the answer of pragmatism to our present problem, I refer to the view of Poincaré. The distinctive feature of this view is the large rôle given to hypotheses which are merely conventions, and the suggestion that mathematical doctrines are conventional tools for organizing the facts of experience. Mathematical reasoning holds true of the external physical world because that mode of reasoning which enables us to deal most conveniently with physics has been selected by the process of evolution from the different forms of thought potentially existing in our minds.

It is hazardous to introduce the essentially vague concept of evolution or even the more definite but disputable doctrine of natural selection as an explanation of why mathematical reasoning is such a help in the understanding of nature. Obviously it would be absurd (and unjust to Poincaré) to take this argument literally. Biologic evolution or natural selection operates through differences between birth rates and death rates, and there is no evidence that mathematical intelligence—or any intelligence—is thus favoured by nature. Newton, the greatest representative of modern mathematical physics, left no heirs of his body. But in a metaphoric sense there is some struggle between different ideas, and those that can be mathematically demonstrated have had some advantage in the actual conflict of opinion in the last few centuries. But why ideas that are not expressive of the nature of things should be intellectually more convenient in dealing with physics, the biologic analogy does not inform us; and the metaphysical bases of this doctrine as presented by Poincaré are essentially obscure. He seems to maintain an agnostic realism—insisting that we cannot know the nature of things themselves, but are limited to a knowledge of their relations. He does not explain to us, however, how we can know the relations between things by any principle like mathematical induction, which he bases on the mind's ability to intuit its own power.

(4) CONCLUSION. The answer that seems to me to be suggested by the progress of mathematics itself is that the relational structure which is the object of mathematics is just as objective—in whatever sense we take that term—as the physical entities related. The laws of mathematics are applicable because they are the laws according to which all objects or realities *can* be combined.

Boundless confusion has resulted from the fact that the laws of logic have been spoken of as the laws of thought. If the laws of thought mean the laws according to which Jones and Smith and others actually think, in those rare moments of their lives when they do, it is hard to see what such psychologic laws have to do with logic. Smith reads Hertz's *Mechanics* and finds it dull and unintelligible. Jones reads it and is charmed to find that he comprehends it in a flash. Neither of these facts, though descriptive of the character of Jones's and Smith's thoughts, has anything to do with the logic of Hertz's *Mechanics*—no more than the question of the stimulants or anodynes which Hertz may have used to keep his thoughts on his beautiful demonstrations and to overcome the painful consciousness of the disease which was dragging him to an untimely grave. These considerations are obvious enough, and yet the prevailing tendency is to regard logic as a part of psychology, and to view mathematical operations as merely mental.[28]

There is, of course, another sense in which the laws of logic are spoken of as the laws of thought, viz. as the laws according to which

---

[28] Thus Lipps (*Grundzüge der Logik*, p. 2) argues that logic is part of psychology "as surely as knowledge occurs only in the psyche." One might as well argue that astronomy is part of physiology "as surely as vision occurs only in or through the eye."

we ought to think if we are to apprehend the real world. The astounding fact, however, is that even the three traditional so-called laws of thought which adorn our logical text-books say nothing at all about thought, but rather make affirmation of existence: whatever is, is; nothing can both be and not be; everything must either be or not be. Would it not be better to call these propositions invariant laws of being or existence? [29]

So far as the nature of anything is a subject of inquiry it includes a pattern of relations or order of transformations which like the form of a crystal, river, or organism is constant relative to the flux of matter which assumes this form. The laws of crystallography, physiography, or biology are, however, invariant only for limited fields of objects. We appeal to the laws of physics to explain them, because such laws of physics (if we find them) are invariant in regard to all bodily existence. But physics does not exhaust all possible being, and the invariant laws of all possible being are laws of logic or pure mathematics.

Instead, therefore, of assuming an alogical nature which somehow or other obeys laws in somebody's mind, would it not be simpler to start from the observed fact that the laws of logic and mathematics do hold of nature, and proceed to inquire what are the other characteristics of nature which follow from or are connected with this fact? Such a procedure involves what the critical philosophy calls dogmatism, viz. the assumption that we do have knowledge. But readers of the history of philosophy know that two such widely different thinkers as Fries and Hegel were each able to point out that the critical philosophy itself is not free from that assumption.

The assumption that numbers and mathematical or logical laws are mental is due to the even more widespread notion that only particular sensible entities exist in nature, and that relations, abstractions, or universals cannot have any such objective existence—hence they are given a shadowy existence in the mind. But this is a shabby subterfuge: for these numbers or relations are also *numbers and relations of things*, and any assertion with regard to these abstractions is either true or not. Now truth, whatever it is, is not a quality which inheres in a proposition

[29] See my paper on "The Subject Matter of Formal Logic," *Journal of Philosophy*, Vol. XV (1918), p. 673, and cf. Woodbridge, "The Field of Logic," International Congress of Arts and Sciences, Vol. I.

simply because it is mental, but a proposition is true because of factors other than the fact that I now think this proposition. If, therefore, abstractions had no existence except in the mind making them, no assertion into which they entered could possibly be true—except the assertion that I now think such and such a proposition.

The vulgar prejudice against the reality of universals is really due to the fact that we cannot point to them and say: *here* they are,—that is, they cannot be localized in space. But for that matter, neither can our civil rights, our debts, or our philosophical misunderstandings and errors; and yet no one has seriously doubted the real existence of the latter. The truth seems to be that there are different kinds or modes of existence. But on pragmatic grounds, at any rate, there seems to be no reason why ratios, percentages, or velocities should be considered any less real than the bed-posts or tables which are held up to us by ungenerous brethren as the only genuine types of existence.

Fruitful mathematical procedure as well as sound metaphysics require of us the greatest care not to forget that universals are abstractions or relations of order among possible existents and not themselves concrete existents. Every concrete existent is the intersection of innumerable relations, and the mathematical study of any field must isolate its subject-matter and disregard all other phases of existence as irrelevant for its purpose. Ordinary mathematics has found its greatest obvious success in physics where the isolation of the object of study is practically feasible. Thus we can solve the problem of the attraction of two bodies because we can take the sun and the earth, or the earth and the moon, and analyze their motion as if they were the only bodies in the universe. The actual departures of these motions from the results of our first mathematical analysis in terms of the theory of gravitation, can themselves be progressively analyzed as due to the influence of other planets operating according to the same law. And although there is always a remainder, we can still maintain the ideal of mathematical analysis, always looking in the deviations between our observations and the implications of our theories for those mathematical threads of identity which bind new laws and new fields of experience to our problem.

Logic and pure mathematics, then, apply to nature because they describe the invariant relations which are found in it. When we consider natural objects purely as the embodiments of such relations we are said to idealize these objects, or to consider them as ideal limits. But such idealization gives us the essential conditions of what truly exists. And only in the light of ideals can we distinguish between what is relevant and irrelevant to any natural transformation. Thus pure mathematics not only extends our knowledge of nature, by enabling us to follow illuminating perspectives, but in insisting on relevance it helps to maintain the conditions of sanity.

Chapter Two

MECHANISM AND CAUSALITY IN PHYSICS [1]

FROM the days of Aristotle to those of Descartes and Kant the fundamental questions of physics supplied much of the stimulus and substance of philosophical reflection. But the development of physics as an experimental science demanding advanced mathematical knowledge—the lack of which discomforted the acute and powerful mind of Hobbes—led philosophers to follow the easier path opened by Locke, which led to the identification of philosophy with mental science. Though Kant himself began as a mathematical physicist, the influence of his theory of knowledge reinforced the influence of Locke, and cosmological questions almost disappeared from philosophy in the latter part of the nineteenth century. But the discovery of radio-activity, the experimental work which has led to the opening up of the world within the atom and to non-Newtonian mechanics, the Einstein theory of relativity, and the newer quantum mechanics have aroused great popular interest which philosophers can no longer ignore. Yet the new situation is not without danger. It is certainly futile to appeal to the "method" of physical science unless we take the trouble to become familiar with it as it actually operates, and it is hazardous to accept the "results" of a science unless we know how much unconscious, but none the less antiquated, metaphysics has entered into their make-up. It is a distinguished physicist who has lately reminded us that a metaphysics is no sounder because it is held unconsciously or professed by one who is not professionally responsible for it.[2]

As the term *mechanics* has been freely used and abused, a few dis-

---

[1] This chapter was written in 1911, read before the American Philosophical Association that year, and printed in the *Journal of Philosophy* in 1918 (Vol. XV, p. 365). At the time I wrote it Einstein had not yet developed the generalized theory of relativity, and when it was printed his leading papers on that topic were unavailable in America on account of the war. As the chapter, however, deals with relatively permanent issues I have left it substantially as originally written but have added an appendix.

[2] Maclaurin, *The Theory of Light*, p. 7.

tinctions at the outset may clarify the discussion. In the first place, we must distinguish between the mechanical and the physical. The term *mechanics* as used by physicists [3] denotes that branch of physics which studies the motions of masses (considering equilibrium as a special case or limit of motion). Now there are physical phenomena such as light, magnetism, etc., which are not prima facie phenomena of motion, and no physicist claims that all these have, as a matter of fact, been satisfactorily explained on mechanical principles.[4] It may seem altogether superfluous to point out that the belief that with increasing knowledge physics may be completely reduced to mechanics, is a pious hope, that had better be explicitly stated rather than be covertly implied in the use of a term. Yet, failure to keep the distinction between the physical and mechanical clearly in mind has actually caused a great deal of confusion in the discussion of the issue between mechanism and vitalism.[5]

It seems also necessary to distinguish between mechanism and determinism. The changes of a physical system may be treated as a function of a number of variables, the mechanical conditions of the system as expressed in geometric co-ordinates being only one set of these variables.[6] It follows, therefore, that a system may be determined in its mechanical features and physically undetermined, without any breaks or discontinuity in our laws of nature. Moreover, the events in a Kingdom

[3] Continental usage has been fixed in this respect since Varignon's *Nouvelle Mécanique* (1667). In England the term *mechanics* is sometimes restricted to the study of machines, but Thomson and Tait (*Elements of Natural Philosophy*, Art. 1) are not justified in claiming the authority of Newton for this—witness the introductory paragraph of his *Principia*. Besides, it is well to remember that when there were no steam or electric engines "rational" mechanics could only deal with *vis viva* or masses in motion.

[4] Boltzmann and Planck, the most distinguished physicists to defend the mechanical methods of the classical physics, have pointed this out clearly—see *Wiedemann's Annalen*, Vol. LVII (1896), pp. 64, 65, and Planck's *Acht Vorlesungen über theoret. Physik*, p. 64. Boltzmann says explicitly: "The possibility of a mechanical explanation of the whole of nature has not been demonstrated, yea it is hardly probable that we shall completely reach that goal." (Op. cit., p. 70.)

[5] Mechanism as opposed to vitalism in biology asserts the possibility of physico-chemical explanation of biologic facts. Loeb, the illustrious protagonist of mechanistic biology, is as far as Driesch from believing that the phenomena of life can be explained by the motion of particles. In the light of recent progress in physical chemistry, also, it is hazardous to assert the existence of a greater gap or discontinuity between physics and chemistry than between mechanics and other branches of physics such as optics, theory of magnetism, or even theory of elasticity.

[6] J. J. Thomson's *The Application of Dynamics to Physics* embodies this procedure.

of Heaven or the inner life of a Leibnizian monad might be absolutely determined and yet not be, except in an obviously metaphorical sense, mechanical.

A third obvious distinction, which has actually been ignored to the detriment of clear thinking, is that between mechanical phenomena and phenomena expressible in certain kinds of differential equations. It has been widely supposed that, whenever the laws of any branch of physics, e.g. those of electricity, can be expressed in the Lagrangian form, something has been achieved in the way of a mechanical explanation.[7] It is interesting to note that Maxwell, whose procedure in his great treatise on *Electricity and Magnetism*[8] is largely responsible for this impression, had previously been careful to point out that the mathematical form of the relation between different quantities might be the same though their physical natures were different.[9] But mathematical analogies have always proved such a fruitful source of physical discoveries that physicists have been too prone to lose sight of the fact that mathematical analogy does not mean physical identity. This confusion has also been furthered by the ready way in which people confuse logical with historical priority. Thus, it has actually been argued [10] that since Lagrange's equations were first derived from mechanical considerations, they are not likely to be general forms of natural law, and hence everything expressed by them must be ultimately mechanical. The date of derivation is, however, no part of the mathematical or physical meaning of these equations. Like other equations, they state the mutual implication of certain functions of variables, and the physical meaning of these equations depends upon the interpretation or meaning that we attach to the independent variables. Clearly, the general form (and even the method of derivation) of the Lagrangian equations does not demand that their variables should be masses and velocities rather than electric charges and their intensities. All sorts of different phenomena, social, economic, or physical, as well as electrical or thermal, may have their variations expressed by the same equations, precisely as they are

[7] Larmor, *Aether and Matter*, p. 83. Maxwell, *Electricity and Magnetism*, p. vii. Combebiac, *Les Actions à Distance*, appendix.

[8] Part IV, Chs. 6-7.

[9] *Scientific Papers*, II, p. 218.

[10] Combebiac, op. cit., p. 81.

subject to the same laws of the multiplication table.[11] The fact, therefore, that the laws of electricity can be made to assume the same form as
the laws of mechanics no more proves the primacy of the mechanical
than it proves the primacy of the electrical.

§ I. MONISTIC MECHANISM

(A) ONTOLOGICAL ARGUMENT

IT IS one of the unfortunate results of Ward's *Naturalism and Agnos-
ticism* that it has strengthened the unhistorical notion that mechanism, i.e. the mechanical interpretation of physical nature, is inconsistent with ontologic idealism. While it is true that mechanism has frequently been developed in the interests of physical monism or materialism, it must not be forgotten that the mechanical view of nature was
fashioned by the founders of modern idealism, Descartes, Leibniz, and
Kant; and today, it is idealists of such diverse schools as Wundt, Aliotta,
and Fullerton, who contend that the mechanical point of view is necessary for physical science.

It is precisely this supposed necessity that is in need of critical examination. Why must all physical phenomena be viewed as ultimately
so many different forms of motion? It is to be observed that the classic
science of mechanics is a deductive system of propositions, all deducible
from Newton's Three Laws of Motion or D'Alembert's Principle in
its Lagrangian or Hamiltonian form.[12] But an examination of Newton's
laws and D'Alembert's principle, or the principle of least action in its
Hamiltonian form, does not reveal any of them to possess inherent
logical necessity; nor has any valid a priori reason ever been adduced
why all events in nature should be deducible from these laws. The
attempts of philosophers like Descartes, Kant, or Wundt, or even of
physicists like D'Alembert or Playfair, to prove these laws, hardly need
any refutation.[13] Careful examination of them readily shows that they

---

[11] Petrovitch, *La Mécanique des Phénomènes fondée sur les Analogies*, esp. pp.
7-20. For an illustration from the realm of economics see the *Comptes Rendues de
l'Académie des Sciences*, 1911, p. 1129.

[12] In *Crelle's Journal*, Vol. IV (1829), p. 233, Gauss has a demonstration that no
other principle will ever be necessary.

[13] Descartes, *Principia*, II, Art. 23. Kant, *Met. Anfangsgründe*, Pt. III. Wundt,

either move in a circle, taking for granted the very principles which they pretend to prove, or else appeal to principles which are no more self-evident (whatever that may mean) than those they wish to prove. But it is not necessary to examine these a priori proofs, since we are in possession of experimental facts tending to show that these principles are not at all universally true, but are only first approximations, i.e. true only within certain limits. Thus, the Newtonian assumptions of the constancy of mass and the proportionality between force and acceleration are now regarded as true only of tangible masses at ordinary velocities (ranging up to the paltry 18 miles per second with which the earth moves in its orbit). When we come to the small particles which compose the cathode rays or the β rays of radium, moving with velocities comparable to that of light, recent experimental physics has been forced to assume that the masses no longer remain constant but vary with the velocity. Thus, even apart from the Einstein-Minkowski relativity theory—the only one that explains the Michelson and Morley experiment—there is evidence for believing in a superior limit on possible velocities. Hence, the principle of the composition of velocities, or that acceleration varies directly as the force, is no longer of universal application. At any rate, there can be little doubt that the question, What fundamental principles of mechanics are actually true? cannot be determined a priori but only by examining the experimental evidence—which involves elements of contingency.

Leaving out of account the specific axioms or laws of mechanics various attempts have been made to prove a priori that all physical phenomena must ultimately consist of the motions of material particles.

The gist of Wundt's argument is that it contradicts our perception to assert that an object can change and still remain the same—except in the case of spatial change.[14] With all due respect, I must urge that this is sheer dogmatism. Our perceptions certainly do not contradict the assertion that an object can be now hot and subsequently cold, or that the same piece of soft iron can be at one time magnetic and subsequently not so—certainly there is no more contradiction here than in saying

Prinzipien der mechanischen Naturlehre. D'Alembert, *Dynamique*, pp. 7, 64. Playfair, *Outlines of Nat. Philosophy*, p. 26.

[14] *Prinzipien der mechanischen Naturlehre*, pp. 179 ff. In substance the same argument is repeated in all his other works. Cf. *Logik*, II, p. 225 ff.; *System*, p. 423.

that the same object can be now in one place and now in another. The contradiction in saying that a house can remain the same though the colour of its roof has been changed, is a contradiction which exists not in perception but only in a conceptual system which arbitrarily defines the identity of an object to consist in the maintenance unchanged of all its possible properties except its spatial co-ordinates. If an object can change its location and still retain its identity, why may it not similarly change its colour, its thermal or its electric properties? The assumption that the only possible changes of reality are spatial is simply the mechanical dogma over again in a different guise, and we have here no genuine proof but a *petitio principi.*

The same logical fallacy of supposing that facts of qualitative change are ruled out from reality because they contradict an arbitrary definition of identity, underlies the remarkably learned and charmingly written book of Meyerson, *Identité et Réalité.*[15]

Remembering, however, that good causes are frequently defended by bad arguments, we ought to consider whether we cannot find a better reason for the belief in the primacy of spatial change, a belief which has persisted since the foundation of modern physics. Such a reason, I believe, is to be found in the historic fact that only by reducing physical changes to phenomena of motion was it possible for the men of the Renaissance to overthrow the scholastic physics of illimitable occult qualities and to build up instead a quantitative physics capable of fruitful mathematical development. This was reinforced in the minds of men like Kepler and Galileo by the Neo-Platonic doctrine that the body of nature was composed in purely geometric terms. It is under the influence of the latter that Galileo brought forth, in his *Il Saggiatore,* the modern doctrine of the distinction between primary and secondary qualities.[16] If only extension and motion are truly existent in nature, and colours, tastes, temperatures, etc., are mere subjective products, then a true physics can be had only by reducing all phenomena to those of motion. The remarkable rapidity with which this doctrine was at once adopted from Galileo by men like Kepler,

---

[15] Pp. 98-99. A similar argument was adduced by that lonely thinker, Spir, *Denken und Wirklichkeit,* p. 424.

[16] Galileo seems to have developed this independently of the older Democritean doctrine used by Bacon.

Descartes, and Hobbes, shows what a fundamental need of the time it met.

Nevertheless, it is to be noted that the only fairly consistent attempt to banish all qualities from physics, viz. the Cartesian attempted reduction of physics to geometry, broke down under the criticism of Gassendi, Newton, and Leibniz. Atomists, Leibnizians, and Newtonians, in turn, postulated besides space and matter, primitive qualities, forces, and the properties of repulsion and attraction, respectively. Moreover, as the number of our instruments of measurement has increased, and as our mathematical methods have developed, changes in all sorts of qualities, such as illumination, elasticity, or electric charge, have become just as capable of mathematical development as changes of distance. Hence, the motive for reducing everything to spatial properties is no longer a living one. Doubtless scientific physics always endeavours for technical and aesthetic reasons to reduce the number of fundamental qualities to a minimum consistent with the known diversity of facts. But this is distinct from the pretended a priori proof that all changes must ultimately turn out to be only spatial.

Against any attempt to prove all physical phenomena to be ultimately mechanical we have the electrical theory of matter which derives the phenomena of mass from the laws of electricity. In support of such a theory it is not necessary to maintain that matter itself is less fundamental than electricity. Indeed, we can define electricity only in terms of the activity of natural bodies; and the notion that modern physics dispenses with such bodies in favour of electrical fields is based chiefly upon a conviction that somehow the abolition of matter leaves more room in the universe for God and man. But though matter is not eliminated as an essential element in all physical phenomena, there is no reason to suppose that the most fundamental transformations of matter are purely spatial rather than electrical.

Aliotta in his *Idealistic Reaction Against Science* has argued that mechanics is an inherently simpler branch of physics than electricity. But though the concept of space seems simpler than electricity (and indeed involved in the latter) there is no difficulty in seeing that descriptions of electrical phenomena, e.g. Maxwell's equations, may explain more of the physical world than do the laws concerning the motion of masses.

MECHANISM AND CAUSALITY IN PHYSICS

(B) HISTORICAL ARGUMENT

There are, however, philosophers who distrust dialectic a priori arguments and even reject the distinction between primary and secondary qualities, who yet believe, as does Professor Fullerton,[17] that all that takes place in the world must be explicable according to mechanical laws. Professor Fullerton frankly admits that the world is not known to be such a system, but the vision of it, he says, is revealed to the eye of faith.[18] This faith, I suppose, is based on a popular impression that the mechanical view has been making steady progress towards a complete explanation of the physical universe, and that it is, therefore, reasonable to hope that the hitherto unconquered fields will in the course of time yield to the sway of mechanical explanation. This is, however, a view which finds no support in any actual history of physics. Indeed, the most competent historian of physical science arrives at the very opposite conclusion.[19] Even if we do not share Duhem's view as to the final bankruptcy of the mechanical view, there can be no doubt that to the conscientious reader of the history of physics there is no such continuous progress towards a mechanical millennium as is pictured in the popular myth. It is easy to show that throughout the history of physics there have never been wanting fruitful researches carried on in utter independence of the mechanical hypothesis: the foundation of thermodynamics by Fourier, of electro-dynamics by Ampère, and the phase rule by Gibbs, are striking and well-known instances. The history of mechanics also shows a perpetual see-saw between those who are partisans of the conflicting claims of motion, the atom, or force, as the primary and all-sufficient category. Thus, the purely kinematic view of mechanics, with its ether and vortices, which seemed to have died with Descartes, was revived by the vortex-ring hypothesis of Helmholtz and Kelvin, by Larmor and others in their attempts to derive matter from ether, and in a different guise by Hertz in his brilliant but uninfluential *Mechanics*. The atomic hypothesis, brought into modern physics by Gassendi, Huygens, and Boyle, was eclipsed by the physics of forces

---

[17] *System of Metaphysics*, pp. 147, 226.
[18] Op. cit., p. 227.
[19] Duhem, *L'Évolution de la Mécanique* (1905). See also his *La Théorie Physique* (1906); his *Essai sur la Notion de la Théorie Physique* (1905); *Le Mixte* (1904); and *Introduction à la Mécanique Chimique* (1903).

213

of Newton and Leibniz (united in Boscovich), and was revived again by Dalton and Avogadro in the early part of the nineteenth century. It suffered some eclipse in the latter part of the nineteenth century—witness Berthelot, St. Claire Deville, and Ostwald—and is now to the forefront again in the form of the electron theory.[20] Nor has the Newtonian dynamics had an unchecked career. Its great triumph in astronomy made its immediate ascendency irresistible, and for over a century and a half all physical phenomena were viewed as those of miniature astronomical systems, governed by central forces. Laplace's treatment of capillarity in the tenth book of his *Mécanique Céleste* is perhaps the most characteristic product of this attitude, in which the attractive and repulsive forces of non-extended points in empty space were regarded as the key to all the secrets of nature. Yet the opposition to the Newtonian concept of gravity as a property of matter—witness the works of Euler and Bernoulli—never completely died out. When Laplace confidently announced the permanent completion of the Newtonian system by his explanation of the double refraction of light, a large part of that structure had already been undermined by the labour of Young, Fresnel, and Faraday, which brought back the ether and contact forces and banished action-at-a-distance. But the multiplicity and complexity of the various models of the ether—elastic, labile, solid, fluid, irrotational, gyrostatic, adynamic, etc.—soon made physicists weary and brought about a reaction, so that good physicists now prefer to go back to something like an emission theory of light rather than lose them-

[20] It is of course only analogically that the present electron theory may be called atomic. In one sense, however, it is an emphatic refutation of the old conception of the *atom* as absolutely *indivisible*. The basis of the present electron theory is not any a priori or philosophic necessity, but the empirical discovery that many physical facts involve multiples of a certain amount of electricity. As to what physical fact corresponds to this mathematical unit, it would be hazardous to assert with any assurance in the present state of our knowledge. I may, however, add that the phenomenalistic view that the physical atom is a mere symbol or mental figment ignores the vast mass of empirical evidence which makes the existence of atoms (i.e. physical indivisibles) as probable as the existence of King David, Croesus, or the man Shakespeare. It is only in imagination that we can go on dividing matter indefinitely without changing its specific qualities. This is the case because the imaginary process of division soon gets to a point where the imaginary division is only a duplication of the small magnitude supposed to be divided. In physics, however, we find a preponderance of evidence to indicate that matter is not indefinitely divisible but that there is a limit to this process below which the breaking up of matter, e.g. uranium, water, or wood, results in radical changes in its specific properties.

selves in interminable seas of hypothetical mechanisms, beside which the Ptolemaic cycles and epicycles were simplicity itself.[21]

(c) PSYCHOLOGICAL ARGUMENT

A third type of argument, the psychological, is represented by Abel Rey's recent book, *L'Énergétique et le Mécanisme*. The substance of M. Rey's contention is as follows: There can be no thought without images, and mechanics is best suited to provide images or models of physical phenomena. Energism, or mathematical physics may formulate the knowledge we have, but it cannot serve as an instrument of research. The laboratory physicist must work with the mechanical hypothesis in mind. This argument can be supported by many quotations from Lord Kelvin and other British physicists to the effect that to understand physical phenomena means to be able to form mechanical models of them. This, however, is not a statement of a universal law. It is true only of a certain type of mind, of those who, when they calculate the forces between the heavenly bodies, "feel their own muscles straining with the effort." But as far back as 1870 Maxwell,[22] the most illustrious representative of this type of mind in physics, had recognized the existence also of the abstract and mathematical type, and that "the tenuity and paleness of symbolic expression" had equal rights in science with "the robust and vivid colouring of physical illustration." There is not a single diagram in Lagrange's *Mécanique Analytique*, and a careful reading of it shows that Lagrange had few physical images before his mind as he wrote it. If there are minds that can dispense with diagrams in geometry and mechanics, why not minds that can dispense with mechanical models of physical phenomena? Mechanical models certainly have not as much relevance to physical inquiry as diagrams in geometry, since it can be shown, as Poincaré[23] has done, that whenever a mechanical model is invented to explain physical phenomena, an infinity of other models is possible.

Nor is it true, as a matter of fact, that the mathematical type of

[21] Campbell, *Philosophical Magazine*, Vol. XIX (1910), p. 181. Trowbridge, *Am. Journal of Science*, Vol. XXXI (1911), p. 51.
[22] *Scientific Papers*, II, 220.
[23] *Électricité et Optique* (1890), preface.

mind is impotent to produce great physical discoveries. From the discovery of the laws of planetary motion by Copernicus and Kepler, or of universal gravitation by Newton, to the discovery of the laws of thermal and electric conduction by Fourier and Ohm, or the pressure of light by Maxwell, a long list of most impressive physical discoveries by purely mathematical methods can be drawn. Physicists, like others, are not always the best judges of what is going on in their own minds when they are working, and many who speak a current language of mechanism really carry on their researches by mathematical methods. Did not Maxwell himself arrive at the electro-magnetic character of light by the purely mathematical analysis of the dimensions of the ratio between the electrostatic and electromagnetic unit? [24] The same is true of many of Lord Kelvin's discoveries in thermodynamics.

I pass over M. Rey's argument for mechanism based on the ground that knowledge must proceed from the simple to the complex. Surely the various strains and stresses in the ether or the lateral vibrations in polarized light are not psychologically simpler than the phenomena of light which the mechanistic hypothesis attempts to explain.

§ II. PHYSICAL PLURALISM AND THE PROBLEM OF CAUSATION

ALL the arguments for the mechanical dogma thus turn out to be vain. But our analysis suggests that a priori arguments *against* mechanism would similarly prove ineffective. The present actual decline of mechanical explanation in physics may render the full revival of such explanations unlikely but not impossible.

It is a curious and noteworthy fact—worthy of greater attention than it has yet received from those interested in the drama of human thought —that philosophic criticism of physical procedure has almost always gone entirely unheeded. Apparently valid arguments by men like Stallo and Ward to the effect that the mechanical hypothesis was inconsistent with itself and inconsistent with the facts, have failed to exert any noticeable influence on physics.[25] The reason for this is that physi-

---

[24] *Scientific Papers*, I, pp. 577 ff.; II, pp. 137 ff.; and *Electricity and Magnetism*, IV, Ch. 19.

[25] Thus, Stallo, and Ward after him, have argued that, on the kinetic theory, atoms must rebound when they clash or else *vis viva* be lost and the laws of mechanics no

cists do not use hypotheses as definitive summaries or pictures of existence but as anticipations which promote research. They are, therefore, willing to use different promising hypotheses in different contexts and to postpone the question of adjusting their inconsistencies. Thus physicists use the atomic hypothesis which regards matter as discrete and then employ integrations which presuppose material bodies to be continuous as regards the space they occupy. Both hypotheses are too useful to be abandoned, and the ironing out of their mutual inconsistencies is effected by conceding certain theoretical differences between our physical states and the mathematical equations which explain them.[26]

Nor is an apparent contradiction between a theory and facts necessarily fatal. For theories are flexible and facts can be supplemented by hypothetical elements. Thus if the facts of radiation do not fit in with the law of the conservation of energy, an ether can be invented and endowed with just those properties which will make the law true. If, therefore, physicists of a certain type of mind find that illustrative models based on a mechanical hypothesis help them to visualize their problems, it is as vain to argue against them as to argue against their religion or political affiliations. The effective thing in the long run is always the elaboration of the possibilities of some alternative method of explaining all of the facts with less hypothetical elements. It is thus that only the advent of the relativity theory brought about the abandonment of the "ether."

Now, the fundamental postulate of mechanism, as we saw, is the as-

longer prevail; but if they do rebound, elasticity becomes a fundamental property of matter and the atomic theory no longer offers any explanation of it. (Cf. Kroman, *Unsere Naturerkenntniss*, p. 315.) Neither horn of this dilemma, however, can be considered fatal. If the atomic theory does not explain elasticity, there are many other facts like the diffusion of gases which it does explain. On the other hand, recent physics has taught us that it is not necessary that the laws applying to ponderable masses should apply in the same way to molecules or atoms. We must guard against the naïve assumption that the laws observed to hold within the limits of the pressures, temperatures, masses, etc., actually observed, must necessarily hold below or above these limits. Maxwell shocked his contemporaries, even the agnostic Huxley, by asserting that the law "two portions of matter cannot occupy the same space" has no application to molecules (*Scientific Papers*, II, 33). Yet it is clear that the law in question is not a priori necessary but founded on the simple empirical observation that ordinary solid matter has the property of impenetrability. If we had been as familiar with the diffusion of gases, or even with the interpenetration of water and alcohol, impenetrability as an absolute dogma would never have acquired its vogue.

[26] See W. Voigt, *Lehrbuch der Krystallphysik*.

217

sumption that there is an ultimate structure of things which it is the primary business of the physicist to discover, and even where it has not yet been discovered, he must still be sure that it consists in nothing but the hidden motions of particles. An alternative to this realistic monism of motion has, as a matter of fact, always existed in physics since the days of Ptolemy [27] and Archimedes. But it has only recently been able to obtain sufficient philosophic backing to make it self-conscious and respectable. Since the days of Kant and Comte, physicists need not be ashamed of admitting that their science is not a means of piercing the veil of phenomena and grasping the ultimate reality behind it, but only a method of extending and organizing our knowledge of these phenomena; and the recent revival of pluralism supports those who refuse to believe that all possible changes of the physical universe must be reducible to just one kind of change: namely, motion. It is interesting to note that Comte's views in this matter were determined by Fourier, whose preliminary chapter in his *Théorie Analytique de Chaleur* contains the essence of the matter. Expressions of it may be found in the writings of the founders of mechanics itself. Thus, Galileo states explicitly,[28] "It does not appear to me at present worth while to investigate the causes of natural motion, concerning which there are as many different opinions as there are different philosophers. Some refer them to an attraction towards the centre, others assign them to repulsion between the small particles of a body, while still others would introduce a certain stress in the surrounding medium which closes in behind the falling body and drives it from one of its positions to another. But it is not worth while examining all these fantasies. All that is needful is to investigate the properties of accelerated motion and define it in such a way that the momentum of the body increases uniformly in simple proportionality to the time." The same attitude was expressed by Newton in his famous adage: "*Hypotheses non fingo.*" By hypothesis, we must remember, Newton meant an explanation not directly derived from phenomena. Thus he says in the concluding *scholium* to the *Principia:* "Hypotheses, whether metaphysical or physical, whether of occult qualities or mechanical, have no place in experi-

---

[27] Duhem, *Essai sur la Notion de Théorie Physique* (1908). See also Delambre's article on Kepler, in Michaud's *Biographie Universelle.*
[28] *Discorsi e dimonstrazione intorno a due nuove scienze. Opere* (1811), VIII, p. 256.

mental philosophy. In this philosophy particular propositions are inferred from the phenomena, and afterwards rendered general by induction. Thus it was that the impenetrability, the mobility, and the impulsive force of bodies, and the laws of motion and gravitation were discovered. And to us it is enough that gravity does really exist, and acts according to the laws which we have explained, and abundantly serves to account for all the motions of the celestial bodies and of the sea."

An even more explicit statement of this we have in Rankine's paper on the *Science of Energetics* (1855). Rankine clearly distinguishes two methods of constructing a physical theory, which he calls the hypothetical and the abstractive. The hypothetical method consists in starting with some hypothesis about something which is not the object of direct perception, and deducing from this supposed constitution the empirical properties. All mechanical theories of physics, e.g. the kinetic theory of gases, illustrate this method. The abstractive method, on the other hand, is described as follows: "Instead of supposing the various classes of physical phenomena to be constituted in an occult way of modifications of motion and force, let us distinguish the common properties which these classes possess and define more extensive classes denoted by suitable terms. For axioms let us frame propositions containing as particular cases the laws of the particular classes of phenomena comprehended under the more extensive classes. So shall we arrive at a body of principles applicable to physical phenomena in general and which, being framed by induction from facts alone, will be free from the uncertainty which must always attach even to those mechanical hypotheses whose consequences are most fully confirmed by experiment." [29] It is to be observed that while mechanical theories of physics are illustrations of the hypothetical method, mechanics, as a branch of physics studying the laws of motion, is itself an illustration of the abstractive method.

Though Rankine was one of the founders of modern thermodynamics and the author of classical treatises on the steam engine and ship building, this paper received very little attention. It came in the heyday of mechanical models, when every one was trying to derive the principles of energy from the principles of mechanics. These efforts,

---

[29] Rankine, *Miscellaneous Papers*, p. 245. Cf. Whitehead, "The Mathematical Concepts of the Material World," *Transactions Royal Society of London*, 1906.

however, soon came to a standstill. The kinetic theory of gases struck a rock in the problem of the equipartition of energy, being unable to harmonize the theory with the behaviour even of diatomic gases.[30] More particularly it was soon realized that the principle of entropy or degradation of energy—the general fact that physical phenomena proceed in one direction and are irreversible—could not be explained on mechanical principles. (Thus, two gases will diffuse themselves one in the other, but will not conversely separate themselves spontaneously.) Maxwell and Gibbs realized this, and introduced the notion of a statistical as opposed to a mechanical knowledge of physical phenomena. Imagine an indefinitely large number of particles moving at random with various velocities, and one can compute on the basis of the various degrees of freedom and the principles of statistical probability what the total effect will be. The famous example of the sorting demon was introduced by Maxwell [31] to show that the second law of thermodynamics was not mechanically necessary but had only statistical certainty. This has recently been reinforced by M. Gouy's investigations on the Brownian movements, which indicate that what we ordinarily call thermal equilibrium, i.e. stable, uniform distribution of temperature, may prevail in sensible volumes, but not within microscopic volumes, so that the second law of thermodynamics may not be applicable within the latter.

As the laws of thermodynamics are empirically verifiable and independent of all atomic or other hypotheses as to the ultimate structure of things, the enormous success which followed the introduction of this method into physical chemistry by Gibbs, Van der Waals, and Van't Hoff, gave support to the empirical or descriptive theory of physical science upheld by Kirchhoff, Avenarius, Mach, and Duhem.

Before examining the philosophic significance of this theory, it is well to note what it has actually meant for physics. No one can compare the prevailing tone of physical theory today with that of a generation ago without noticing the greater recognition today of the provisional, empirical, pluralistic, and yet thoroughly mathematical character of physics. No one asserts nowadays, as did Maxwell, that atoms never change and are today as fresh as when they came out of the

---

[30] See the various papers of Rayleigh and the preface to Gibbs' *Statistical Mechanics.*
[31] *Treatise on Heat.*

hands of the Creator. Formerly we used to be told that when hydrogen and oxygen combine to form water, the two substances, as represented by the H and O atoms, remain the same, though most of the properties of the H_2O molecule in no wise follow from those of the H and O. Now reputable authorities on physical chemistry, like Ostwald and Duhem,[32] find it possible to return to the Aristotelian conception of change and to suppose that when an electric spark causes the H and O to combine, the H and O both disappear and a third something, namely, water, takes their place. Modern physics has learned to be suspicious of eternal substantial forms, and is not awed by the scholastic dogma *ex nihil nihil fit, nihil in nihilo.* We no longer think that because the white light that enters a prism issues in the form of many colours, it follows that the white light actually contained all the various colours [33] and we are careful not to say that when a cool body is brought into the presence of a warmer body, the heat gained by one is precisely or identically that which is lost by the other. But perhaps the most striking illustration of this point is to be found in the current statement of the law of gravity, which has so long served as the typical law of nature. Instead of asserting dogmatically that every particle of matter attracts every other particle precisely as the product of the masses and inversely as the square of the distance, careful physicists, like Poynting and Thomson, point out that astronomic observation is by no means decisive on this point and that all we can say is that when we take large masses like the planets, the mean results fit in with our formula.[34]

From this point of view the classic notion of uniform laws of nature holding with absolute accuracy for the smallest atom as well as for the largest star cluster can be replaced by the more modest doctrine of empirical or statistical constants holding in limited fields. Our knowledge of physical phenomena is in some respects like that of social phenomena when studied through such facts as marriage rates, birth rates, tables of exports and imports, etc. But while statistical averages in the social field show large fluctuations from year to year and as we pass from one locality to another, physical laws do not seem to have changed within the time open to scientific verification and are uniform

---

[32] Duhem, *Le Mixte*, p. 165. Ostwald, *Lehrbuch d. Allgem. Chemie*, II, pp. 5-9.
[33] Wood's *Physical Optics*, Ch. 21.
[34] Poynting and Thomson, *Properties of Matter*, p. 46.

with a remarkable degree of precision in different parts of space. Part of this precision may be due to the fact that in the human field individual variations obtrude themselves, while in the physical realm the constituent individuals or atoms are for the most part beyond our range of observation.

There are philosophers to whom the slightest suggestion of contingency in the physical world or any doubt as to whether everything does happen absolutely in accordance with universal laws, is an atrocious and unpardonable blasphemy. But whatever may be said for the sublime faith back of this attitude, it surely is not necessitated by the experience of the physicist who works with instruments of precise measurement. Laboratory workers know how difficult it is to get phenomena to repeat themselves even approximately, i.e. within the range that we call the limit of probable error, and they will readily subscribe to the statement in Chwolson's great international text-book, that when we study physical phenomena more closely we can convince ourselves that there is almost no physical law which can be exactly verified.[35]

We need not here press the hypothesis of C. S. Peirce that there is a domain of radical indeterminism, that besides the variations due to errors of observation there are variations due to the fact that our physical laws do not express with absolute accuracy the actual behaviour of things. But modern physics is beginning to recognize more and more the point made by Poincaré[36] that the simplicity of Newtonian laws may be the result of averaging large numbers of very complicated phenomena, in accordance with the well-known fact that the larger the number of cases considered, the simpler the expression of the prevailing type. Spectrum analysis and other evidence as to the structure of matter suggest that an atom of sodium may have a structure as complicated as that of a piano or stove, and the variation in the behaviour of the atom may consequently be as great as that of these somewhat capricious objects. But when we remember that the number of atoms in a pin-head is greater than that of all the human beings now alive,

[35]*Traité de Physique*, I, p. 29. Cf. Poincaré, *Value of Science*, Pt. III, Art. 4-5, and Thomson and Tait, *Natural Philosophy*, I, Ch. 3.
[36] *Thermodynamique*, p. vii.

we can readily understand why any tangible piece of sodium behaves so like any other piece.

The principle of uniformity of nature is usually stated thus: like causes produce like results. But in physics, as in social science, we never have the entire identical situation repeating itself. What we observe is that when the antecedent situations are alike, the sequences are also alike. Now likeness is a matter of degree, i.e. it depends on the fineness of our classification. When we say water freezes at $32\,^{\delta}$ F. we regard all samples of water as alike, and the result is approximate enough when measured by an ordinary thermometer. We may, however, treat our water as consisting of different samples having, e.g. different degrees of density, and notice slight variations in the reading—provided our thermometer is adequately graded.

§ III. CONTINGENCY AND MATHEMATICS IN PHYSICAL SCIENCE

THE considerations involved in the last section are so well established in the daily procedure of scientific investigation, that any form of rationalism which puts its face against them is bound to come to grief. The old rationalistic conception that the principles of mechanics are a priori self-evident axioms which will never be successfully attacked,[37] has been effectively disposed of by the rise of non-Newtonian mechanics. The primary laws of mechanics, as of any other branch of physics, are now seen to be logically contingent, i.e. they are not to be derived from the non-temporal laws of logic.[38] Their contraries are possible hypotheses. There is nothing illogical or inconsistent in supposing the Newtonian laws of motion or the formula of gravitation to

[37] This used to be the basis for preferring physical deductions from the laws of mechanics rather than from the laws of thermodynamics. (J. J. Thomson, op. cit.)

[38] Similar considerations hold true of the mixture of logic and psychology which passes as epistemology. To derive the fundamental laws of physics from the laws according to which the mind operates, does not really remove this contingency. There is no reason to suppose that any known laws according to which the mind works are absolute constants, and the only evidence we have as to the way in which the mind operates at its best is the changing body of actual science. There is a close and suggestive parallel between the attitude of modern epistemology to physics and that of the older theology. The older theology tried to derive the truth of physics from the will of God. Kantian epistemology tries to derive it from the ways in which the mind operates. The physicist may believe as much as he pleases in "the ways of the mind" as he does

be inaccurate. Not a single established law of physics but its absolute accuracy has now been shaken by recent experimental work in Brownian movements, radioactivity, the phenomena of radiant energy, etc. But while all this fortifies the view that all known natural laws are contingent or expressive of relations which are slowly changing in time, it seems to me an unseemly intellectual haste to jump to the conclusion of the positivism or phenomenalism of Mach, Pearson, and their followers, who assert that the world simply is, and that all necessary relations are fictions or mental products.

We may grant at the outset that the positivists are right in regarding the popular use of the word *cause* as embodying remnants of primitive animism. When we popularly speak of a thing's causing something else, we undoubtedly tend to attribute to the thing something analogous to human compulsion, something of muscular tension or the feelings of activity and passivity when we wilfully push or are pulled contrary to our will. Such animism is out of place in modern scientific physics. The Humian analysis of causation and its replacing of the ideas of production, of power and force (as synonyms of compulsion) by the idea of regular sequence, was the *coup de grâce* which modern thought administered to the scholastic physics of occult qualities and powers. If we do sometimes find authoritative physicists still speaking of the operation of forces in an anthropomorphic way, or lapsing into the popular manner of speaking of heat or gravity as causes, we must remember the great difficulty of freeing ourselves completely from prevailing popular use of words, and the even greater difficulty of expressing ourselves vividly without the use of metaphors, of which anthropomorphism supplies the bulk. Technical and mathematical language, however, is surely, if slowly, replacing expressions of causal relations with mathematical functions or equations, which are neutral to all anthropomorphic hypotheses. In formulating Newtonian laws of motion in popular language, physicists may still use such phrases as: bodies acted on by forces, etc. But when the physicists' actual deduction from these laws is carefully examined, we find actions replaced by changes in certain physical co-ordinates or parameters, and "force" denotes that part of the change that

in the "will" of God. But he must not introduce them as principles of physical explanation, for the simple reason that they are not principles of determination. We have no scientific way of telling the way the mind or the will of God works except by examining the results.

varies as *mA*, the product of the mass by the acceleration. Mathematical expressions like *mA*, which keep on recurring, are usefully denoted by some name; and the conferring of a name unfortunately always tends to reify or hypostatize that which is thus denoted. But the whole tendency of modern experimental as well as mathematical physics is to eliminate the metaphysical notion of matter as an *ultimate* substance, and to find the element of permanence—without which there would be no science—in the mathematical relations. Thus Helmholtz, who in his youth thought that "the final aim of physical science is to find the ultimate unchangeable causes of the processes in nature," became satisfied later that the principle of causality meant nothing more than that natural phenomena happen according to law.[39]

Not only must we, then, admit with Hume that conscious analysis does not show any single event to necessitate any other event, but modern physics suggests that the laws of nature which do correlate these events are themselves contingent, in the sense that they are known to be true only within the limits of observation, and may perhaps not prevail outside of the infinitesimal portion of the universe whose surface we have scratched. We cannot be sure that these laws held true in the distant past any more than we can be certain that they now hold in the more distant parts of space not available to our instruments.[40]

Empiricism breaks down, however, in failing to account for the fundamental assumption underlying all scientific procedure: namely, that the logically necessary relations which hold between mathematical expressions hold of natural phenomena themselves. No physicist for a moment doubts that all the unforeseen logical consequences of a true physical hypothesis must necessarily hold for the physical universe in which that hypothesis is true, and that, if any of these consequences turn out to be false, it must be due to the falsity of our original assumption and not to the fact that nature fails to behave in accordance with the rules of mathematical deduction or computation. So long, therefore, as the laws of logic and mathematics are applicable to the physical universe, necessity of a certain kind, namely, the necessity which connects ground and consequent, must be predicated of it. It would not be diffi-

[39] *Wissenschaftliche Abhandlungen* (1882), p. 68.
[40] For a vigorous refutation of the easy assumption that the known laws of physics must hold for the whole universe see Chwolson in *Scientia*, 1910.

cult to show that this is precisely the necessity which common sense and physical science actually attribute to the causal relation. A stone thrown up *must* fall down after its upward velocity is spent and it has thus become a free body, if we assume, as we do, the law of gravity. If carbon combines with oxygen and thus burns, any substance like paper, made of wood pulp, *must* burn. The consequences in both cases are necessary and physically explained, though the major premises are contingent. If the law of gravitation or that of valency could themselves be deduced from another law—for example, some law of electro-magnetism—the realm of physical explanation would be widened and greater unity be introduced. But the logical character of physical explanation would remain unaltered. Actually, the search for physical causes or explanations is, thus, a hunt for appropriate major premises or middle terms. The principle of causality (as distinct from particular causal laws) is thus simply the general maxim that physical phenomena are connected according to invariant laws. While this maxim is properly a postulate or resolution of the scientific understanding to look for such connections, it can be maintained only because the world of physics is full of universal elements or relations which repeat themselves indefinitely.[41]

The significance of this obvious truth, that logical or hypothetical necessity holds of nature, has been obscured by a number of powerful dogmas in modern philosophy which are certainly not the outgrowth of reflection on the nature of modern physics. These dogmas are: (1) that logical and mathematical relations and abstractions generally exist in the mind only, while physical phenomena exist in the external world; (2) that strict or deductive reasoning is a series of tautologies which cannot extend our knowledge; (3) that science deals only with the actual existing world, and (4) the monistic or organic view of truth which fails to note that approximations or partial truths are still truths.

Without attempting a complete refutation of all these dogmas, I may suggest how dubious or in need of radical revision they appear, when we deal seriously with the fact that after all nature does behave in conformity with logical and mathematical principles. Consider, for

---

[41] Poincaré, "*Les equations expressent des rapports et si les equations restent vraies, c'est que ces rapports conservent leur realité.*" *Congrès des Phys.*, 1900, *Comptes Rendus*, I, p. 15.

instance, the following statements of Mach: "No one will fancy that vibrations in themselves have anything to do with circular functions or the motions of falling bodies with squares." "These mental expedients have nothing to do with the phenomenon itself." Here clearly Mach, the monistic sensationalist or empiricist, is at one with the metaphysical dualism of Descartes and his dogma that universals and principles are in the mind only, while the physical world of extension lies outside of it.[42] But this fails to explain why phenomena seem to occur as if the law of gravitation with its inverse squares were true, or why the properties of circular functions have proved most potent instruments for the discovery of important facts in almost all branches of physics. Doubtless, equations are not vibrating strings; but is it not straining the dualistic dogma to assert that they have nothing to do with each other? Do not let us be misled by the term expedient or invention. A map or chart is an expedient or invention. Yet if it fairly represents its object, is it not because certain relations between its parts are precisely those between the corresponding parts of the object represented? Mach admits that it is easier to deal with natural phenomena when the relations between the quantities investigated are similar to certain relations between familiar mathematical functions. But what does that similarity here mean, if not identity of mathematical relations? No one denies the suggestive value of popular analogies, such as that which speaks of certain social phenomena as periodic, rhythmic, or typifying the swing of the pendulum. Such analogies are dangerous, because popular language does not indicate clearly where the differences begin and where identity of relations ceases. When, however, different processes are expressed in analogous equations, the extent of the identical elements is unmistakably indicated, and deductions made from these equations are applicable to all the possible regions in which exist relations such as expressed in the equations before us.

Grant that the law of identity asserts that there are diverse elements which remain identical; grant, for example, that the relation represented by the ratio 2:1 may hold between all sorts of different entities; and most of the artificial problems of the classical epistemology disappear. At any rate, it becomes rather easy to explain the seeming paradox that physical laws may be true and yet physical phe-

[42] Mach, *Mechanics*, pp. 492-494. Descartes, *Principia*, I, Art. 23.

nomena show departure from them. For physical entities may have invariant relations between their parts and yet, being complexes, not have their entire character expressed by simple laws or mathematical formulas that contain only a small number of factors or operations. Consider, for instance, Newton's first law of motion: "A body not acted on by force would continue in a state of rest or uniform motion in a straight line forever." If we adopted the mode of argument prevailing among certain positivistic philosophers, we should say: This is a foolish statement. No one has ever seen a body not acted on by any force, much less verified its motion forever. Indeed, no such body does exist, if Newtonian laws of universal gravitation be true. Yet, though no single physical body acts as if the law of inertia were the only law, the law of inertia is still an indispensable part of the Newtonian mechanics—which, with all its limitations, is still one of the most accurate descriptions of nature that the human mind has produced. Again, consider Boyle's law that the volume of a gas varies inversely as the pressure. There is not a single gas that conforms to it exactly, but it is not therefore false. It is a true first approximation, rendered more adequate when we introduce additional factors, as was done by Gay-Lussac and later by Van der Waals. As our instruments of precision and control over nature increase, we attempt to eliminate more and more of the residual variations. This seems an endless task, but the physicist must always assume that phenomena measurably depend not upon an infinite but upon a very limited number of factors.

Note that dependence upon a limited number of factors means independence of all others. The organic point of view, or Mill's notion that the total state of the universe at any one time is the cause of the total state of the universe at the next moment, ignores this element of independence which our physics is constantly asserting. (For example, the friction in a gas is independent of its temperature, etc.) It seems to me also quite clear that a principle such as Mill states would be inconsistent with the principle of causality as actually employed and as explained by a physicist like Maxwell. As stated by the latter, it is the following: "The difference between one event and another does not depend on the mere difference of the times or places at which they occur but only on differences in the nature, configuration, and motion of the bodies concerned." [43] If we did not eliminate from our consideration the particular moments of time or points of space at which events occur, physics would remain impossible. To speak of an event at all, there must be kinds or classes of them. They must be capable of repetition. And these repeatable time intervals (seconds, years, etc.) and space intervals (yards, etc.) rather than instants and points enter into the causal relation of actual physics. But the principle of causality as thus formulated carries us further and determines our choice of space reference and time measurement, as well as the form of our mathematical equations. Suppose I measure the length of a certain rod and find that it varies irregularly. The maxim of causality means that such variation is due not to the mere difference in time or place at which the measurement was made, but to other factors, such as heat, pressure, etc. Similarly, if I notice that the tide rises higher on some days than on others, the principle of causality means that it is not the time at which it occurs but certain factors such as winds, nearness of moon, etc., which must be taken into account. No single law of physics would have meaning if everything depended upon everything else. If the freezing of water depended upon an infinite number of factors, there would be no sense in saying it depends upon temperature and pressure or that one of the latter can be varied while the other is constant. We can speak of water at all only because certain qualities or groups of qualities maintain their identity and keep on repeating themselves while other things change. The particular gunpowder whose explosion fires a particular shell will never again do so, but the elements of mass, units of force, velocity, etc., will repeat themselves indefinitely.

Without the assumption of the existence of identical elements, all common sense, as well as scientific assertion, becomes not simply false but meaningless. On the other hand, unless physical nature behaves according to the laws of diversity (excluded middle and contradiction) not a single mathematical principle could be applied to it, and it might, as far as physics is concerned, be one big blooming buzzing confusion. In all physical operations where addition is applicable, we see operations which are essentially independent of each other. It is not necessary, for purposes of physics, to believe that nothing ever can happen except what is adequately described by the actually known laws of physics.

[43] Maxwell, *Matter and Motion*, p. 31.

It is not necessary to believe that the world exists solely for the satisfaction of our scientific ideals. But it is certainly most reasonable to suppose that the relations expressed by physical laws are actual constituents of the world, and that large domains or aspects of the latter are as described by our physical science.

To sum up: mechanism as a formulation of the laws of masses in motion is a basic branch of physics but not an adequate account of the whole of it. An adequate analysis of the latter bears out the contention that not ὕλη, formless matter, or blind sensation, but mathematical and logical relations form the intelligible substance of things. But the world must contain more than this form if the concepts and procedure of physics are to have meaning.

APPENDIX. THE PHILOSOPHICAL SIGNIFICANCE OF PHYSICAL RELATIVITY

THOUGH the essentials of Einstein's theory of relativity were known to the scientific public as far back as 1905, it was only after the astronomical confirmation of its law of gravitation in 1919 that it captured the world's imagination, and that people began to ask what "message" it had for philosophy in general, what light it threw on the nature of man and the physical world, on mind, and perhaps on the fate of civilization.

Clearly, however, if we draw (as we must) some distinction between physics and philosophy, and regard the latter as dealing with the more general or constant phases of truth about nature, any theory which successfully predicts specific physical phenomena is in that respect a physical rather than a philosophic theory. But while the bending of a light ray is a physical fact, in itself of no greater philosophic importance than any other fact, the theory of relativity has radically transformed our ideas of time, space, matter, and gravitation, and has so revolutionized our classical cosmology that no adequate philosophy can ignore it.

We need, however, to be warned that as philosophic theories are wide and flexible, different philosophies may accommodate themselves to the same set of facts. Einstein himself put it somewhat ironically when he said that all true philosophies agree with the established results of physical science.

(A) METHODOLOGIC BEARINGS

From the point of view of scientific method the theory of relativity is a great triumph for the abstractive method—all the more notable because Einstein himself, far from being, like Mach, unsympathetic with the hypothetical method, has indeed made notable contributions to our understanding of the Brownian movements and of the theory of quanta of action, by framing suitable hypotheses as to the invisible structure of matter.[44] The special theory of relativity may long serve as a model of what can be achieved in physics with a minimum of simple assumptions and a most rigorous mathematical development.

Nineteenth century physics had succeeded in systematizing our knowledge of inorganic nature into two divisions, mechanics and electricity, the former governed by Newton's laws, and the latter by Maxwell's equations. But no effective bridge or connection between them was established. The effort to construct mechanical models of electricity and ether failed utterly, and electro-magnetic theories of matter and gravitation found difficulties in the fact that while electro-magnetic forces are polar (positive and negative), can be screened, and spread with the velocity of light, gravitation cannot be screened, acts on all bodies alike, and seems to act instantaneously through distance. Moreover, the failure of numerous experiments, that of Michelson and Morley being the most illustrious, to detect the motion of the earth relative to the ether, lay as a cloud upon the state of scientific physics. In this situation Einstein succeeded in introducing a remarkable degree of order and unity by uniting the electro-magnetic principle of the constant velocity of light with an extension of the principle of relativity that is involved in classical mechanics.

From the Galileo-Newtonian principle of inertia it follows that no mechanical experiment on a body (or physical system) can tell us whether it is as a whole at rest or moving uniformly in a straight line.

[44] It is interesting to note that in the volume *Physik* in *Die Kultur der Gegenwart* (1915) Einstein wrote not only the chapter on Relativity but also the one on Atoms—thus showing himself a disciple of Boltzmann as well as of Mach.

If we can extend this to the electrical field and say that no experiment of any kind, optical or otherwise, can tell us whether a body is at rest or in motion relative to the ether, the negative result of the Michelson-Morley and similar experiments will become necessary. But how can we combine this extended principle of relativity with the fundamental equations of electro-magnetism in which the velocity of light is a constant? Einstein succeeded in doing this by analyzing the nature of measurement and showing that it involved the relativity of simultaneity as well as of distances and of temporal durations. This correction of our ordinary ideas as to time and space has upset many who have found various paradoxes in it. But these paradoxes are due to our trying to think of the new ideas without abandoning the old ones which are inconsistent with them.[45] The mathematical development of these new ideas and the way they have reconciled and explained most diverse physical phenomena guarantees them a validity greater than the opposing uncritical assumptions as to time and space.

But the original theory of relativity was not altogether satisfactory. Restricted to uniform motion, it could not explain the starting of any motion, which involves acceleration. Moreover, on a priori grounds there seems no reason for restricting the principle of relativity to uniform rectilinear motion. Every observable motion means a variable distance between one body and one or more others; and mathematically or kinematically it ought to make no difference where I take my centre of measurement or system of co-ordinates. The invariant laws or truths of nature ought to be the same to all observers, no matter in what relative motion they are to each other. Thus observers on Mars, if gifted with as much scientific power as we, ought to be able to learn the facts of astrophysics as well as we do, and to the extent that both our and the Martian observations are true, it ought to be possible to frame transformation

[45] Robb, Krauss, and others have alleged that the theory involves self-contradiction. In our daily life motion is associated with conscious effort or power and there seems to be an absolute difference between resting and acting. But physically that is irrelevant since one resting with reference to the earth may be in motion with reference to the sun. In daily life also we so often refer to the earth as the centre of reference for rest and motion that it is not necessary to mention it. Hence we get into the habit of regarding statements such as *A is at rest* and *A is in motion*, *A has a velocity of 16* and *A has a velocity of 17 miles per hour* as mutually exclusive. But if we remember that such statements are incomplete expressions so long as they do not specify the thing with reference to which the rest, motion, or velocity is asserted, we shall recognize that these statements are not full expressions and therefore not subject to the law of contradiction.

formulae which will enable us to pass from one set of observations to another. The original restricted theory of relativity provides such a formula only for the case where the observers are in uniform rectilinear motion to each other. This restriction seemed justified on the experimental ground that accelerated and circular motion produced special effects. Thus, the fact that the earth revolves around its axis, it was urged, can be found without reference to any other body—by merely measuring it and finding that it is flattened at the poles.

Thus the argument stood since the day of Newton, who argued that rotation involved absolute space. But it did not satisfy Einstein, who, following Mach in this respect, could not see why an unobservable entity such as absolute space should be invoked as the cause of a particular phenomenon such as the flattening of the earth at its poles, or the curved shape of the water surface in a revolving bucket. After all, these bodies are known to revolve not in respect to absolute space but in respect to surrounding objects on the earth or to the stars. Why may not the observed effects be due to the interaction between the rotating body and the earth or fixed stars?

In the effort to show the possibility of a purely kinematic representation of gravitational "force," Einstein developed the tensor calculus as a wonderful instrument for the formulation of physical laws that hold for all co-ordinate systems, i.e. for all observers moving relatively to each other in any definite manner.

To those engaged in physical research the new theory justifies itself by the various phenomena, such as the bending of light rays or the spectrum shift, which it predicted, and by other phenomena which it explained better and with fewer hypothetical elements than any other theory. But as a method of studying nature it justifies itself by making clearer what has always been implicit in the ideal of physical science from the days of Thales and Archimedes to those of Galileo and Newton, and that is that the true laws of nature must hold for all observers.

(B) COSMOLOGICAL BEARINGS

Einstein's fundamental discovery is the fact that the spatial and temporal dimensions of physical objects are not independent variables, but that both depend on the velocities of these objects and on the masses in the physical universe. Can we dismiss this as of no importance to metaphysics? If we do, what shall we say of time and space and matter, when all their measurable qualities are removed?

(1) TIME. Turning more specifically to the light which the theory of relativity throws upon the nature of time, we may consider, first, the location or date of an event, second, the length or duration of an event, and third, its relation of precedence or sequence to other events.

(i) *Temporal Location*. As to the date, consider the instant at which some event occurs, begins, or ends. It is popularly supposed that any two such events in different parts of space either do or do not occur at the same time. The chief novelty of Einstein's theory is the conception of the relativity of simultaneity. Two events which will be recorded by accurate physical instruments as having occurred at the same time will be recorded as having occurred at different times by physical apparatus that are in motion relative to the first set of instruments. This divergence as to whether two events are or are not simultaneous involves no illusion but is a necessary consequence of the principle of relativity and the nature of measurement as applied to events in different parts of space. It is physically true that two clocks which are synchronous to measuring instruments on the sun are not synchronous to similar recording instruments on the earth. Of course, if we look within us introspectively, it seems impossible that the two events which are simultaneous to us should not be simultaneous to every one who observes them truly. But the properties of things are not to be determined by introspection but by objective measurements. Now there are cases of subjective or apparent simultaneity which can be corrected by taking account of the time it takes for the signal from a distant event to reach us. Thus, at the instant when I see my watch mark one o'clock I may hear a distant bell strike. These two events appear simultaneous to me. But on taking into account the time required for the sound of the bell to reach me I find that the two events are not really simultaneous, i.e. would not appear simultaneous in all possible positions which I might take. The divergences as to simultaneity to which Einstein calls our attention are not of this apparent kind, but are found to hold after all corrections as to the time it takes the signal to reach us. Subjectively each one of us is an absolute clock for the entire physical universe, but objectively there are many different clocks, all running true and yet diverging as to their synchronous character according to a definite function of their velocity relative to other bodies. This liberalization of our notion of simultaneity is as consistent with ancient truth as the discovery that each one of us has the zenith directly over his head.

(ii) *Duration*. If we grasp the relativity of simultaneity, there is little difficulty in seeing that the measurable physical duration (or elapsed time) of any event depends upon the velocity of the centre from which it is measured. By *time* men have always meant days or years and their fractions or multiples. We are now accustomed to regard the rotation of the earth around its axis or around the sun as the measure of the days or year. These rotations as physical events are convenient measures of time, if by *time* we mean the correlation of all events in one series. But any other repeatable event can serve as a measure of time if two events would synchronize with each other. Our ordinary life and our classical physics have taken for granted that an hour is an hour throughout the universe, i.e. that if a certain physical operation coincides throughout its extent with a physical process that is our standard time measurer, it will coincide with all other processes in nature that are coincident with that standard event; but this is approximately true only in physical systems which move relatively to each other with velocities like those of our aeroplanes or even of planets in their motion around the sun, which are negligible in comparison with that of light. However, when we take our fastest-moving planet, Mercury, or the motions of electrons that begin to approach the velocity of light, then Einstein's formula describes the facts more accurately than the classical view. Nor need we be surprised at this if we remember that a well-constructed clock will go at a different rate at the equator from what it does in the polar regions.

(iii) *Temporal Order*. Time involves a unique order which is irreversible. The past can never be restored though it may be intuited. This has been connected in the minds of modern physicists with the law of entropy, i.e. that all physical changes are in the direction of making energy unavailable. Now the analysis of our physical concepts by the theory of relativity leaves entropy as an invariant, which guarantees it a certain mathematical reality and thus secures a certain order of succession in physical events. The theory of relativity not only takes for granted the irrevocability of the past, that the status of events as past is unalterable, but in making the velocity of light a maximum it makes vision or other communication with the past impossible. But it is at first rather surprising to learn that of two events in distant parts of space, one may precede the other in one physical system and follow it in the measurable determinations of another system that is moving relatively to the first. This seemingly paradoxical situation, that event A may as truly be said to precede B as to follow it, depending on the different referents, is limited by the finite velocity of light as a maximum. This also limits the application of the principle of causality between events in different parts of space. In order that any event A should be the cause of or influence event B, the time-space interval between them must be such that the light signal from A will reach B at the instant of or previous to its beginning. In thus eliminating action at a distance it makes for contiguity in the causal relation.

(2) SPACE. Little need now be said about Einstein's discovery that the length of a body is not the same when measured by different observers that are in relative motion to each other. But the relation of space to matter suggested by the general theory of relativity is not so clearly understood. In asserting that time-space measurements are influenced by matter or gravitation, Einstein seems to involve himself in a vicious circle. Space can be measured only by rigid bodies or by light rays of constant velocity. But in the general theory of relativity there are no absolutely rigid bodies and the velocity of light is not a constant. It would seem then that we cannot build up a geometry of our space on the basis of rigid bodies nor can we measure the actual dimensions of bodies on the basis of pure space. This dilemma points back to the more primary difficulty as to the reality of space. Either space is prior to matter and exists as a vacuum or it is merely an abstract phase of matter. If we take the former alternative there seems no reason why its geometric traits should depend on the presence of matter. If we take the latter alternative there seems no meaning in the assertion that the presence of matter affects our measurements. This dilemma, however, arises from a failure to get close to the facts of the case. The matter of our physical universe occurs in various densities. When these densities are arranged in a mathematical series, we have at one end the ideal or perfect vacuum and at the other completely space-filling matter. Neither limit seems actually attainable. But the variation in the ratio of the two

elements is so great that in certain situations we can neglect the influence of matter and study the behaviour of metric systems apart from such influence. Compared for instance with the sun, the earth may, for certain purposes, be regarded as having a negligible mass. We may therefore study the behaviour of bodies on the earth and then compute how these phenomena would be modified if they were transferred to the sun. Thus a ray of light which may be regarded as a straight line uninfluenced by the relatively small mass of the earth becomes somewhat bent when passing very near the large mass of the sun. Naturally a geometry which is so closely connected with physics cannot be absolutely accurate. Ordinary Euclidean geometry involved in Newtonian mechanics has an accuracy sufficient for all ordinary phenomena. Non-Euclidean geometry has to be brought in only when the masses exceed certain values.

To Einstein as to Plato, space is not mere negation. The Eleatic objection that there can be no empty space because we cannot say that non-being *is* can be answered by insisting that space is a certain potentiality of motion of matter or occupation by matter. This potentiality is definite and not altogether indeterminate. As such it has a certain determinate being.

Einstein's view that a finite mass in the total physical universe necessitates a finite space follows consistently from his fundamental assumptions. A finite amount of matter radiating energy into infinite space would sooner or later dissolve into nothing. But in a non-Euclidean geometry a ray of light sent into the void would after a certain period come back to the material universe whence it started. The assurances, however, that the amount of matter in the universe is finite rest in part on empirical considerations which in the nature of the case cannot be indubitable.

(c) ULTIMATE PHILOSOPHIC ISSUES

What suggestion, if any, do these considerations as to the theory of relativity throw on the ultimate nature of the physical universe, and more particularly on the issues between nominalistic or subjective empiricism and realistic or objective idealism? Here Einstein's and Minkowsky's conception of the essential unity of time and space in existence and the idea that the laws of nature must be absolute or invariant in all spatio-temporal transformations is the capital point to keep in mind. This will enable us to dismiss at once the popular impression which, confusing physical relativity with the relativity of knowledge to mind or consciousness, sees in Einstein's work a new support to anti-materialistic subjectivism. In none of the phenomena explained by relativity is there any special reference to mind. When we speak of the length of a body depending upon the state of the observer we do not refer to his consciousness or subjective feeling but to the velocity of the physical system in which he might set up yardsticks and clocks to measure physical processes. Some colour to the subjectivist interpretation is lent by the fact that the same body can be said to contract from the point of view of one observer but not from that of another. But this is not a question of what the observer thinks or feels [46] but of an objective consequence of the nature of measurement.

Traditional psychologic relativists like Petzoldt [47] use Einstein's work to back up the contention that "the world is to every one as it seems to him." But physical relativity like any other scientific theory helps us to correct the illusions of mere seeming. It deals not with what happens to seem to any one but rather with what must seem to any one who makes precise physical observations. Indeed the essence of the theory of relativity is a formula from which all these diverse appearances can be deduced, just as from the length of a rod we can deduce all the angles which it subtends at all possible distances. The search for laws or mathematical relations invariant for all observers has always been at the heart of science. Einstein is in the tradition of Kepler, Galileo, and Newton in the conception of nature as revealed in mathematical form, in ideal entities like rigid particles, light rays, instants, and points which are not direct objects of sense-perception but limiting concepts which enable us to order series of natural objects.

There is, however, a sense in which the conception of invariant mathematical relations brings together Plato and Protagoras, Newton and Mach. For Einstein more successfully than Leibniz has combined the principle of observability with the principle of mathematical continuity. We may regard the general theory of relativity as a mathematical triumph of Mach's ideal that the motions of any physical body

[46] Bergson, *Durée*, etc., pp. 102 ff.
[47] *Das Weltproblem*, p. 203.

should be described in terms of measurable qualities or relations to other bodies rather than with reference to unverifiable "absolute" space or occult "forces."

To those who have realized the polarity of unity and diversity there is nothing paradoxical in combining the empiricism of Mach (as embodied in the principle of observability) with the Neo-Platonic mathematical conception of Galileo and Newton. The wonder rather is that the two were ever separated. Popular difficulties with the theory of relativity show us that naïve rationalism conceives the absolute or the real object as a thing in itself which cannot be different in different relations. But if we hold on to the reality of universals—not as additional terms but as identical relations between many different terms—then there is no reason why the same object should not have different relations to different other objects. Its identity consists in a certain constant pattern of relations embodied in all its sensibly different aspects.

We can see this in the relation of the Lorentz equations or transformation formulae to the various co-ordinates which these formulae relate. Einstein, and Eddington even more so, sometimes speak as if the fact that our co-ordinates are arbitrarily chosen deprives them of all reality. But if this were the case, if there were no element of truth in the way each set of co-ordinates describes the nature of things, the law which unites all these descriptions would have no significance from one point of view. A cat may look at a king, and the latter may be unaffected thereby in all of his kingly functions. But it is of the essence of his reality (or of the world of which he is a part) that he should present the specific view which in the cat's perspective he does. The fact is that the specific numerical relations indicated by any arbitrarily chosen co-ordinate and the law uniting the infinitude of such relations are each elements in the logical analysis, and both are phases of the physical world.

Chapter Three

LAW AND PURPOSE IN BIOLOGY

ANTI-INTELLECTUALISTS have argued with some plausibility that the rationalism of the Enlightenment was due to its preoccupation with mathematical physics which made such rapid strides in the seventeenth and eighteenth centuries, and that the progress of biology in the nineteenth century has now turned men's minds from mechanical to organic considerations. This half-truth seems especially convincing to those to whom biology means popular Darwinism of the Spencerian or Neo-Lamarckian type, a biology which looks on struggle and adaptation as the primary fact of life. It is well, however, to remember that the emphasis on the organic is already strong in Kant to whom natural science is primarily mathematical physics. Also, if we think (as we should) of biology as an experimental science, its modern career is intimately connected with the rationalism typified by Descartes, Harvey, Borelli, and Fr. Sylvius, and with the revolt against the abuse of final causes which led great idealists like Spinoza, Leibniz, and Kant to advocate universal mechanism. Biologists themselves trace the modern beginnings of experimental biology to the two memoirs on breathing and on animal heat by Laplace and Lavoisier—two typical representatives of the Enlightenment. None of the founders of modern biology, Redi, Hooke, Malpighi, and Leeuwenhoek were much concerned with teleology. However the physical synthesis of many substances formerly not known outside of the living body, and the tremendous progress in the discovery of physico-chemical influences on life, led some of the nineteenth century writers to over-hasty claims that life is nothing but motion of physical particles and chemical change. Naturally enough the recent romantic reaction against science and its conception of a world determined according to laws, has led to a crusade to redeem biology from the domination of mechanistic ideas. Vitalism as a fundamental faith in an anthropomorphic world has, of course, never been completely eliminated. But its present resurgence raises issues that have important general bearings.

§ I. MECHANISM AND VITALISM

IT WILL help to remove misunderstandings if we note at the outset that mechanism in biology is not the same as mechanism in physics. No responsible biologist today claims that the phenomena of life can be explained by the laws of mechanics which govern material particles in motion. Extreme mechanists like Loeb are at one with vitalists like Driesch in rejecting as inadequate the older attempts of Liebig and others to explain such action as that of fermentation by the vibration of atoms. The physical results of the vibration of atoms doubtless hold true in living as in non-living tissue; but to explain the nature of life by the laws of motion is like explaining the peculiar disposition of Kant by the fact that he was a vertebrate.

This will enable us to dispose of Huxley's version of the mechanistic doctrine of Laplace: "An intelligence, if great enough, could from his knowledge of the properties of the molecules of the [original cosmic] vapour have predicted the state of the fauna in Great Britain in 1888 with as much certitude as we may what will happen to the vapour of our breath on a cold day in winter."

If the properties of the molecules that Huxley refers to are strictly mechanical ones, i.e. co-ordinates of position, masses, and instantaneous velocities, then the failure of the attempt to reduce physics itself to mechanics hopelessly weakens its claim in the more complex field of biology. The progress of electricity and chemistry has made us less certain than were Laplace and Huxley that all aspects of physical phenomena can be deduced from the merely mechanical co-ordinates of the molecules, or that these molecules themselves are absolute constants that have suffered no change since the primal mist,—assuming the very questionable nebular hypothesis to be a statement of fact.

This decline of strict mechanism does not decide the question whether biologic phenomena can be reduced to physical and chemical elements. But it puts our discussion on a more empirical basis. The illusion that we can know a priori what physical laws must govern the natural world, is easier to entertain when we restrict ourselves to a simple, fairly complete, and deductively organized science like mechanics. The Newtonian laws of motion and even that of gravitation have become so familiar and simple-sounding that it is tempting to argue, as so many have, that thought or the nature of mind tells us, in advance of experience, that nature *must* be so constituted as to be governed by these laws. But when we come to the more complex and incomplete branches of physics and chemistry it becomes more obvious that our laws are often but generalizations of empirically observed results and not necessary rules which could have been deduced in advance of actual observation. Thus, that the vapour of our breath will form a little cloud is something which we know from actual experience of seeing vapour condense in lowered temperatures. But could any intelligence of which we can form any idea have predicted the effect of such a cloud on the human eye before the existence of eyes?

Consider the matter more closely. We believe the laws of mechanics to be universal and applicable to all material bodies. They must therefore form a *necessary part* of the explanation of all physical phenomena, but they are not on that account necessarily *sufficient* to explain the characteristic of every particular group of physical phenomena, e.g. the magnetic properties of soft iron. To explain electro-magnetic phenomena we must assume new laws, additional to and not derived from those of mechanics. Similarly, why may not the particular group of natural phenomena called biologic, while dependent on physics and chemistry, be governed by additional laws which are specifically biologic? The view that finds difficulty in this conception is based on the widespread but groundless assumption that the properties of any combination of elements can always be obtained by adding the properties of the elements in isolation. The progress of physical chemistry may indeed enable us to deduce *some* of the properties of compounds like water from those of their constituent elements. But there is no reason to suppose that *all* the properties of compounds are but simple additions of the properties of their separate elements,—any more than that the amount of good or evil that two persons can produce together is necessarily the same as the sum of the amounts that they can produce separately. In the terse language of mathematics: all things or phases of nature do not necessarily form addition groups. Only experience can decide which processes do in fact form addition groups. We have been taught this in connection with the principle of the composition of velocities in the Einstein Theory of Relativity, and similar considerations hold in biology.

The question before us is therefore an empirical one: what evidence is there that biologic phenomena are governed by laws that are not deducible from those of physics and chemistry?

A. *The Case for Mechanism*

The case for the mechanist may be put in the form of three arguments: (a) that since life on the earth was at one time impossible it must have been formed out of inorganic matter, (b) that the actual progress of biologic science along physico-chemical lines demonstrates the validity of that type of explanation, and (c) that any other type of explanation would be beyond the scope of natural science. Let us examine each of these arguments more closely.

(A) THE ARGUMENT FROM CONTINUITY

This popular argument assumes that geology and geo-physics have demonstrated that our earth was once at such a high temperature that life as we know it could not possibly have existed on it. If this be granted, and there is some reason for doing so even if we reject the questionable nebular hypothesis, we seem compelled to admit that life on our earth must have evolved from originally simple inorganic material. To be sure, neither in nature nor in our laboratories have we ever observed the synthesis of life from inorganic matter alone. But this objection can be readily disposed of. For the earth is admittedly no longer in the condition it was many million years ago when the original synthesis of like took place. It no longer has as much solar, radio-active, and other forms of energy. Its atmosphere, velocity, gravity, etc., are no longer the same. Perhaps also the synthesis of life from the inorganic requires a longer period of time than any of our modern laboratory experiments have been able to devote to it, just as the radical, geologically indicated changes of species required a longer time than the period in which direct scientific observation of nature has been going on. Time has refuted so many prophecies as to what science will never be able to discover or produce that our failure so far to make living protoplasm artificially cannot be taken as a decisive argument against its possibility.

Those who wish to obviate the force of the argument from geology and geo-physics that life was produced out of inorganic material contend either (1) that life has always existed on our globe, or (2) that it has been brought over here from other parts of the universe.

(1) That life has existed even in the enormously high temperatures which have prevailed on our earth, is not absolutely ruled out by the fact that it completely disappears today even at the relatively low temperature of 300° C. For admittedly the early period of our globe differed also as to solar energy, radio-activity, pressure, etc. Nor is the possibility of such life completely ruled out by the negative argument that it left no traces in the earliest or igneous rocks. But there is no *positive* evidence whatsoever for believing that life did always exist on our globe, and all that we know about fusion of rocks points against it. It is clearly the kind of argument that no one advances except when anxious to believe the conclusion for other and extraneous reasons.

Let us note, also, that those who maintain the existence of life while the earth was in an incandescent state, really abandon, for that period at least, any empirical or phenomenal difference between living and non-living matter. In general the assertion that matter has the germ or potentiality of life in it does not contradict our mechanistic argument which maintains the continuity of organic and inorganic nature. Between panpsychism which asserts that even matter has the elements of life, and materialism which asserts that "in matter and motion is the potency of life," there is really no theoretical difference. They are equivalent statements of the postulate of continuity from different points of view and with perhaps different emphases.

A genuine denial of our present mechanistic argument is the view of MacDougall[1] that if life appeared on this earth after inorganic matter "it was due to the co-operation of a new factor." If this "co-operation" means something more than new physical circumstances, it amounts to the introduction of a new supernatural creation; and all the objections against the view that the world was created at any given time hold against it.

[1] *Mind and Body*, p. 233.

(2) The idea that life came to this globe on meteors has appealed to many. But all that we know of the conditions of life points against the possibility of its being able to maintain itself on the surface of meteors in empty space throughout the enormous ages which it would require for a meteor to reach us from even the nearest star-system. As meteors go through our atmosphere they become incandescent with a heat sufficient to destroy all possible life on them.

Arrhenius has tried to avoid some of these difficulties by supposing that life came to this earth in microscopic organisms propelled through inter-stellar space by the pressure of light. This is not free from all the objections against the meteoric hypothesis and has the additional difficulty that micro-organisms cannot live in the ultra-violet rays of the sunlight in the upper regions of the atmosphere. At best it is an hypothesis *ad hoc* with no positive evidence in its favour.

We must conclude, then, that our evidence so far points to the actual synthesis of life out of inorganic matter in the remote past, and therefore, to its general possibility. From this, as we saw before, it does not follow that living processes can be explained *entirely* in terms of physico-chemical laws. To prove the latter, another argument is necessary.

(B) THE ARGUMENT FROM THE PROGRESS OF SCIENCE

The second argument for mechanism is based on the past progress of science. Is it not reasonable, it is urged, in view of the steady progress in the actual explanation of vital phenomena by physico-chemical considerations, to expect that sooner or later every vital phenomenon will be so explained? This argument will always seem persuasive to many, but it is by no means conclusive. We must remember that progress in science depends on human enterprise in which arguments from the past to the future are always hazardous. The history of science frequently shows examples of modes of explanation which make headway for a time and then come to a standstill, e.g. the attempt of E. DuBois-Reymond and others to explain physiologic phenomena on the basis of electricity, or the attempt to explain all physical phenomena on the basis of mechanics. It is therefore not surprising to see this argument from the progress of mechanistic explanation directly denied by a physiological chemist, Bunge, who has himself contributed to the progress of the science by showing the synthesis of hippuric acid in the kidneys. "The more thoroughly and conscientiously we endeavour to study biologic problems, the more are we convinced that even those processes which we have already regarded as explicable by chemical and physical laws are in reality infinitely more complex, and at present defy any attempt at a mechanical explanation." [2] Thus it is usually cited as a great triumph of mechanism, that we have been able to explain the absorption of food from the alimentary canal by the laws of diffusion and osmosis. But a closer study of the facts, according to Bunge, shows that the epithelial cells which cover the intestinal canal do not act as dead membranes, but take up the food by active contraction of protoplasm in the same way as independent unicellular organisms do. They exercise selective functions, absorbing fat globules, but rejecting finely divided pigments and all sorts of poisons soluble in the gastric and intestinal juices.

While such counter-arguments show certain weaknesses in the logical armour of the argument for mechanism, they cannot destroy its persuasiveness to those familiar with the overwhelming mass of biologic phenomena that have definitely yielded to physico-chemical explanations. Moreover this argument of Bunge does not deny that to explain the phenomena of life in terms of the wider laws of physics and chemistry serves the interests of science in reaching the simplest possible explanation. The fact that increased study often shows phenomena to be more complex than we had previously supposed them to be cannot put a stop to the persistent effort at simplicity. The faith of those who believe that sooner or later all biologic phenomena will yield to physico-chemical explanation is not therefore to be thus denied, though opponents will remain unconvinced so long as the mechanistic ideal—like other ideals—remains unachieved.

Another way of meeting the argument from the progress of mechanical explanation is to assert, as Von Baer did, that the undoubted advance has taken place only in respect to the purely physical aspects of organisms, not in respect to their life as such. Thus, the circulation of the blood as a physical fact must have a cause in accordance with physical principles. Shall we, however, say that the rustling of leaves or the movement of the pollen when blown by the wind from stamen to pistil

[2] *Lectures on Physical and Pathological Chemistry*, pp. 2-7.

is any more a biologic fact than the floating of a piece of paper in the air? Motion is physical whether caused by the pressure of air (wind), or by the sunlight which induces the wind, or by latent chemical energy into which past sunlight has been converted. However, if we carry this contention to its logical conclusion, it becomes doubtful whether there *are* any phenomena peculiar to biology. Cannot the pressure of an inverted image on the retina or the spatial changes which we call the accommodation of the lenses and the muscles of the eyes, also be regarded as physical facts?

This would leave nothing to biology itself except possibly psychic elements such as sensation, etc. Biology would thus maintain its independence as against physics only by losing it to psychology.

(c) THE ARGUMENT FROM THE POSSIBILITY OF VERIFICATION

The third argument for biologic mechanism is one that appeals primarily to laboratory workers. If biology is to be a natural science we must assume its phenomena to be subject to causal laws in such a way that under determinate observable conditions certain determinate results may be expected, so that if these results do not in fact happen our assumed law is denied. To make our conditions observable and our laws verifiable we must restrict ourselves to objects in time and space (including the relations which hold between them) that are perceptible to all observers under the given conditions. Biologic phenomena can, therefore, be the objects of science only to the extent that they involve physically verifiable elements. This is not a direct denial of the existence of psychic or other non-physical elements in the organism; but it denies that there can be a coherent natural science of biology if these non-physical elements can produce physical effects that thus have no physical causes. It is no answer to this to point to non-empirical hypothetical entities like atoms or electric fluids in the physical sciences. If atoms were not veriable—in fact we count them and measure their masses, electric charges, etc.—they would have been eliminated from physics as the school of Mach and Ostwald advocated. As to electric fluids or the various models of the ether, few physicists now regard them as anything but symbolic conveniences. No physicist can claim them to be genuine existences and yet hold them to be unverifiable. Certainly there can be no experimental science of such entities.

Neither is it a refutation of this argument to urge that it involves a dogmatic denial of the possibility of an experimental science of psychology. The mechanist might well contend that a *purely* introspective psychology is a dialectic science but not a natural or experimental one. Actual experimental psychology in fact deals with perceptible physical expressions of organic bodies. Lest this contention be hastily dismissed as eighteenth century materialism or new-fangled behaviourism, I hasten to add that this is precisely the position of Plato's great successor, who, in a classic treatise *On the Psyche,* clearly distinguished between the physical and the dialectic aspects of a psychologic phenomenon such as the emotion of anger. (If the inclusion of Aristotle in any group except animists or vitalists sounds absurdly incongruous, the incongruity is due to our translating *Psyche* by *soul* and confusing the latter with *mind*. But Aristotle is clear in excluding only the *nous* or rational mind from the field of natural causes and making it the field of purely dialectic considerations. The vegetative and animal "souls" are clearly organizations of natural bodies.)

Surveying the three arguments for mechanism we can see that they assert three distinguishable doctrines: (a) that the subject-matter of biology offers no inherent discontinuity with the subject-matter of the physical sciences, (b) that the laws of biology must ultimately be deducible from laws of inorganic physics (including chemistry), and (c) that the method of biology must be the same as that of physics or continuous with it to the extent of postulating determinate causal laws that are verifiable in perceptible bodily phenomena in time and space. Those who call themselves mechanists do not take the trouble to distinguish these doctrines, because they usually hold the three together. Subject to certain reservations as to (a) and (b), which we shall discuss later, the mechanists in general are actually in possession of the field, so far as experimental biology is concerned. The vitalists are protestants, and as such naturally divided into divergent sects using varying arguments. It is, however, worth while to make a systematic survey of their chief arguments.

B. *The Case for Vitalism*

The vitalist attempts to meet each of the three foregoing arguments by asserting (a) that there is a radical discontinuity between vital
and non-vital phenomena, (b) that biologic laws are different in character from the laws of the inorganic world, to which they add the fundamental category of purpose, and (c) that the method of biology must
differ essentially from that of physical science, showing a greater resemblance in several respects to the method of history.

(A) THE IRREDUCIBILITY OF LIFE

What is common to all vitalists and gives force to their attacks on
mechanism, is the unmistakable and overwhelming difference between
the living and the non-living. The difference between life and death
does not merely strike the eye. It provokes in us fundamentally different attitudes and emotions. Even those who have no particular interest in domestic or wild animals can seldom accept without repugnance
the Cartesian view that animals are merely mechanical automata devoid
of feeling.[3] It is vain to point to the many new types of growth artificially produced by Loeb, Carrel, and others. All these depend on the
presence of previous life. It is also of little avail to point as Leduc and
others have done to the many similarities between mechanical and vital
phenomena: to the way inorganic mixtures will simulate the circulation
of protoplasm in the cell, to the parallelism between the growth
and regeneration of crystals and that of organic forms, to the apparently free and capricious movements of fine particles in solution,
etc.[4] Not only has no machine ever been constructed that could simulate all the characteristics of the simplest living creature, but there
is a very deep conviction among some vitalists that even if a mechanical
contrivance could imitate all the usual external manifestations of life,
growth, self-repair, and reproduction, it might still remain a lifeless
automaton. Life can be no more identical with the imitations of its outer

[3] To be sure we do not feel the same about plants, grasses, vegetables, or the seeds
of trees. But that is not of great moment except to indicate the shortcomings of our
feelings as a clue to external nature.

[4] Driesch, *Archiv für Entwicklungs-Mechanik*, Vol. XXIII (1907), p. 174; Hoff-
mann, *Annalen der Naturphilosophie*, Vol. VII (1908), p. 63.

manifestations than my own spoken words can be identical with their perfect echo.

On this view, life is an absolute something that has diverse or outwardly visible physical manifestations but is not to be identified with these. This way of putting the case is convincing to common sense or to its ancestor, the old metaphysics that believed in substantial entities quite apart from their appearances. However, no modern science like biology can concern itself with such non-phenomenal and therefore unverifiable realities. You may say: So much the worse for such science; if it explicitly limits itself to the study of objective phenomena it is surely in no position to deny the existence of the non-phenomenal. But we are not concerned with denying the existence of the non-phenomenal. We are denying that there is any reliable knowledge whatsoever of realities apart from their phenomenal manifestation.

The denial that we can have any knowledge of non-phenomenal absolutes, is countered by Bergson with the claim that our own conscious existence is such an indubitable absolute, known introspectively. Analogously it is argued that we know life absolutely by inward intuition of its impetus. Bergson to be sure does not explicitly claim that such intuition without empirical observation of outer phenomena can be adequate for a natural science of biology. But the popular failure to note that psychology would not be an empirical natural science unless mental phenomena were located in organic bodies lends support to the view that biology may similarly be concerned with life in itself. In any case what may be called the psychologic school of vitalists begins with a more or less unavowed assumption that life is essentially psychical, i.e. identical with consciousness in some elemental form of feeling and desire.

Against such identification of life with conscious existence various objections are worth considering.

(1) In the first place the life of our own organism is not always or entirely conscious. We are not conscious during sound sleep, or during fainting spells—certainly not in our prenatal existence. Nor are we, in our normal activities, conscious of the organic processes going on in the kidneys, glands, etc. To claim that we *are* conscious during sleep or of the activities of our separate organs seems very much like asserting that we are unconsciously conscious. The charge of verbal contradic-

tion might be obviated by a distinction between consciousness and self-consciousness. But there is a real difference between being conscious of pain and not being so conscious, between being conscious of an organic process like the contraction of the heart muscle and not being conscious of it. No definition or theory can deny this difference, or that the non-conscious organic phenomena exist.

(2) Other difficulties with the notion that all life is characterized by consciousness arise when we consider the union of two or more organic cells to form a new individual. The two cells being alive are therefore individually conscious before their union. What happens to their consciousness after they unite? Any answer to this question involves the difficulty of adding two entities which are not quantities at all, creating a new consciousness, and other difficulties made classic by William James in his critique of the mind-stuff theory.[5]

(3) The adherents of the doctrine that all life involves consciousness may object to the foregoing distinction between the conscious and the unconscious. The consciousness that knows that it is conscious, they maintain, ought to be called knowledge or self-consciousness. Why deny that there are degrees of consciousness below the level of knowledge? Here again the real issue is whether the suggested meaning of the word consciousness in this contention denotes anything definite, and if so, what evidence there is that what it denotes has any existence. If consciousness and life are made identical by definition there can be no question of evidence. We are then dealing with a verbal convention. But if consciousness denotes something recognizable as distinct from life and yet always a part or trait of it, it is difficult to see what evidence can prove its existence in all animate objects, and not in fire, wind, or sea.

For consider the position of the vitalist. He bases his case in the beginning on introspective evidence. At what is ordinarily called the conscious level we certainly can discriminate degrees of clarity, from definite knowledge to very vague awareness. But the moment we venture to speak of consciousness below the level of introspection, we can no longer distinguish between its different degrees introspectively. We are thrown back on some objective test connected with physical phenomena which you may call the manifestations of life. It then becomes doubtful, however, whether the consciousness that is evidenced by in-

[5] James, *Principles of Psychology*, Vol. I, p. 158.

trospection, and that which is not so evidenced, are really homogeneous. What is there in common between the consciousness which feels difficulties and the unconscious growth of tissue which can never be produced by consciousness directly? No one except an absurd amateur like Samuel Butler supposes that consciousness is necessary to oxidize the blood. Why should it be necessary for its circulation, and kindred phenomena? The vitalist can answer only by pressing the principle of continuity to bridge the phenomenal discontinuity. But then his case becomes indistinguishable from that of the mechanist who as we saw argues from continuity to a denial of the sharp distinction between the living and the non-living.

(4) Is not the idea that life always involves some amount of consciousness largely due to the pathetic fallacy which attributes some element or ghost of prevision, choice and effort, to the acts of animals that in any way resemble ours? We shall examine under the next heading the question whether the category of purpose is not still abused in biology as it used to be in physics. Here I wish to urge only that while men are undoubtedly often conscious of what they are doing and while some animals may also be so, there is no reason to suppose that conscious prevision always accompanies every act which can be interpreted as a means for the attainment of an end. It is characteristic of naïve rationalism to assume it as universal not only in human beings but in animals as well and sometimes even in plants. The analogies of human conduct and our nervous structure, however, are against this assumption. When the gastric juices destroy certain noxious germs or when plants and animals reject certain poisons there is no reason to suppose any awareness of consequences. As we are not conscious except when the brain-nerve system is affected, it is hazardous to attribute consciousness to animals without such a system. Moreover, when we observe how certain reflex acts go on in frogs and dogs after the removal of the brain it seems absurd to suppose that prevision or consciousness is involved in these acts.

Vitalistic writers generally appeal to certain stock examples of complicated delicate and marvellous adjustments whereby bees, wasps, or eels manage to perpetuate their existence; and the question is asked, "Can these adjustments be conceived as the result of blind chance?" To which we all say, of course, "*No!*" For the word *chance* is ambiguous and denotes the absence of cause as well as the absence of design; and few are willing to assert that complicated processes can regularly recur without any cause. But if the question assumes that the conditions of life cannot be the result of non-conscious forces, we are dealing with an assumption as questionable as it is widespread. Few indeed of the processes (or "adaptations") which make life possible are the result of conscious effort, though consciousness *may* supervene to perfect their development after a certain stage. It is certainly not by prevision or conscious effort that the unborn or even recently born infant develops his body and all its various functions, or that the hair continues to live for some time after the rest of the body is dead.

To naïve rationalism is due the popular Lamarckian doctrine that unconscious adjustment is the outcome of conscious effort that has become habitual. This is but partly true of those of our own movements that involve the voluntary muscular system. The feeling of effort and tension and the consciousness of adjustment do tend to diminish (though they do not quite disappear) as we master the arts of walking, talking, writing, etc. But if this is made a general biologic principle, and the evolution of species is attributed to such effort, we are landed in the Lamarckian doctrine of the inheritance of acquired characteristics, for which the best that can be said is, that it does not cohere with such well known facts as the development of worker-bees that certainly do not transmit their characteristics.

There can be little doubt that many of the acts of animals seem to us purposive because our imagination about them is not controlled by experimental tests. If you watch the movements of a moth as it alternately dashes itself against and recoils from a lighted lamp, you may see in them a parallel to the way in which humans fly into noxious excesses and struggle in vain to escape. But scientifically planned experimental study of animal life is decidedly inimical to the view that prevision precedes *all* animate processes or that the latter always involve purposive adjustments that can truly be said to be conscious.

When vitalists recognize the difficulties of identifying life with consciousness, they set up non-psychologic "principles" of life. Such are the unconscious of Von Hartmann, the entelechy of Driesch, the *élan vital* of Bergson, the determinants of Reinke, etc.

While these principles all function as if they were psychological, they are not directly knowable either subjectively or objectively. They are entities invented to explain life, but no definite laws or phenomena are really deduced from them, any more than from the "dormitive principle" of opium in Molière's caricature of scholasticism. These vital principles serve only as sign-posts to emphasize the undoubted fact that the phenomena of life are different from those of non-living nature. But wherein the exact difference consists is an empirical question to be answered by a consideration of the biologic laws which sum up what we actually find life to be. This brings us to the second issue in the mechanist-vitalist controversy.

We may, however, at this point sum up the truth as to our first issue, as it emerges from a consideration of the arguments of both mechanists and vitalists. The vitalists are unquestionably right in insisting against over-hasty mechanists that the phenomena of life are different from those of inorganic physics. The difference is at least as great as that between the phenomena of chemical affinity and those of mechanical motion. The vitalist is therefore sound in rejecting the suggestion that this difference is illusory. A difference between phenomena remains real even if we can find a theory or system of laws which explains those common characteristics. Chemical affinity will remain a distinct phenomenon even if it should be explained as the effect of the laws of mechanics, operating under special conditions. Similarly the phenomena of life will remain distinct if we should succeed in deriving them from physico-chemical laws operating under special conditions. The vitalist, however, needlessly prejudices his case by insisting on certain discontinuities in nature which the mechanist denies. There is no necessity for assuming that life appeared on this earth by a special or supplementary act of creation distinct from all that went on before it. If we suppose that only elements existed before a given time, and that their compounds were formed, the present reality of the latter is not thereby impugned. Similarly the actual distinctness of living organisms is not impugned by the mechanist's assumption that at one time they were formed out of inorganic matter and that they may be so formed again if the proper conditions recur or even that life may be arising today in the sub-microscopic realm out of inorganic materials. Whether the laws which express the behaviour of the new combinations can be derived from the laws governing their separate elements, or whether additional laws are needed is, as we have seen before, another question.

To the extent that vitalists believe in the possibility of a science of biology they assume that there are certain general truths about living organisms and that these truths cannot be reduced to or derived from physico-chemical laws. Now the historic fact is that many general features of life have actually been explained by physico-chemical laws, e.g. that animal heat depends upon the oxidation of carbon, or that organic motion involves kinetic energy governed by such laws. It is surely impossible to understand protoplasm without remembering that it is fluid and subject to all the laws of gravitation, viscosity, and solutions generally. As biophysics, however, is a relatively young science, and has so far explained only very elementary processes, its great progress can readily be minimized by pointing to an impressively large number of phenomena or phases of life which it has not yet explained. This argument is psychologically powerful because we all tend to think that what has not yet been done is impossible. It is, however, logically inconclusive. The question whether certain methods will or will not enable us to solve that which is as yet unknown, is largely a matter of faith. Vitalists, however, claim for their position something of greater cogency than the argument from the inconceivability of what has not as yet been achieved. They claim to demonstrate their position by analyzing the nature of life and the nature of physico-chemical explanations and showing that the latter can never be adequate to the former. Let us consider their outstanding arguments:

(1) VITAL PHENOMENA AND THE LAW OF ENTROPY. The argument has been advanced that while in the inorganic realm all energy runs down and is dissipated in the form of heat-radiation "the processes of organisms seem to be exceptions to this law." [6] The best that can be said for this argument is that it has no merit whatsoever. The facts of the case are not so. The food that enters the animal body contains highly available chemical energy which is converted in the body into the energy of motion and heat, so that the remains are at a lower level of energy-

[6] MacDougall, *Mind and Body*, pp. 244-245.

potential. Surely this is perfectly in conformity with the law of the degradation of energy. It is true that green plants make the synthesis of carbohydrates out of carbon dioxide and water and this product has a higher chemical potential than water. But here again there is nothing contrary to the law of increasing entropy. For the process takes place only with the aid of external energy derived from the sunlight. We cannot only duplicate this in the laboratory, but we daily see the radiant energy of the sun raise the water of the seas from a lower to a higher gravitational potential in the clouds.

Professor Johnstone, who states the facts of the case accurately enough, nevertheless jumps to the conclusion that to *retard* the degradation of energy is distinctive of life.[7] That he should draw a conclusion so little supported by the facts is probably due to the misleading anthropomorphic suggestion of the idea of available energy. The fact however is that all substances which absorb heat and convert it into kinetic energy seem to that extent to increase the store of available energy. This can be seen in the process whereby winds are created or surface waters are converted into clouds which as rain and rivers are sources of kinetic energy. Biologic phenomena are not in this respect distinctive.

(2) DEPENDENCE OF VITAL PHENOMENA UPON HISTORY. Bergson, Driesch, E. S. Russell,[8] and others have made current the idea that biologic phenomena are distinguished from physical phenomena in depending upon the past. Ice, for instance, will melt at 32° F. no matter what its previous history; but the conduct of a dog or a salmon does depend upon its previous experience or life.

That this argument is over-hasty and not supported by any careful analysis of the case ought to be evident from the fact that no one questions that the history of our earth or of an automobile may throw light on its present state. Thus it is commonly known that the working of a physical machine depends on its age. The parts gradually wear out, lose their elasticity, or suddenly break under a tension heretofore easily bearable. Multitudes of similar illustrations can be brought from the realm of electro-magnetism and chemistry. The theory of elasticity and of the strength of materials and the theory of hereditary mechanics or

[7] *The Mechanism of Life*, p. 57 and Ch. 2.
[8] *Scientia*, Vol. IX (1911), p. 338.

of hysteresis are concerned with such phenomena. In these, not only is something in the past necessary to explain the present, but the history may cover a considerable period and the reaction of our physical system to a given pressure, magnetic force, etc., may be different in different periods of its history (although it may seemingly not have changed its structure). On the other hand historic considerations are practically negligible in the study of the behaviour of the lower organisms. All this is so obvious that it is instructive rather to inquire as to what it is that has misled the vitalists. I shall confine myself to the contentions of Bergson and Driesch.

The motives for Bergson's position may be fairly stated as follows:

The laws of physics in their ideal mathematical form are expressed in terms of instantaneous states. But a single instant or moment without any duration is an abstraction and is no more time than a point is a line. In this sense then, it may be rightly claimed that physics eliminates time. This conclusion, however, moves in a certain atmosphere of confusion that surrounds such limiting concepts as instantaneous velocity (ds/dt)—a confusion inherited from the old ideas about infinitesimals which used to be treated both as quantities and as absolute zeros or denials of quantity. But if we recall the true nature of mathematical limits, the issues become clear. Every existing physical state or process, e.g. velocity, involves physical duration, or a time-interval. Velocity without time is meaningless. But the limiting concept of instantaneous velocity denotes, not a single physically existing velocity, but a real relation or law, according to which all possible velocities can be arranged. It is this which enables us to pass from the differential to the integral form, which our equations assume when applied to actual situations. In the integral form, duration or the time-interval as measured by some clock becomes explicit.

The identification of duration with a time-interval measured by a clock is rejected by Bergson, who identifies duration with the introspectively felt flow of events. Bergson may, of course, define *duration* in any way he pleases. But he cannot by merely defining time as subjective prove the non-existence of the physical time that is measurably homogeneous and indefinitely divisible. His denial of objective physical time does not seem to me to be worthy of very serious attention. For the arguments by which the characteristics of subjective time are derived will all be found to assume the existence of some clock which measures objective homogeneous time. The argument, for instance, that a process such as the melting of a piece of sugar will be felt as longer by one person than by another, or by the same person on different occasions, presupposes a common objective time. Biology as a natural science certainly lends no support to the substitution of intuitive *durée* for objective physical time. For the latter is involved in the recognition of the periodicity of biologic phenomena. That the life of an organism occupies a certain length of time, that the various processes going on within it occupy so many seconds, hours, or days, and that some processes are more rapid than others, are matters of fundamental importance. And what are days, hours, and seconds but correlations between phases of life and the time intervals of the rotation of the earth about its axis?

The actual difference between physical and biologic phenomena in regard to time is the greater homogeneity of the former. Physical phenomena seem to have fewer critical points, or points of discontinuity. The parts of the motion of a planet are still motions. But when a process such as digestion is divided at critical points, the different parts are likely to have different names. Yet in all this we deal with a difference of degree of complexity, and largely also with a difference in intensity of interest. Thus chemical processes formerly treated as uniform are now regarded as consisting of different stages.

The maxim of physical causality, as we saw in a previous chapter, depends on eliminating mere locality and mere date as causes, and substituting for them repeatable magnitudes of length and time-intervals. As in biology we deal with more specialized configurations, indefinitely repeatable time and space relations are not so often observable. This lends colour to the view of Driesch that biologic acts are specific in a sense absolutely different from physical reactions. This absolute distinction, however, cannot be maintained if we remember that we know of no biologic act apart from physical expression, and that any description of biologic acts must be in terms of repeatable aspects if it is to be at all intelligible. What can be more specific and historical than the reaction between a given mountain and the rain, winds, and rivers which are wearing it down? The action will be different at different times, depending on which particular stratum of its formation happens to be uppermost. Driesch rejects this as a parallel to the historic character of organic action because the mountain cannot properly be said to act at all. But if to act is to be defined so as to apply only to conscious or animal acts, there will be no inorganic acts according to this definition. Arbitrary definition, however, cannot remove the actual parallel between geologic and organic action in respect to time. The real reason why Driesch refuses to see the parallel is that he thinks only of the difference, that the organism is more or less conscious while the mountain is not. But this difference need not prevent resemblances in respect to temporal aspects. I think it will be found that *all* differences asserted by vitalists to hold between living and non-living nature *assume* that the former are teleologic and more or less conscious, and that the latter are not. In the light of that assumption, various differences seem apparent which are not evident otherwise. Let us then proceed to examine that fundamental difference.

(3) THE PURPOSIVE CHARACTER OF LIFE. The usual proof of the argument that life involves processes of adjustment and discrimination which the laws of inorganic systems can in no way explain, consists in the elaboration of lists of the most diverse biologic phenomena which seem prima facie too complicated for the purely material modes of explanation open to physical science, but which seem readily intelligible as the expression of purpose.

We may, following Driesch, group these under three heads, viz. phenomena of restitution, heredity, and organic movements. It will be more convenient for our purpose to reverse his order and state his arguments in less technical terms.

(i) *Organic Movements.* Many phenomena of animal motion are so like our own purposive acts that it seems impossible to escape the conviction that they are properly described in terms of purpose. But there are undoubtedly some movements for which a physical explanation seems very fitting, e.g. the contraction of certain muscles when stimulated by an electric current, or the constant turning of plants toward the light. Obviously an admission that the complicated conscious conduct of higher animals is an outcome of the simple reflexes of the lower forms, would be an admission of the *possibility* of explaining all conduct in terms of increasingly complex mechanisms. Otherwise, we should have a discontinuity in the evolution of life, and we should have the problem: at what stage in the evolution of life is the new type of organic motion suddenly created?

The thoroughgoing vitalist, who believes that all life is conscious, can avoid the difficulty here, by denying that even the simplest organism can have its motions explained mechanically. Thus Uexkull, MacDougall, and others insist that even the amoeba shows choice in accepting or rejecting food, in escaping from a larger amoeba, etc., and that such spontaneity is not explicable by physical laws. The difficulty, however, of putting conscious choice at the basis of reflex and instinctive movements, or of the motions of lower organisms that show very low variability or modifiability of conduct is felt by even as radical a vitalist as Driesch, who is too thoroughly trained in biology not to recognize the arbitrary character of naïve pseudo-psychology. Also while the process of accepting food and rejecting other materials is not yet fully explicable by physical laws, physical analogies, such as the way in which a solution of chloroform will accept and digest shellac and reject the glass which it covers, weaken the force of the dogmatic assertion that physico-chemical explanations of digestion will forever be impossible.

At the other extreme of organic movement we have the conscious motions of *homo sapiens*. That these are inherently incapable of being explained by physical laws is a very deep and perhaps invincible conviction. But such a conviction is not necessarily infallible, and we are, at any rate, only considering the logical force of the argument advanced in its favour. For it is well to remember that where a deep feeling of certitude prevails the logical quality of supporting arguments is apt to be bad. These arguments are likely to be but after-thoughts and not the real basis of our convictions. May it not be that because the question of mechanism is generally held to touch the dignity of human personality, our arguments on the subject are somewhat hasty? Certainly not everything which seems to support a sound doctrine is itself sound. Let us then consider rather closely the arguments for the view that conscious movement can never be explained in terms of physico-chemical laws.

The nearest approach to a rigorous proof of this is to be found in the work of Driesch. Physical laws, he maintains, presuppose a quantitative relation between the stimulus and the reaction; but in conscious action the response is specific and out of relation to the quantity of the stimulus. Busse and he have made current a striking example:

A father receives a telegram. The reading of the telegram may cause him to jump for joy and start a complicated series of preparations for receiving his son, or it may throw him into utter despondency and even kill him if it announces the death of the son. The difference between the two telegrams is but a few markings on the paper (in German it would be the difference between *an-* and *um-gekommen*).

The logical weakness of this argument is its assumption that a physical and chemical explanation must assume the immediate external antecedent to be the sole and sufficient cause. On this ground one might as well argue that there can be no physical explanation of the explosion of a powder magazine or the starting of a vast forest fire by a single spark. The smallness of a stimulus need not prevent it from releasing vast stores of energy, and the human organism certainly contains stores of physical energy.

Similarly the fact that the same telegram will not so affect a stranger to whom it is delivered by mistake, or even the father so long as he cannot understand its language, in no way proves the impossibility of physical explanation. For past differences of organic behaviour in respect to the expected son, or even such minute organic adjustments as are involved in learning a given language, can readily be supposed to create differences in the present structure of the cerebro-nervous system. A small drop of moisture in a given mixture often determines whether a chemical transformation shall take place or not. Driesch apparently forgets that *small* is a relative term, and not the same as *inexistent*. A minute variation in the mechanism of a gun's adjustment can decide whether a very distant powder magazine and the huge ship containing it shall explode or not.

Nor is the impossibility of physico-chemical explanation proved by Driesch's further argument that the same effect can be produced on the father by two physically different telegrams, e.g. one in French and one in English. The contention that a physical system cannot respond in the same way to two different stimuli can be met by the illustration of the explosion, which may be caused by a blow or by a match.

These physical analogies doubtless sound absurdly far-fetched. For the father *is* conscious and the gunpowder is not. It is no part of my contention to deny this. But it is one thing to assert that consciousness is present, another to assume it to be the sole cause, and still another to insist that no physical explanation of the actual result is possible. The last is what Driesch tries to *prove;* and the physical analogies are sufficient to show the inadequacy of this attempted proof.

MacDougall has advanced the argument that organic movements are characterized by persistence combined with variability. The falling stone stops dead when it strikes the ground, but organic movements persist. Here again it is the belief in vitalism that makes the argument seem convincing rather than the argument that proves vitalism. Outwardly studied, persistency is certainly a character of physical pressure; and even as to variability inorganic nature offers parallels in the actions of the wind, fire, and the sea. If like the ancients we attributed purpose to the latter we could readily see instances of persistent and varied efforts to destroy certain structures or to eat away certain lands.

A much more popular argument proceeds by taking some supreme work of human genius, e.g. Shakespeare's *Hamlet,* and asking: Is it conceivable that any physical system like a Laputan machine can turn out such a product? If the question assumes that such a machine can never produce a composition like *Hamlet* such assumption is demonstrably false. For given a finite number of letters or words to begin with, the number of *all possible* combinations of words equal in length to the number of words in *Hamlet* is necessarily finite. Hence given sufficient time a Laputan machine will certainly turn out the play of *Hamlet* among its products. It may be objected of course that the required number of combinations is so enormously large that the odds against the actual play of *Hamlet* being so produced in historic time are overwhelming and that no rational person can entertain such a possibility. The reader may be assured that I do not for a moment entertain the hypothesis that the brain of Shakespeare *was* such a Laputan machine. But I find this argument from probability logically inadequate, as indeed are all arguments from probability as to what has occurred only once in infinite time. *Every* actual arrangement of nature at any moment of infinite time has infinite odds against it; and there is nothing in the nature of possible physical explanations which necessitates that a play like *Hamlet* should occur more often or less often than it actually has. What justly renders so absurd the idea of any resemblance between Shakespeare and a Laputan machine is precisely the fact that a Laputan machine is so constructed that to it as to some ultra modernistic "poets" all sorts of combinations of words are equally possible. But to a human being who has grown up in a definite community, who has become habituated to associate words in certain orders, all sorts of combinations that might be otherwise possible become actually ruled out; and the more we know of Shakespeare's constitution, his sensitiveness, etc., the more classes of abstract possibilities are ruled out. If the workings of Shakespeare's brain are to be viewed as determined by physical laws these laws must necessarily be so much more numerous and complicated than those of a Laputan machine that the difference in the two cases is, from the point of view of our present knowledge (and possibly of any future finite intelligence), an unbridgeable gulf. Nevertheless we have here no valid proof that there is anything in the workings of Shakespeare's brain contrary to physical law. The best we can say is that while physical laws are universal, in the sense that we know of no exceptions, those that we do know are certainly inadequate to explain the more complicated phenomena of organic behaviour.

Our examination of the vitalist argument so far shows that it can at best prove the inadequacy of the physical laws now known. It has not proved the impossibility of physical explanations in the field of animate nature. The importance of this distinction will be more fully evident if we examine the positive side of the vitalist's claim, namely that in the category of purpose we do have a simple explanation. Few confusions in human thought are as mischievous theoretically and practically as the view that teleology to the exclusion of mechanism can serve as an adequate explanation.

Description of animate nature in terms of purpose is usual and legitimate to the extent that it points to actual analogies to purposive human conduct. But all arguments which assume that a purpose explains anything in biology where a mechanical explanation is impossible or even absent, are certainly based on a false dilemma.

The relation of means and ends coincides with that of cause and effect to the extent that both means and causes are antecedents which precede ends and effects as temporal consequences. The difference between the two relations is largely in the fact of valuation. In the relation of purpose the value of the antecedent is subordinated to that of the consequent, while in the causal relation the antecedent is viewed as determining the nature of the consequent. Hence where no causal relation can be established (because no invariant relation between antecedent and consequent can be found) we have no warrant for speaking of the adaptation of means to ends. An animal act that is sometimes followed by vital advantages (and sometimes not) cannot be regarded as a means for the advancement of life.

We may put this more positively either as a question of evidence or as one of possibility.

Assume, for argument's sake, that animals in their movements are always more or less conscious of purpose. Such consciousness cannot be an explanation of these movements unless there is an invariant and not merely occasional sequence between them. But how can the biologist tell that a determinate mental state is the invariable antecedent of certain movements? Obviously only if the mental state which we call purpose always has a determinate observable expression in the external world. This physical expression will, therefore, invariably be correlated with the organic movements which are assumed to be the consequence of the conscious state. Hence an outer or physical causal relation will thus always have to exist before we can tell with any degree of scientific certainty that a purpose has really been effective.

Moreover, assume that a purpose is known to exist even apart from its physical expression. Can it possibly explain how a physical occurrence comes to pass unless we also know of some mechanism used for the purpose? Let us take the simplest case. A smaller amoeba moves in such a way that the observer ascribes to it the purpose of running away from a bigger amoeba. This description is intelligible enough because it is put in terms analogous to those in which we habitually describe a familiar human situation. Do we, however, have here an adequate scientific explanation? Do we mean that the amoeba is conscious of danger or of wishing to run away and that this consciousness by itself is sufficient to cause the consequent action? Even in human beings we do not regard the mere psychic element of fear as the adequate cause of running away. Apart from the Lange-James theory of emotions which makes fear but a psychic translation of the incipient act of running away itself, it seems fairly well established that an adequate explanation of the running must involve a physical mechanism. The vitalist may maintain the debatable position that the psychic element is independent and necessary to set the mechanism in motion. He has, however, no evidence for the view that the kinetic energy of the organic motion has no physical cause but is created *ad hoc* by the mental element. Such a view is indistinguishable from a belief in magic. It is not coherent with biologic science.

Against the foregoing test of the existence of purposes the vitalist may argue that it is too narrow, that it is based on what is characteristic of human conduct only, and that even human beings frequently achieve their purposes without any clear conscious prevision. The doctrine that purposes may be unconscious is the basis of the philosophy of Von Hartmann, who is not very popular today, possibly because he shows that vitalism can be closely connected with pessimism. His arguments, however, have become current coin and are freely used by contemporary vitalists. Freud's unconscious, Bergson's *élan vital*, Driesch's entelechy, and Reinke's determinants are all Von Hartmann's unconscious over again.

If the notion of purpose could be cleared of all reference to consciousness, unconscious purpose might denote the fact that certain activities have results which are advantageous to the organism. Conscious purpose might then be a special case whose scope is easily exaggerated. Reflection shows that we seldom have an adequate notion of what our purpose is before it is actually realized. But vanity makes us attribute to ourselves prevision. Rarely do we have a *complete* view in our mind of the picture we set out to paint, of the story or drama we wish to write, or of the statue into which we wish to shape our block of marble. We find our purpose only by actually working, in the course of which we must thus rely on unconscious forces. We have to wait till some hidden unconscious force brings our ideas to us. We let our pen, or pencil, or fingers on the piano, run ahead of our conscious ideas before we achieve our ends. Indeed we learn to walk or talk consciously only when apparently blind impulses are successful in achieving our ends, so that repetition becomes more easy. In all this the essentially romantic philosophy of the unconscious (which dates back to Schelling) has brought us a much-needed correction to the old, narrow and anthropomorphic rationalism. But its main interest has been the emotional liberation of life, which becomes insufferably cramped and deadened when it is entirely dominated by conscious human purpose in the manner of worldly or utilitarian rationalism. The philosophy of unconscious purpose has not, therefore, been very critical of its own positive concep-

tions and has consequently fallen into a naïve anthropomorphic teleology of its own. A typical instance of the latter is the argument of
Von Hartmann that the unconscious "deems special venereal organs
necessary in the case of beings whose consciousness is more highly developed—for the greater the risk of its thwarting the demands of instinct, the more desirable does a bait become to entice to the performance of instinctive action." [9]

Waiving the mythologic character of the facts assumed here, does it
not seem that Von Hartmann, who denies the adequacy of conscious
purpose, nevertheless makes the unconscious work in exactly the same
way as the halting, fallible human consciousness works when it qualifies
the general purposes to which it is committed? Is not this rather typical
of the attempt to stretch explanations in terms of conscious purpose and
yet avoid the responsibility of proving the existence of such purposes?
In any case we have here no explanation as to why the unconscious
should develop consciousness in such a way as to make it likely to
depart from its instinctive or unconscious ends and then create baits
(which do not always work) to correct that result. In point of fact
the whole explanation is mythologic.

Unconscious purposes cannot explain biologic processes any more
than conscious or semi-conscious purposes can. Take any of the wonderful organic or instinctive acts of the spiders, wasps, or young chicks.
Do we get any light on these acts if we call them the expression of
unconscious purpose?

It is easy to be convinced of the vitalistic explanation if you are certain of the fact that all organic acts *are* purposive—and not merely purposive in general, but adapted to a specific end, viz. to prolong the life
of the organism or its offspring. Is this, however, true?

Let us recall some very elementary facts. Many of the acts of animals
held up as models of adaptations to perpetuate life are continued under
conditions which make these acts useless or injurious, even to the very
destruction of life. Abysmal stupidity, ineptness, and positive maladaptations are natural facts not confined to the human species. The study not
only of worms but even of mammals like horses and sheep leaves no
doubt of this. Growth is a typical manifestation of life, and it is certainly not true that all forms of growth or regeneration further the

[9] *Philosophy of the Unconscious*, I, p. 222.

continuance of the organism or of its species. It is not necessary to enumerate instances of utterly useless forms of regeneration or of ordinary pathologic growth. The indisputable universality of death is conclusive proof of the wide extent of the maladaptations of plant and animal organisms. Unpleasant as the contemplation of death and decay in nature may be to popular philosophy, we cannot deny them to be essential biologic facts. Perhaps it is the social admiration for the prudent law-abiding citizen which makes us cling to the old naïvely rationalistic and anthropomorphic doctrine that nature does nothing in vain. But it is also possible to see in animate nature a wild profusion of aimless growth surging like the storms and volcanic eruptions of the inorganic realm. In any case the almost inconceivable over-production and terrifically lavish destruction of life make it impossible to identify the processes of animate nature exclusively with purposive adaptations to prolong the life of the individual organisms. Nor is the matter mended by urging that the species rather than the individual is the end of every biologic act. This is simply not true. Many forms of growth are of no use for the survival of the species and many others are injurious. Many species have in fact perished, and no one is in a position to tell how long any existing species will continue to exist.

> "So careful of the type?" but no,
>> From scarpèd cliff and quarried stone
>> She cries, "A thousand types are gone:
>> I care for nothing, all shall go."

Vitalists, like Bichat, too serious to overlook the fact of death, have tried to define life exclusively in terms of the forces of the organism which resist its disintegration. In the realm of abstract science it is as possible and worth while to distinguish life and death as to distinguish positive and negative magnetism. But it does not follow that either has any separate existence in nature; and in fact the concrete biology of organisms is unintelligible apart from the processes which lead to death. Anabolism is unintelligible apart from catabolism. Since the days of Virchow pathologists have given up the attempt to look upon the processes of disease as foreign to the processes of life. Pathology is an integral part of physiology. Hence, though the values of life and death are most sharply antagonistic, we cannot explain the actual phenomena of biology by reference to life purposes only.

It may be objected that our criticism of biologic teleology ignores the fact that individuals and even species are but eddies in the great stream of life which has kept up for many millions of years and may continue for many more millions of years if this earth of ours lasts. But biology, as a natural science, must explain these eddies and we surely cannot explain the facts of plant and animal physiology as all adaptations to prolong this great stream of life. Does it really explain the annual production of so many millions of eggs by a single fish to say that it is necessary to prolong the species? The truth is that purpose is never an explanation of the origin of any phenomenon unless we know the mechanism by which the phenomenon was produced and can prove that it was a conscious intention that set the mechanism in operation. Without conscious intention the teleologic relation of adaptation is but the causal relation viewed retrospectively. As such it is of no particular aid to the vitalist. For in this sense we can speak, as Ruskin does, of the earth-crust being adapted to support the weight of rocks, or with Prof. Livingston of pebbles being adapted to be washed down by river floods.

In general we may conclude that while the obvious and undoubtedly significant and important differences between organic and non-organic behaviour predispose us psychologically to reject the possibility of physico-chemical explanations even of the external movements of animate bodies, yet the vitalist has so far failed to prove the impossibility of such explanations. True, the actual progress of biology has so far succeeded in describing in terms of physical law only the most elementary aspects of life; but the vitalist cannot legitimately refute the possibility of indefinite progress in this respect, by asking the rhetorical question: "Can you imagine a physical machine performing the functions of the organism?" There is no reason for assuming that the laws of physics and chemistry apply only to the simple machines that we can imagine on the basis of our present knowledge. Too many phenomena formerly regarded as beyond the scope of purely physical laws have yielded in the course of time to the inexorable advance of purely naturalistic explanation. If, as we have suggested in our consideration of applied mathematics, it is an empirical question whether any natural events or things fall within the addition theorem, there is no reason to suppose that all the properties of the complex we call an organism are merely the sum of the physico-chemical properties of the separate elements. But to discover what in fact are the additional laws needed to describe the properties of the new combinations we need a much more positive attitude to mechanism and a more critical attitude to teleology than the vitalists have so far shown. If this seems true in the discussion of animal motion it is even clearer when we consider the more vegetative aspect of life, to wit the life cycle of growth, heredity, and regeneration, on which vitalistic arguments have been based.

(ii) *Growth, Reproduction, and Heredity.* The vitalist asks us to fix our attention on the marvellous way in which every animal and plant begins as a single cell and goes through a definite cycle which we call its life history, in which it produces other cells each of which in turn goes through the same cycle. Is it conceivable that any mechanism can bring this about? We certainly know of no machine that can duplicate all these phenomena.

This however is not a logical proof that physics and chemistry will never be able to explain the phenomena of life. There are enough physical analogies to enable us to dismiss as a piece of loose rhetoric the assertion of Dr. J. S. Haldane that "the idea of a mechanism which is constantly maintaining or reproducing its own structure is self-contradictory." Atoms, crystals, our solar and sidereal system do *maintain* their own structure. The maintenance of a certain form together with the splitting off of a part which forms a copy of the original is physically illustrated in the phenomena of vortex rings. That there are prima facie enormous differences between biologic phenomena and these physical analogies, no one will deny. But their presence may serve to remind us that while life *may* involve something beyond all physics and chemistry the proof that it does so has not yet advanced to the stage where we can justly say that every other alternative involves self-contradiction. Neither Dr. Haldane nor any other vitalist has as yet offered any logical reason for being certain in advance as to where the progress of mechanical explanations will stop. Recent chemistry *has* made enormous progress in enabling us to understand the elementary processes of protoplasm—no scientific vitalist now denies that the cell is a Gibbs system and subject to the laws of solutions. Under the influence of Boveri and Mendel—the latter, by the way, trained as a physicist—some notable steps have recently been taken in the direction of what may be called a mechanistic explanation of heredity. In any case we have made enough logical progress to realize that any statement that the embryo is "prompted by psychological impulse" or is really "resolved to acquire a certain form and structure," [10] are mythologic or at best metaphoric. Indeed all vitalists practically admit that such statements have no scientific value in explaining anything.

When we recall that fifty years ago the actual function of the sperm cell in fertilization was not yet established, and that a great embryologist like Von Baer could think of the spermatozoa as parasites in the seminal fluid, the progress made in recent years is most encouraging. When we face the infinitely larger region of the unexplained, no candid mind can deny that we have but touched the edge of the problem. This need not, however, prevent us from recognizing that the supposed absolute obstacles which the vitalist sees in the progress of biophysics are by no means insuperable. They are largely based on a systematic underestimate (α) of the influence of the environment, (β) of the complexity of the cell, and (γ) of the kinetic implication of the term *structure.*

(α) The relative constancy of organic forms, the almost invariable way in which each generation goes through the same cycle as its ancestors despite most diverse conditions, readily generates the feeling that the determination of the whole process of development is in the organism itself and more especially in the germ-cell. This is the tacit assumption which makes plausible the supposed parallelism between the embryologic development of the individual and the history of the whole race. But recent studies by Loeb and others and the actual stimulation of artificial growth and parthenogenesis by physical means, have shown that the actual influence of the environment if properly analyzed is more decisive in determining growth than was formerly conceived possible. The biophysicist need not, therefore, look upon the organism exclusively as a self-controlling and self-perpetuating machine. He can view it also as a reaction system [11] (i.e. a set of reactions between one part of the world and the rest).

A body in stable equilibrium, e.g. a properly loaded ship or a gyroscope, seems to refuse to be disturbed out of its own state, and we are

[10] MacDougall, op. cit., pp. 242, 244.
[11] Bechold, *Die Kolloide*, p. 31.

likely to think its state independent of the changing environment. In fact, however, the notion of stability unites the polar or contrary categories of independence and dependence. The ship or gyroscope will not be disturbed out of its state by certain pressures because it is adjusted to and by something cosmically large. So also the organism as a reaction-system remains in constant or continuous equilibrium so long as the *relevant* factors in the environment remain the same.

From this point of view we can understand and overcome the difficulty which Driesch and other vitalists feel in the query: How can a single and relatively simple cell, if it is merely a physical system, determine such a complicated process as the perpetually repeating life-cycle? The difficulty was really removed when biologists rejected the old theory of preformation or evolution in its original and still popular sense. If we think of growth not as mere unfolding of what is already preformed, but as a process of modification through the interaction of the organism and the environment, the original relative simplicity is not an insuperable objection to subsequent complexity. Nor is there any reason why physical processes should not be cyclical.

(β) The admission that growth is a progress in complexity should not be confused with the view that the germ cell is inherently simple. Driesch has made a great deal of the fact that during the cleavage stage of certain eggs we can remove a part without interfering with the production of the normal form of the organism (excepting its size). This, he argues, shows that the potency of development is up to a certain point the same in all its parts; and how, he asks, can a machine function if it is the same in all its parts? We may answer by pointing to the more recent experimental work of Boveri which largely destroys the force of this argument by showing that the egg is not equipotent in all its parts, but that the nucleus plays a determining and differentiating rôle. Also while the process of differentiation is going on in the egg, the various parts are not completely isolated from each other, and there is no inherent difficulty in conceiving that even when somewhat specialized they still retain up to a certain point their original germinal potency. We have here the helpful analogy of a bar-magnet. Though half is south pole and half north pole, each part becomes bipolar when the bar is broken.

For our general purpose, however, it is more significant to point to the inherent complexity of the single cell. Not only has it a complicated chemical composition and a diversity of physical parts, but there is a good deal of reason for supposing that the smallest visible parts of the cell are complex structures of units intermediate in size between molecules and the optically minimal parts of the cell. We must remember that there may be more molecules in an ordinary cell than there are human beings in all the lands of the earth. True, the infra-microscopic units formerly suggested by Darwin (gemmules), Nägeli (micellae), Weissman (biophors), De Vries (pangens), Hertwig (idiosoms), and others have not proved verifiable. But that the chromosomes, so influential in heredity, are determined by certain smaller units generally called biogenes seems now acceptable to most biologists. As careful an investigator as Nilson admits that "this conception has rendered important service in physiologic analysis."

We may, therefore, regard the cell as a complex of the second order in respect to its constituent molecules. This means that the motions and arrangements of simple molecules can *directly* explain only the intermediate structure and not the distinctive traits of the activity of a single cell.

(γ) Since chemical composition is determined in the molecules, it follows that if the cell is a complex of the second order it will be determined by its intermediate structure and not purely by its chemical composition. In fact it seems fairly established that the chemical composition of protoplasm is somewhat variable. Hence mechanists admit that in biology we have to deal with another element besides chemistry or molecular structure. This is variously called form, organization, or structure. Even as extreme a mechanist as Loeb insists that "without a structure in the egg to begin with no formation of a complicated organism is imaginable." [12]

The word *structure* may have a static and a kinetic suggestion, i.e. it may refer to a purely spatial arrangement of different parts or to a spatio-temporal pattern of parts and their rhythmic or cyclical processes. There is nothing in the mechanistic position to rule out the latter meaning. Just as the modern chemical doctrine of isomerism has made us familiar with the idea that two substances composed of equal numbers of the same kind of atoms may yet differ radically because of different

[12] *The Organism as a Whole*, p. 39.

spatial arrangements of these atoms, so do recent physical chemistry and electro-magnetism, in their account of the structure of the atom, show us anew the rich possibilities in the old idea that different systems may be explained by the various periods of the processes or motions of the component parts. The simplest form of a pattern of activities may be illustrated by any relatively permanent physical form that has a changing content, e.g. a waterfall, a flame, a vortex, a wave, etc. But in the living cell, which Wilson has aptly called "a delicately balanced moving equilibrium," [13] we have a most complicated system of systems. Add to this the fact that a relatively simple aspect of life, its chemistry, has recently made enormous progress in the explanation of vital phenomena by the study of hormones, and it certainly appears a narrow policy to set up premature bounds to the progress of biophysics.

In truth, if the vitalist were satisfied to insist that the older and simpler ideas of physics and chemistry will never be adequate for biology, and that the latter needs an additional category such as structure, form, or organization, there would be no serious issue between him and the mechanist. An issue arises when the vitalist insists, as does Driesch, not only that heredity is a process having direction, but that there is in *addition* [14] some *director* of the process denoted by the term *entelechy* (or some other name) but essentially unknowable. Not a single aspect of heredity can be really explained so long as these mythical or shadowy entities are admittedly unknowable. We only have the additional mysteries how entelechy manages to control the physical process, how as a qualitative unit it can divide itself into many entelechies and direct the various offspring entelechies when the organism dies. Driesch is fond of appealing to the authority of Aristotle. The latter, however, has once for all pointed out that when the abstract name of a process is made into an additional thing, we can have no genuine explanations but only additional difficulties.

(iii) *Regeneration.* The phenomena of regeneration—the restitution by growth of destroyed parts of the organism—is one of the most striking characteristics of life.

A good illustration of its apparently purposeful character is seen when we cut the tail of a fish obliquely. The new tissue does not grow at a

---

[13] *The Cell*, p. 637.
[14] Driesch, op. cit., Vol. II, p. 53.

uniform rate all over the cut surface, but grows more rapidly where more material is needed to restore the original form of the tail. An even more remarkable example is the way the eye lens of a triton is regenerated even when the old lens is not destroyed but only rendered useless by being pressed into the vitreous humour. The development starts not from where the lens originally begins in the embryo, but from the iris, so that its black pigments have to disappear by absorption in order that the new lens should function. The vitalist seems, therefore, on impregnable ground when he asks whether any machine can thus regenerate a special part from some other special part. Yet biologists who have studied these phenomena closely and systematically refuse to be converted to vitalism. Their reasons have readily appealed to their fellow students.

It is easy to emphasize the teleologic aspect of many cases of regeneration but there are also instances in which regeneration serves no useful end and may even be injurious. Instances of superfluous regeneration may be found in the two or more lenses developed in the eyes of newts or salamanders, or two tails in lizards. The regeneration of a new head after the removal of the anterior end and reproductive region of an earthworm cannot possibly serve any useful end to the species. On the other hand, the development of one organ to take the place of a different one, e.g. of a tail instead of an eye, can certainly not serve any useful purpose if we suppose the organism to be properly equipped for life in its original uninjured form. Natural selection may explain why some (not all) useful forms of regeneration have survived. But it is not true, as is often claimed, that the power of regeneration is greatest where liability to injury is greatest. Where regeneration is useful, the fact of utility will not explain how it originates.

While the cause or causes of regeneration are obscure, and no simple physical or chemical explanation can be expected, physical analogies are not lacking. Crystal growth offers one. If we observe closely how organic material appears at the surface of any organic lesion or injury, we cannot escape the suggestion that as the organism is a system in which there is equilibrium between the expansive forces and inner tensions, any new opening for the protoplasm to flow out must produce a readjustment of the whole plastic material of the organism. As the equilibrium of organisms is remarkably stable, it is natural to expect that when equilibrium is restored the organism will in most cases resemble its original form. T. H. Morgan has developed a theory of regeneration along this line which certainly has the advantage of being stated in terms which admit of experimental verification. The older mechanical attempts to explain regeneration by the reserve of food near the injured surface have broken down in failing to explain the polarity of growth [15] or its determinate direction. It looked to the material parts and not to the whole organism. But expansion in the line of least resistance is not necessarily teleologic. Fischel,[16] who has studied these phenomena closely, concludes that they show only that an organism reacts "always in a manner that corresponds to its limited possibilities without regard to a teleologic principle. A planarian, for instance, responds to a stimulus and makes a new head, even when it possesses one or more already; a tubularian produces a hydranth at its basal end, if this end is freely surrounded by water; an octirian forms a new mouth on the side of its body, etc. . . . . Working blindly, without respect to the consequences as far as they concern the whole, the one thing only is produced for which the conditions are present that bring about its formation in the cells."

Our long and perhaps needlessly laborious discussion brings us to the conclusion that explanations in terms of purpose cannot have a very high value in a natural or experimental science of biology. For purely descriptive purposes it may be convenient to avail ourselves of the analogous and more familiar (but unexplained) facts of human conduct. It may be even difficult to avoid using the analogies so habitual in ordinary discourse. When however we wish to formulate general truths or invariant relations between biologic phenomena, causal relations of the type exemplified in physics become necessary, though they may not be sufficient. For though the teleologic relation involves the causal one viewed from the reverse direction, it thrives only where our knowledge is of a low degree of determination, so that we can speak of a plurality of means or causes to bring about a given effect. As our knowledge grows, we learn what is the common element in the various causes that lead to the same result, and that common element uniquely determines

[15] Child, "A Dynamic Conception of the Organic Individual," *Proceed. Nat. Acad. Sci.*, Vol. I (1915); Wilson, *The Cell* (3d ed.), p. 107.
[16] *Archiv für Entwicklungs-Mechanik*, Vol. VII (1898), p. 557, quoted by T. H. Morgan, *Regeneration*, p. 207.

the result in accordance with the ideal of the causal relation. When teleology is thus viewed as a low degree of causal determination, it has, as Kant somewhat confusedly insisted, a heuristic value in suggesting the search for the adequate cause. But pure teleology, which dispenses with all physical causation, has not even the heuristic value of empirical laws that suggest the search for the relevant intermediate circumstances which rationally unite antecedent and consequent. Final causes by themselves are still, in the words of Bacon, "vestal virgins devoted to the gods but barren."

On the other hand, our discussion has also justified some of the positive motives that lead vitalists to reject the cruder forms of mechanism. To the extent that the phenomena of growth, metabolism, reproduction, etc., demand explanations in which the particular form, organization or structure of the organism is a factor, they involve something more than those physico-chemical laws that have been observed to hold only of the separate elements which constitute these organisms. But the "something more" is additional to the physical laws already known. There is no evidence for the assertion that in the more complex natural configurations or kinetic structures called organisms, physics, and chemistry cease to be 'true. Hence extreme mechanists can be, and are, at one with vitalists in this emphasis on the influence of the organism as a whole. All the observable laws of biology can therefore be related to each other in the same way by both schools. What really differentiates the two schools is a metaphysical interest in ultimate interpretation. The vitalist generally wishes to interpret not only animate nature, but the whole world, on an anthropomorphic model. The mechanist wishes to conceive of animate nature and the whole world on the model of a physical machine. We shall perhaps see this more vividly by considering the difference between mechanists and vitalists in respect to the general meaning of scientific methods in biology.

(c) THE INDEPENDENCE OF BIOLOGIC METHOD

(1) THE INDIVIDUALITY OF THE ORGANISM. One who has studied physics for any length of time and has become familiar with its usual abstract language is likely to feel that he is in a new intellectual atmosphere when he turns to a book on biology. Treatises like those in the

Cambridge Natural History do not seem to utilize methods different from those followed by good historians in describing ancient peoples or by intelligent travellers dealing with modern primitive tribes. This is true even if we should take books like Darwin's *Descent of Man* or *Domestication of Animals*. While in the latter we have an attempt to establish general laws, the method of reasoning seems less like that of mathematical physics and much more like that of history or those which a conscientious referee or judge follows in a court of law. This situation lends colour to the view that biology, like history, is a more concrete science than physics and nearer to reality. Some followers of Bergson like the present Master of Baliol have tried to use this to prove that biology and history, though empirical, give us absolute knowledge of things in themselves, not subject to those limitations of phenomenal and mathematical knowledge that were pointed out by Kant.

Apart from the dubious character of the metaphysical arguments that the abstract is unreal and that mathematical physics eliminates real time, it ought to be clear that this attempt to vindicate for biology a superior type of knowledge different from that of physics, is based on a flagrant neglect of some elementary facts in the case.

One of these is that physics and chemistry, like biology, *begin* with what might be called the natural history stage, i.e. with mere description. But as our knowledge increases, we are able to introduce analytic and experimental methods, and to formulate the relations between phenomena in ever wider laws. Thus the descriptive study of the motion of individual planets gives rise in the course of time to the mathematical theory of astronomy, and so the study of the strength of different materials gives rise to the theory of elasticity, the study of the shape of crystals to mathematical crystallography, etc. So likewise the descriptive stage in the study of atomic weights is followed by Moseley's law in which they are all united by mathematical relations. Is biology an exception to this rule? Not at all. The branches of biology which have made most progress, e.g. biochemistry and genetics, are assuming more and more an experimental and mathematical character like that of other branches of physics. That as we thus progress in the fullness and accuracy of our knowledge we fall behind in the knowledge of reality, is an amazing claim. We can understand its seeming plausibility only if we remember that as our description becomes more scientific, it becomes more removed from the familiar level of common sense. Obviously to the extent that habitual and familiar descriptions seem more "real" to us, all progress in science is at first away from reality. But it is to be observed that the common sense level is not one of primitive metaphysical innocence. The language of common sense is full of animistic, ancient, and scholastic metaphysics; and theoretic science arises not only to satisfy practical needs, but also as a way out of intellectual dissatisfaction aroused by the perplexing contradictions which infest the realm of common sense.

In the second place no test of individuality has ever been proposed that does not accord some individuality to planets, crystals, or electrons, or that will apply without any difficulty throughout the organic realm. Bergson himself is aware of these difficulties and mentions some, e.g. the difficulty of distinguishing between a member of a colony and what is merely a part of an organism. Is the germ cell a separate individual or a part of the body? Bergson tries to avoid these difficulties by assuming only one true individuality, the whole of life, of which all existing individuals are branches. But whatever interest or value may inhere in such a revival of the ancient idea of a cosmic life or a world-soul, Bergson is certainly unfortunate when he tries to use such a metaphysical concept as the whole of life to explain empirical biologic facts such as the supposed resemblance between the eye of a scallop and the vertebrate eye. Apart from the unfortunate circumstance that he is mistaken as to the fact, Bergson's logic in the case will certainly not stand scrutiny. Granted the fact of resemblance, it will not follow that physico-chemical explanations of such resemblance are impossible. Why cannot similar circumstances in different groups of animals produce somewhat similar results? Even if all mechanical explanations were proved impossible, the idea of the whole of life acting on both molluscs and chordates to produce a resemblance is a myth or a metaphor, not a biologic explanation of how or why anything happens. A scientific explanation of such facts of convergent evolution can be stated only, as Willy has done, in verifiable elements according to a method not substantially different from those of physics and chemistry.

(2) THE INFLUENCE OF THE ORGANISM AS A WHOLE. Admittedly biology must always consider the influence of the organism as a whole. This has seemed to some an admission that thereby biology differs in

method from the physical sciences that proceed by the summation of the influence of the various parts. As a rough characterization of an outstanding difference, this seems unobjectionable. The attempt, however, to find here an absolute difference between the organic sciences which proceed from the whole to the parts, and the mechanical sciences which proceed from the parts to the whole, does not agree with the facts of the case. In the first place, it is not true that all physical science operates on homogeneous systems only. The Phase Rule for instance is a very important scientific law which deals explicitly with heterogeneous systems. In such systems no part of the effect is fully explained without taking account of the total state of the systems. This becomes obvious when chemical action takes place between the parts or elements of a system. But in the second place even in what may be called a homogeneous system, e.g. the masses of the sun and the planets, the influence of the total whole is necessary to explain the movement of any actual part. It is only because the influence of the sun figures so predominantly in the bulk of the effect that we forget that no part of the system is simply passive. The physical law of action and reaction means therefore that every part of a mechanical system is also reciprocally cause and effect just as in an organism. There is doubtless such an enormous difference in degree of complexity here that it seems vain to expect that the mathematical methods of homogeneous continuous quantity will ever be applied to biology. But the mathematics of discontinuity has a place in physics as well as in the more developed portions of biology.

(3) THE BIOLOGIC LEVEL OF ANALYSIS. A seemingly promising compromise between mechanism and vitalism as to method is the suggestion that as science studies the relational structure of things, it may analyze different entities in the same way, and the same method may prevail in biology as in physics, though the subject matters are on different levels. While the units of electricity are electrons, of chemistry, atoms, of mechanics, molecules, why may not the units of biology be living cells? Will not this grant the irreducibility of life and yet the applicability of mechanistic methods? Do we not in fact speak of the mechanism of life?

The difficulty with this compromise is that biologic phenomena cannot all be adequately explained in terms of cells as units; and even if such explanation were possible, most vitalists would still be dissatisfied with the compromise. For at bottom, vitalists are opposed to the determinism which is characteristic of scientific methods in physics.

(4) THE SPONTANEITY OF LIFE. No one can well deny that organisms have a greater observable complexity and variability of motion than inanimate systems. If this variability is called spontaneity and spontaneity is identified with indeterminism, then the behaviour of organisms is indeterminate within the limits of our observation. This is a conclusion which vitalists from Bichat to Driesch and Bergson do and must accept. Bergson, more interested in introspective psychology than in biology, is radical in his indeterminism; but even Driesch allows his non-physical entelechies and psychoids to suspend (to what degree?) the operation of physical laws.[17] This fits in very well with the common impression summed up by Bichat that while physical motions are constrained and thus subject to law completely, biologic phenomena involve an element of freedom, and therefore depart more or less from law.

This view is generally rejected by biologists as incompatible with the character of their method. It is the very business of science to bring apparently anomalous phenomena under the rubric of universal law; and it cannot accept any a priori limit as to what natural happenings it may not hope to describe in terms of law. When we look at the facts closely, we see that the greater complexity and observable spontaneity of the organic realm is no new type of obstacle in the presence of which scientific method should disarm itself. Men have been impressed with the fact that the wind bloweth where it listeth, and the weather is proverbially uncertain. Yet we have little faith in the dogma that it is hopeless to seek for explanations here in terms of physical law. What can be more erratic than the seemingly irregular movements of the bodies called planets (i.e. wanderers)? Yet by persistent study their motions have been reduced to law with a degree of accuracy that is far beyond the power of the ordinary imagination to realize. There seems little reason to doubt that a good deal of the popular resistance to determinism in biology is grounded in the anthropomorphic idea that invariable uniformity means inability to resist the compulsion of external law. But to the scientific observer, the motions of a physical system are not imposed upon it by any observable agency external to it; and the

[17] Op. cit., II, 125, 136, 178, 181, 254, 336.

"laws" of motion do not express any compulsion at all. They are rather the forms of abstract mathematical description. Why, then, should not that method of description be extended to biologic phenomena?

The pretended proofs of an essential inapplicability of deterministic physical method to biology can be dismissed as sheer logical fallacies. One may describe biologic phenomena in terms analogous to those of human purpose, but one has no right to turn around and use the possibility of such description as an argument against the possibility of a mathematical-physical type of description. So also we can characterize as fallacious all of Driesch's arguments to the effect that an organism, unlike anything physical, can respond differently to the same stimulus and in the same way to different stimuli. These arguments depend upon taking "same" and "different" in an arbitrary absolutistic sense, not warranted by the character of phenomenal inquiry. Following his analysis, we could also say that the same bar of iron will act differently to the same objects (according to whether they are magnetized or not) and in the same way (e.g. in the way of attraction) to a number of different objects. In the absence of knowledge of magnetism, Driesch's argument would prove that these phenomena are inherently beyond the possibility of physical explanation.

But while the methodologic determinism of physical science is an irrefutable postulate necessary for the undertaking of experimental tests, it is after all only an ideal and there is no logically conclusive proof that this ideal is always attainable. Nature may not be constructed altogether for the convenience of the experimental scientist. Actually we do observe a higher degree of variability in biology than in physics; and the observed regularities may be likened to the regularity with which very large numbers of men learn the language of their parents, earn their living, marry, honour the dead, etc. If a fundamental physical principle such as the law of the conservation of energy has not been experimentally demonstrated with absolute accuracy, why insist that some departures from it are impossible in biology? The positive reasons for rejecting this suggestion are in the end based on faith in the future progress of science.

This faith need not be in the least blind. We may grant that physics and chemistry have not explained and may never fully explain the behaviour of organisms. But this shortcoming of our knowledge is not cured by adding unverifiable imaginary entities. Vitalism may help us to dramatize nature and give it an anthropomorphic familiarity. But while a traveller through a dark forest may manage to keep up his courage by filling the darkness with guardian angels, he cannot thereby change the nature and direction of the right road through the woods. Vitalism can no more increase the corpus of our knowledge than imaginary delicacies can provide sustaining food for our actual bodies.

(5) CONCLUSION. We conclude, then, that while neither mechanism nor vitalism is free from undue dogmatism, their merits and vices are by no means equal. Vitalism clings to the primitive sense of the mystery of things. It prevents us from sinking into mechanistic dogmatism, and it keeps a window open into the abysmal darkness outside of our little metaphysical kennels. But vitalism cares so much for the sense of mystery that it dogmatically blocks the path of rational physical research and it keeps its door open to arbitrary and wilful dreams. Like other attempts to cling to our primitive feelings, it is delightful but childish and barren. The vice of mechanism in practice is at bottom similar to that of vitalism—it will not open its imagination to the possibility of physically determining factors quite other than those already known. It is a vice of economy which becomes deadly to all intellectual life if it rules out everything except sensible qualities. In the end, however, despite the association of vitalism with a hazy idealism (which is really subjectivism and nominalism) mechanism is much more in harmony with true objective idealism (which insists on the reality of universal ideas). It keeps the essential faith in the rational concatenation of things according to universal law. Not the nominalistic Berkeley but the neo-platonic Newton is the true idealist.

§ II. IN WHAT SENSE IS EVOLUTION A BIOLOGIC LAW?

So CLOSELY is the word *evolution* identified in popular thought with the very essence of biology and so many varied and ambitious metaphysical doctrines have recently been erected on the foundations of this vague word, that a closer examination of the question at the head of this section is one of the intellectual necessities of our age. I can only give a succinct summary and refer the reader to the technical works on biology for the supporting evidence.

(A) EVOLUTION AND THE UNFOLDING OF THE PREFORMED

In its original and still popular meaning, the word *evolution* denotes the unfolding of what is involved in the organism from the beginning. The oak is simply the acorn unfolded, and the great diversity of life is but the unfolding of what was implicit in the original germs of life. Scientific biologists have abandoned this view since the days of C. F. Wolff in the middle of the eighteenth century. But popular philosophy clings to it partly because it makes the creation of the world picturable like the winding of a clock, and partly because it saves us the trouble of actual study as to why any particular form of life succeeds any other determinate form. This in large measure accounts for the popularity of such schemes of cosmic evolution as those of Spencer and Bergson. Having begged omnipotence in the initial assumption—whether in the form of an *élan vital* or universal differentiation and integration—everything else follows without further trouble of thought.

(B) EVOLUTION AND TRANSFORMISM

As currently used by English-speaking biologists, evolution denotes the rejection of the idea that "species" or organic forms have remained constant since creation. This denial is based on two different considerations: (1) empirical evidence such as the facts of paleontology, and (2) the general a priori bias in favour of change as a universal trait of nature.

(1) Few who know the facts are inclined to dispute that the ancestors of many of our existing plants and animals were markedly different from their present descendants. There seems little ground for doubting, for instance, that vertebrate animals have arisen in the course of time through modification of older forms. This, however, is not true of all forms of life. There is little, if any, factual evidence to show that our unicellular organisms have changed their form at all; and even some multicellular invertebrates seem to have maintained themselves without substantial change throughout the ages. Darwin and Huxley

recognized this and explained it on the ground that those forms that are adapted to their environment need not change at all. This throws an interesting light on the meaning of *adaptation:* a form of life which does not change at all throughout the revolutions of geologic time is best adapted to maintain itself. But passing over this parenthetic observation, we must agree with Huxley [18] that "facts of this kind are undoubtedly fatal to any form of the doctrine of evolution which postulates . . . an intrinsic necessity, on the part of animal forms which have once come into existence, to undergo continual modification."

(2) In the light of these considerations we need not pay much attention to a priori proofs that the transformation of species is not merely an observed process but a necessary consequence of the very organization of life.[19] Spencer's attempted proof of this in terms of dissipation of force and integration of matter can be dismissed as a jumble of inadequate and undigested physical ideas. Moreover, if one of the proofs of evolution, viz. the parallelism between ontogeny and phylogeny, is taken seriously, the constancy of unicellular forms throughout the history of life is thereby already assumed. In general, a priori arguments as to the universality of change must assume that there is also something constant, and they cannot decide what empirical parts or aspects of nature have undergone the actual changes.

(c) EVOLUTION AND ORTHOGENESIS OR PROGRESS

Certain biologists, by no means all, are unwilling to limit themselves to the doctrine that (some) animal and plant forms have changed in time. They maintain that these changes must be in a definite direction, generally from the simple to the complex (Spencer). For this, also, an overwhelming mass of illustrative and confirmatory material can be gathered. But our test as to what is simple and what is complex must necessarily be somewhat vague, unless we adopt the popular anthropomorphic idea of life as a unilateral development from the amoeba to man—expressed by Emerson in the lines:

> And striving to be man, the worm
> Mounts through all the spires of form.

[18] *Science and Hebrew Tradition,* p. 82.
[19] Kellogg and Jordan, *Evolution and Animal Life,* p. 4.

While this idea of man as the head of the line of evolution is too flattering to human vanity to be ever completely eliminated, it is no part of the science of biology. From a purely scientific point of view every existing form is the end of its line of evolution. If, however, we abandon resemblance to man as the test of simplicity or complexity, shall we say that the evolution of the horse's hoof from four toes is a development from the simple to the complex? Whatever test of simplicity we set up we shall have to admit that many animals lose complicated organs in the course of time. You may, if you like, call these changes degeneration. But by whatever name you call them, so long as they are facts it is not true that all changes of organic form are from the simple to the complex.

In general it may be true that knowing two or three points in the development of an organic form we may often venture to interpolate the form that must have existed in an intermediate stage of time. But the general history of life has not yet shown us any one formula for evolution which will enable us to predict the future forms of life with any greater certainty than we predict the future of political or ecclesiastical organizations.

(D) EVOLUTION AND COMMON DESCENT

The doctrine of evolution is generally (though not by all biologists)[20] identified with the view that all living forms have a common ancestry. If this ancestry is identified with the relatively simple unicellular organisms, there seems no difference between this and the doctrine that all development is from the simple to the complex. Logically, however, there is the possibility that it is only the limitations of our knowledge or means of exploration that makes us think of all unicellular organisms as alike in their simplicity. If this absolute simplicity or homogeneity is questioned, the assumption of a common ancestry for all forms of life becomes questionable even if we grant that all multicellular organisms are descended from unicellular ones. All sorts of possibilities are then opened. There may then be a certain rough parallel in the way different complex organisms have developed from *similar* beginnings but no common ancestry. There is also the possibility of different earlier

[20] T. H. Morgan, *Critique of the Theory of Evolution*, p. 189.

forms becoming more alike through the mixing of germ plasm by interbreeding. In any case the facts of convergent evolution—of increasing resemblance between different forms—throws some doubt on the a priori argument that all resemblance must be due to a common ancestry. It has recently become a sort of habit to regard all diversity as having arisen in the course of time out of a common simple source. But to deduce all history from this convenient modern habit is not in the spirit of sober science. There is a good deal of evidence that many species, genera, orders and phyla do have a common ancestry. But that is not enough to establish the universal rule.

(E) EXPERIMENTAL EVOLUTION AND GENETICS

The facts of variations and heredity, the way offspring differ and yet resemble their parents, form today the subject matter of the rapidly developing science of genetics. With a fine loyalty to the earlier generation of biologists, this science is generally called experimental evolution. But it is well to note that the present study of the laws of genetics under experimental conditions and in the light of mathematical ideas, actually has little to do with Spencerian evolution or even with the more or less speculative ideas of Darwin. Modern genetics originates in the work of De Vries, and more especially Mendel, combined with the careful microscopic study of the cell begun by Schleiden and Schwann. So long as the explanatory ideas of Darwin were the biologist's interest, Mendel's work was unknown. But with the growth of experimental work Mendel's ideas have come to the foreground because they open up methods of research rather than mere explanations of facts already known. Mendel's law, however, is hardly likely to become popular among those who use biologic doctrines to bolster up what are called "organic" philosophies. For Mendel treats the organism as a bundle of more or less independent elements which should be studied separately (as far as possible); and this will not help sweeping assertions about the whole of reality.

(F) EVOLUTION AND NATURAL SELECTION

As various ideas of Darwin on the causes of the "origin of species," e.g., sexual selection and pangenesis, have been abandoned by modern biology, the principle of natural selection has become the object of heated controversy. Curiously enough this controversy went on for many years before it occurred to some students of biology to undertake experimental investigations to determine whether any such process as natural selection actually prevails. I think it fair to conclude from these recent studies that natural selection is a factor, though not the principal —certainly not the only—factor, in the change or "evolution" of species. But "natural selection" is only a name for a whole group of factors which lead to the elimination of certain organisms embodying certain variations. It does not explain the causes of variation, nor does it at all prove that every trait of an existing organism makes it adapted to its environment. Many traits are indifferent to survival-value and many that are injurious nevertheless persist. This supports the contention of Huxley that natural selection does not mean the survival of the fittest in any moral sense of the word. It should also warn us against the Spencerian assumption that because thought has been evolved in the struggle for existence it must necessarily serve some use or survival-value. Above all it lends no support to the view that the latest product of "evolution" is necessarily the best for human life, or that carnage and brutality will promote any specific human or moral values. Those who use natural selection as a substitute for a benevolent Providence can find no genuine support in the facts of scientific biology.

§ III. BIOLOGY AND THE ARGUMENT FROM DESIGN

FROM time immemorial men have been profoundly impressed by the wonderful adaptation within organisms and between organisms and their environment. The view, therefore, that animate nature cannot but be the creation of an intelligent and benevolent cause has appealed powerfully to all generations. Is it conceivable that all these marvellous adjustments are the results of blind mechanical forces? Of course in a literal or logical sense the question can readily be answered in the affirmative since there is no logically conclusive proof of any logical contradiction in the idea (which has actually been held, e.g. by Epicureans) that the complicated vital adjustments do arise out of purely physical forces. But as the creation of life by exclusively non-living forces is unknown in human experience, the positive analogy of the way human beings create purposive arrangements naturally appeals to the popular mind as the most plausible account of the origin of life. If you see a watch, you conclude that some intelligence created it; and if you see the even more complicated and delicate adjustments in living creatures, should you not conclude that a much greater intelligence is back of them?

It is generally believed that the theory of natural selection has broken the force of this argument. But there are enough occasional but vigorous denials of this to make worth while a reconsideration of this vital issue.

Let us note at the outset that even before the advent of Darwinian natural selection, the weakness of the argument from design had been recognized by philosophers from the ancient Greeks to Kant.

In the first place, while arguments which conceive an unknown cosmic creation on the analogy of human activity are psychologically vivid, they are analogies of little logical cogency. Human activity is such a negligibly infinitesimal part of the natural world that there is no reason to suppose that the latter only repeats the human pattern. Let us take a concrete example. It doubtless sounds very unreasonable to suppose that mechanical molecules can of themselves combine to form a watch. It is inconceivable because all the watches we know of are made by men and we have never observed them as natural formations. Let us, however, take another illustration. What is the probability of a sample of uric acid being an artificially manufactured product? A century ago no one would have hesitated to answer: None at all. Uric acid is an organic product and it is impossible for chemists to make it. We now realize the limitations of our previous experience. Yet nothing can be more certain than that our present experience also is limited, and that it is therefore most hazardous to base on it arguments as to what is cosmically impossible.

In the second place the popular idea of creation involves us in such insuperable difficulties that no philosophers—except as they have been subject to theologic influence—have maintained it. The scientific study of nature since the Greeks has always analyzed natural production as a transformation which requires previously existing material. Creation *ex nihilo* has no support in such study. Nor does it really explain anything to say that the animate or inanimate world began by an avowedly incomprehensible and supernatural act. If you need a creator in time to answer the question who made the world, you are bound to face the question who made the creator, and so on ad infinitum. It seems therefore intellectually safer to limit ourselves (as regards production) to the infinite chain of natural events and to the relations which we can discover in it.

Despite these considerations, however, the argument from design, based largely on biologic adaptations, continued to figure largely in popular thought and in Anglo-American academic education up to the triumph of Darwinism. Nor need we be surprised at this if we remember that though Kant recognized the inadequacy of the physico-theologic argument, he still characterized as absurd the idea that an organic phenomenon like the growing of a blade of grass could ever be explained on purely physical principles. This concession is, strictly, of no aid to theology, since if the formation of the organic from the inorganic is unthinkable, we shall no more be able to think or understand God's creating life out of previous inorganic matter than *ex nihilo*. But this is perhaps too subtle for popular theology.

The doctrine of natural selection did not directly face the question as to the cause of adaptations. But it effectively weakened the case for an omniscient, omnipotent, and benevolent cause of these adaptations by calling attention to the frightful wastage of life. A creator who has to make so many imperfect or maladapted specimens to achieve the relatively few that survive for any length of time, seems either devoid of love for the imperfect or lacking in power. It is of course open to the adherents of the old theology to say that the misery and suffering of the imperfect is, in some mysterious way hidden to an imperfect intelligence, for the good even of those ill-adapted creatures. But this really abandons the case. One can similarly argue that a great deal of apparent good may turn out to be evil. The undeniable fact is that nature is full of maladaptations, though the observable number of them is rendered small by the fact that the creatures thus affected are so rapidly and extensively eliminated. When, therefore, you find a creature that seems for a time adapted to its environment, you may think of the countless others that were eliminated, and take the adaptation of the present specimen for granted. You may even argue that it could not be here if it were not in some way adapted to maintain itself. But the latter assertion is hardly more than a tautology. If the present seeds of future destruction are maladaptations, the latter are universal.

Yet the causes of adaptations are not thereby explained; and so long as this is true, revival of the physico-theologic argument from design are bound to recur. Especially is this true if we ignore—and we have plenty of emotional motives for ignoring—the fact of death and the inherent unlikelihood of an infinite duration of life on earth. An interesting current example of such revival of the argument from design is the reversal of the position of Paley and the Bridgewater treatises. Instead of arguing from the fitness of the organism, it argues from the fitness of the environment.

Consider the many seemingly exceptional physico-chemical conditions necessary to make life possible. The odds against such a combination occurring by chance are enormous. Hence, it is concluded, some designing agency is more probable. This argument, however, is extremely unfortunate. For even if we were to grant the validity of its mathematical reasoning, the latter could prove nothing more than that in a chance universe spread through time and space, the occurrence of life should be extremely rare. That, however, is precisely the actual case. Life as we know it is a relatively recent episode in an infinitesimal part of space. Even on our tiny globe it occupies a minor part. We do not find it a few miles below the surface of the earth and it disappears a few miles above it.

Nor is the argument for design improved if we rely on empirical evidence. A favourite example of evidence for design is the fact that water unlike most substances expands before reaching the freezing point. This prevents the ocean and the rivers from freezing to the bottom and thus makes it possible for fish and other organisms to continue active life. Here, however, it is only the apparent exception to what we expect to be a law of nature that lends colour to the view of a special intervention to make life possible where otherwise it could not be—just as a man who reaches his destination just before the downpour begins is tempted to think that Providence has held back the rain for his sake. When, however, we are caught in the rain, we do not generally think this has any connection with Providence. Suppose that we lived in a world in which marine life were not possible, on account of water continuing to contract below 39° F. Would it have been legitimate to use this as an argument against design? Actually we do live in a world where life is impossible above certain temperatures. Is this a legitimate argument against design? All we can say with certainty is that life is possible under certain conditions and not under others. Naturally those conditions under which life is possible can be called favourable to life. But this is an analytic proposition and can hardly support a proof that life is the result of design.

The belief or hypothesis that the total universe is the expression of a purpose—even of a definite purpose revealed to us—cannot be disproved. One who holds to that faith can always appeal to the remote future for verification. But neither can we disprove the assertion that the total cosmic process shows no purpose with reference to human life. Theoretically it seems reasonable to suppose that since the category of purpose arises in human affairs, it ought not without adequate justification be stretched to cover the entire universe of non-human relations. It is a common experience that categories applicable in a given realm cause confusion when stretched beyond that realm. But the emotional pull of rival hypotheses does not generally permit of even intellectual neutrality. A universe that is not alive to its core strikes us as cold and bleak and fills us with the almost instinctive fear of the unknown; while the idea that human or quasi-human forces are cosmically dominant produces a satisfaction similar to that of returning home from a lonely trip in a desert.

For the purposes of currently prevailing religion, it is not enough that the world should be merely purposive. It must be purposive in the interests of humanity and in accordance with a definite scheme as to what our best interests are. A purposive world in which the fate of humanity is a mere incident, in which this whole earth of ours plays no greater part than a stray chip from a statue which an artist is perfecting, offers little more support to current religion than a dogmatic materialism. Yet so ingrained is the fear of empty spaces and so strong the human desire for a conscious spectator of our intense but often incommunicable inner strivings, that millions have preferred to believe in a demoniac world, designed to torture all but a few of the elect rather than in a world that indifferently pours its beneficent and destructive rains on the just and the unjust.

Chapter Four

PSYCHOLOGY AS A NATURAL SCIENCE

§ I. FROM COMMON SENSE TO SCIENTIFIC PSYCHOLOGY

I**T SEEMS** paradoxical that science should be both an extension of common sense and also a correction of it. Yet that is only one illustration of the more general fact that all the arts grow out of nature (human and environmental) and yet transform it.

However, there is a real difficulty which the scientist faces when after rising high above the common-sense level he comes back to it (as a human being he must) and tries to give some account of what he did. This difficulty becomes critical in the sciences of human nature and social affairs.

The mathematician is not much embarrassed in this regard. As he builds his road straight up the empyrean of pure thought he charts every step of it in his cryptic but luminous symbols, and when he returns to the common-sense level he does not have to give any further account or justify himself. His proofs are open to all, and every one who is willing to make the necessary exertion can go over each and every one of his steps and rise to the same height.

Though their subject-matter is more concrete, nearer to man's interests, the physicist and biologist can still deal with it in a purely technical way and their departure from common sense is sanctioned not only by the practical utility of their results but also by rigorous reason, fungible specimens, and repeatable experiments. The psychologist is not so advantageously situated. His subject is nearest to men's "business and bosoms" and he has little support in his efforts to escape the common-sense views of human nature. For the latter have resulted from the attitudes developed through the ages and have been embodied in our common speech-habits. Our daily conduct postulates that we not only are alive but also feel, understand, and think. Indeed, the success of our practical efforts and our *amour propre* depend upon our right anticipation of what others will feel and think as a result of our conduct. Common sense cannot be persuaded to doubt the existence of mind or consciousness in ourselves or in those with whom we are connected by ties of love or common enterprise. So certain are we that there is a real difference between conscious and unconscious acts, that we believe even a dog can tell whether he has been kicked intentionally or unintentionally.

At times, indeed, we may believe our neighbour's mind inaccessible, encased as it seems in a separate body. But we do not hesitate to draw all sorts of conclusions as to his or her mental traits and motives on the basis of outer conduct. Actions, we say, speak louder than words. Thus we judge from observation that A's mind acts quickly or slowly, that he is credulous or critical, sympathetic or vain, and so forth.

Of this kind of knowledge some possess more than others, often to a notable extent. Certain writers, let us say Flaubert, Dostoevsky, or Proust, are reputed to have unusual insight into the workings of the human mind and heart. But such insight is not something that can be learned. We regard it as a gift of the gods and we do not generally examine it critically or test whether its compelling plausibility is not just verbal magic rather than verifiable truth.

Our continued biologic existence, however, does not depend simply and entirely on the accuracy of our knowledge. And the very necessities of daily life, our anxiety to please or subdue or to arouse and maintain sympathy, admiration, or fear frequently put our vision out of focus. It is not easy, even if we have the inclination and opportunity, to study human conduct with the same disinterested curiosity with which the naturalist studies the refraction of light, the flowering of plants, the habits of birds, or the effects of rain and wind on mountains. Thus vanity, credulity, the blindness of love and hate, naturally make us subject to even more illusions about human nature than about the physical world. We not only fail to see all that is actually happening before us but we feel certain that we see or have seen things that cannot possibly be. The proverbial illusions of most people as to their virtues or mental powers, or their estimates of their friends and enemies, amply testify to the tricks that our psychologic judgments and beliefs as well as our memories are constantly playing on us. All sorts of idols, traditional loyalties, and conventional taboos prevent us from looking for and recognizing the actualities of our mental life. It is therefore not at all strange that reflection, bent on seeing things as they are in their systematic connections, should find common-sense psychology confused and inadequate.

The confusion and inconsistency of common-sense psychology is perhaps best seen when we ask what it means by the terms *mind, soul,* or *spirit.* The answer will generally be found to contain an unstable mixture of supernaturalism and naïve materialism.

The soul is an ethereal or divine essence so utterly distinct from the earthly body as to be capable of continuing its separate career when the latter disintegrates. But on the other hand common sense cannot conceive of the soul except as some sort of body in space. Minds or spirits are shadowy quasi-corporeal beings like smoke or breath (*anima, spiritus, Geist, πνεῦμα,* etc.) and their thoughts and feelings are unhesitatingly localized in actual bodies, in the head, heart, liver, or other viscera.

It is difficult to liberate ourselves completely from such confusions because they are encrusted in metaphors or other linguistic deposits of past thought, and it is not easy to overcome their misleading suggestions.

When we become aware of these inconsistencies and begin to question the assumptions of popular psychology and seek for methods of testing them we take the first step in the type of reflection called the philosophy of mind. But such a philosophy cannot develop prosperously without additional and more accurate factual knowledge. The beginnings of such knowledge are cultivated in various professions. Physicians, teachers, lawyers, priests, political and business administrators, all depend upon a certain ability to size up what is going on in the minds of those with whom they come into professional contact. They often see the need of more accurate and adequate information than the vague impressions of common sense and they sometimes coin their cumulative insights into professional maxims which they hand on for the guidance of others.

But even such professional knowledge is scientifically inadequate precisely because it is primarily concerned with practical results. The doctor, teacher, diplomat, or statesman must adjust himself to the peculiarities of the individual before him. This adjustment is an art and involves a subtle sense for the weight of many imponderable factors rather than an ability to formulate clearly the law of any one factor. Hence many who see clearly what is demanded in a given case come to grief the moment they begin abstract formulation. The latter is a special art (of questioning assumptions and testing generalizations) and any one untrained in this art or without special gifts for it is apt to cut a sorry figure, as does the successful business man when he lays down rules as to how to attain "success."

When the professional practitioner begins to formulate his wisdom as to man's nature, he is under the influence of popular psychology and this expresses itself in his proverbial utterances. His statements strike us as true until we meet the very opposite assertions. All sorts of opposite psychologic generalizations are thus expressed with boundless assurance and received with equal approval.

Some teachers find that the best way to make children learn is by enforcing strict discipline which develops habits of work. Others find that the only way is to get their pupils interested. Their respective generalizations as to the nature of the child are apt to be as different as their practices. Some moralists and legal or political theorists believe that men can be saved for the good life only by generous doses of fear. They differ in their conception of human nature from those who believe that men are inherently gentle and can be caught as well with honeyed words.

Some medical men work best with surgery and *materia medica*. Their psychology is apt to be physiologic and materialistic. Others work best with hypnosis, suggestion, psycho-analysis, or the like. In their psychology (as in Freud's) the physical body may almost disappear as an influence.

The striking and important fact is that men using the most opposed and diverse maxims are all professionally successful in curing the sick or treating the insane, just as homeopaths and allopaths both undoubtedly succeed in the more definite field of medicine.

Historically the passage from such a common-sense and practical psychology to science begins in Greek philosophy through the introduction of unifying principles or hypotheses and the attempt at a rational organization of the material. The great human service of philosophy is of course in the wide vistas it suggests; but its technical value is to compel our knowledge to submit to the form of a system and thus to have its inconsistencies brought to the foreground and eliminated.

However, we cannot systematize knowledge unless we have a fund of more or less accurate observations, and it is therefore no accident that the founders of the science of psychology—Empedocles, Democritus, the Hippocratic writers, Aristotle, and Galen—were also physicians. At any rate, we have in Aristotle's *De Anima* and *Parva Naturalia* the ground plan of psychology with the topics of sensation, perception, memory, thought, dreams, etc. Both the physiologic and the dialectical methods are illustrated in his account of anger. We owe also the most enduring classification of human types or characters, that into the phlegmatic, sanguine, bilious, and choleric, to the Greek physicians.

The progress of psychology as a natural science was hampered by the fact that the immortality of the individual human soul became a religious dogma. The path of scientific psychology was not made easier when the New Testament canonized certain distinctions between the *psyche,* the *pneuma,* and the *nous,* and gave authority to the notion that certain mental disturbances are the result of possession by spirits. Even more injurious perhaps was the separation of the faculty of medicine (physicist and physician were originally the same) from that of philosophy or theoretic science.

The modern revival of psychology as a science was principally due to two movements. In the first place, Hobbes, Locke, Hume, and their British and French followers re-introduced the experimental methods or attitudes of the natural sciences into psychology through the processes of self-observation or introspection. This led to the development of pure or analytic psychology, according to which our conscious life can be broken up into a number of mental states or ideas bound together according to the laws of association of ideas already indicated by Aristotle, just as the laws of attraction bind together the small particles which compose physical bodies. The disregard of the physiologic aspect of mental life becomes explicit in books like Stout's *Analytical Psychology* (1896). In the second place, the development of psychology from the physiologic side was one of the results of the rapid expansion of physiology, especially the physiology of the nervous system, in the nineteenth century (largely aided by the perfection of the microscope).

The studies of Johannes Müller on the physiology of the sense organs and his doctrine that each nerve can produce only one kind of a sensation irrespective of the object which stimulates it, proved basic for several generations. By developing suitable laboratory methods, men like Weber, Helmholtz, Fechner, and Wundt laid the foundations for a definite science of psychophysics or physiologic psychology. Through the aid of modern machinery and the invention of special instruments for the measuring of sensations of sight, hearing, touch, movement, heat, and pain, they developed experimental psychology. It cannot be gainsaid that as many of the phenomena of psychophysics have become measurable, definite laws or invariant relations have been established as to hearing, sight, pain, sensations of heat and cold, and their physical and psychologic conditions. Definite knowledge has thus been obtained as to the reaction time to diverse kinds of stimuli, the conditions of memory and emotional disturbances, and other rudimentary phases of our conscious life.

The physiologic movement that began in Germany naturally spread to Great Britain and America, and in William James's great book, *Principles of Psychology*, we have the result of the union. Despite his medical training and his approach to psychology through the teaching of physiology, despite the fact that his was the first psychologic laboratory, James was by temperament an introspectionist of the older British school, with a rare genius for fresh inward vision. He chafed somewhat at the minutiae of microscopic or "brass-instrument" psychology and referred to its practitioners as "those new prism, pendulum, and chronograph philosophers" "with their spying and scraping, their deadly tenacity, and almost diabolic cunning," having little of the grand style about them. Yet he admitted the success of their workman-like efforts to eliminate the uncertainties of introspection by introducing quantitative methods and operating on a large scale so as to use statistical means. "In some of these fields the results have as yet borne little theoretic fruit commensurate with the great labour expended in their acquisition. But facts are facts." [1] James proudly claimed that he had treated psychology as a natural science.

James's great book, however, marks the end rather than the beginning of a period. The syncretism which it represents has not had

---

[1] James, *Principles of Psychology*, pp. 192-93.

a prosperous career. Human nature is a field in which it is difficult to experiment and in which it is even more difficult to resist the temptation to unverifiable speculation. The number of human beings who can be experimented upon and the range of feasible experiments upon them limit the investigation at every step. A scientific psychology can recognize these limitations and limit itself to the study of verifiable facts or laws as to the elements of psychologic life in relation to their bodily conditions. But man's interest is greatest in the more complicated facts of his life. He is loath to recognize the difficulties in the way of real knowledge and he is inclined to grasp with greater eagerness than caution everything that promises to lift him above human limitations and solve for him the riddles of existence.

As literature is becoming increasingly psychological it is difficult for psychology to refrain from being literary rather than scientific. In America, especially, the psychologist is under special temptation to assume greater knowledge than he really has, for is he not appealed to by teachers, and anxious parents, by hospitals and social agencies, yea by business itself as to personnel, advertising, and the like? Research by laboratory methods is slow and not always successful. Knack or personal adaptability, aided by the prestige of an imposing technical vocabulary, may make some individual psychologists very successful in their applications. But though he may gain many *aperçus* in such work, he must rely on impressions and opinions and he cannot in this way build up a science of verifiable propositions.

The limited progress of psychology as a science and the temptation of popular expansion and practical success have brought about a profound dissatisfaction with the existing state of scientific psychology and as usual the reformers have broken up into conflicting sects—functionalists, intentionalists, behaviourists, institutionalists, motorists, dynamists, hormists, and Gestaltists, and outside of these, and despising them all, are the psychoanalysts of the Freudian, Jungist, and Adlerist schools. At times the representatives of these warring psychologic faiths seem to realize that they are over-emphasizing their differences—which is quite obvious when we examine the diverse text-books which they produce and note how much they all have in common not only with each other but with predecessors like Wundt, on whom, in their enthusiasm for the new, they nearly all bestow a too generous amount of contempt. As among religious sects, those that seek unity succeed only in creating new sects. For fierce devotion to battle cries has always characterized those who do not have much ground of their own to cultivate. Some indeed are willing to give up the name of psychologists, but like various religious modernists they wish to carry the clerical offices and church property—the chairs of psychology—along with them.

Confronted by this clamour of conflicting voices, it is surely well to go back to fundamentals and reflect on the subject-matter of any psychology that can pretend to be a natural science, and the kind of laws that we can reasonably expect it to establish.

§ II. THE SUBJECT-MATTER OF PSYCHOLOGY

(A) CONCEPTIONS OF PSYCHIC SUBSTANCE

(1) THE SOUL. According to the classical view, still popularly current and embodied in Catholic and other philosophical treatises, the subject-matter of psychology is the psyche or soul. But though the vast majority of men not only have no doubt as to the existence of the soul but believe that it survives the body (and may even have had a previous career) most scientific psychologists have rejected this conception as too metaphysical or even mythologic.

It is instructive to examine the reason for this rejection.

Popular thought starts with the sound perception of and belief in the genuine existence of motion, life, and consciousness. It invokes a supernatural cause for them because in its pre-scientific stage it assumes that material bodies are by nature immobile, lifeless, and incapable of conscious experience.

But scientific physics does not conceive matter as by nature immobile. It assumes rather that bodies will move according to ascertainable formulae, and this is justified by the number of such formulae which it does find. When asked: Is not a creator or world-soul (*anima mundi*) necessary to set the whole of the material world into motion, the reply is that the question rests on an unnecessary assumption, to wit, that there was a time when the material system had no motion, or that it is still intrinsically immobile so that a force outside of it must keep it in motion. Biology similarly answers the argument that we must assume a soul (originally psyche is identical with vital principle) to explain how matter can become alive. Most of the bodies in the universe are lifeless, but some *are* alive and they have the power, under certain limited conditions, of transforming certain non-living bodies into living tissue, just as living tissue is constantly dying or being transformed into lifeless matter. The fact of life is undeniable but there is no more necessity for making a separate substance or thing out of it than out of death. Catabolism or disintegration is just as much a fact as anabolism or growth.

Similarly, in regard to the argument that we need a soul to explain thought and feeling: "Can soul-less bodies think or feel?" The answer is that under certain conditions persons who are organic or living bodies do think and feel, and that with patient study we can find out somewhat more precisely what these requisite conditions are. To speak of the soul as a substance that is the active source of movement, the cause of thought, the owner of the various states or the bond which unites them, etc., is to speak in metaphors, and such metaphors are a fruitful source of confusion if taken literally.

If by calling the soul substance you mean that it is a continuous existent entity other than the body and its observable conscious phenomena, then it is inherently indescribable and beyond our knowledge.

We may and must distinguish between any temporary mental state or event and the general stream of our conscious life. We may call the latter our self and rightly insist that it has a unity of feeling or conscious being that is not of a mere aggregate of atoms or cells. For the feeling of personal identity normally pervades our whole life and all the issues of our personal existence involve it. But the certainty of feeling is an observable fact manifested in and inseparable from our bodies. We do not in fact know any unembodied self or soul, and we cannot tell what would be its essence or inner life apart from the consciousness of body such as gives colour, warmth, depth, and reality to our own consciousness. And since we cannot formulate its nature it cannot serve as a verifiable hypothesis to explain any actual event.

Why, for instance, do we not remember as well at some times as at others? To invoke a faculty or act of the soul simply duplicates the fact to be explained and leaves us no wiser. But if we suppose that remembering depends upon certain bodily conditions, then fatigue becomes relevant and we have a field of investigation before us in which we may acquire more definite knowledge. Why, for instance, should thyroid extract in certain cases stop progressive mental apathy and incapacity? The soul hypothesis is silent or only repeats what we already know. But if mental apathy is an intimate part of, or essentially conditioned by, the physiologic processes of the brain and the nervous system, then we may find out how the chlorine in the thyroid extract affects the bodily and mental functioning. So evident is all this that even as stout a believer in the soul as William James admits that "it does not strictly explain anything," and that what we know directly are states of consciousness with which some corresponding brain states must be assumed.[2]

In thus rejecting the soul as the subject-matter of the science of psychology, we do not discriminate against any known phase of conscious life. But we set ourselves against the fallacy of reification, of supposing that because we can speak of the soul as a noun or subject of discourse it must necessarily be an existing thing in which properties inhere. This fallacy of reification is not avoided if for the word *soul* we substitute any other term such as *the conscious* or *unconscious mind, the non-empirical self, the psychic organism independent of the body,*[3] or the like. Conscious life is a series of events in the history of an organism. It is not a separate non-empirical thing. Some realization of this led William James to reject consciousness as an entity though not as a function.

(2) THE TRANSCENDENTAL EGO OR SELF. Similar considerations enable us to dispose of the argument of those idealists who reify one element in the empirical act of knowing, viz. the "I think" and make a transcendental substance of it, called the *transcendental ego.* Granted that every act of knowing may be accompanied by the implicit recognition that this is what I think, or this is how things appear to me, the "I think" or the "me" is still only an element of an empirical event which can become the object of a perception which is another empirical event.[4]

[2] *Principles of Psychology*, p. 182.
[3] Ladd and Woodworth, *Elements of Physiologic Psychology*, pp. 191, 656.
[4] In the paralogisms of the pure reason Kant denies that we know the soul to be a substance. But elsewhere he is at one with Fichte and Schelling in urging that the mind

If these empirical acts actually have the kind of interrupted continuity which the stream of consciousness or the minds of individual human beings in fact have, we do not thereby get beyond the empirical or phenomenal realm. We can speak of knowledge in general and we can thus speak of the subject of knowledge in general. But this is a logical abstraction, not an enduring existential entity. Idealists like those of the Neo-Kantian school are victimized by their own metaphors when they speak of the growth of knowledge in general as if knowledge could exist without actual empirical beings who know. If I think of the world before the human race appeared or after it will cease, I may imagine a *possible* observer. But the infinitude of such possible observers do not form an actual being. They are at best universal accompaniments of our thought, like language, but not relevant to what we should think of as existing.

(3) THE SUBCONSCIOUS. Another way of reifying mental phenomena, of speaking of mental events as if they were material things having a continuous existence, can be seen in the doctrine of *the* unconscious mind. In current psycho-analytic literature especially, the unconscious self, with unconscious thoughts and purposes, functions exactly as does the soul in popular or mythologic thought.

The doctrine of unconscious perceptions was introduced into modern philosophy by Leibniz in order to justify the view of mental life as a (continuous) substance not dependent on the body. It became basic for the panpsychism of the Romantic philosophy (especially Schelling and Schopenhauer) and it was a logical consequence of the Spencerian view of evolution as a continuous process of unfolding what is involved or latent in the primitive nebula.

Since the publication of Von Hartmann's *The Unconscious* it has become the basis of various theories of instinct, of vitalistic biologies such as Bergson's, and of various popular psychologies which prosper on unverifiable psychic entities.

of man as the subject to which all nature is mere appearance is a thing in itself and not a natural object subject to categories. This rules the mind or the thinker out of empirical psychology and leads to the denial that thinking is itself a natural experience. (See Kant's short essay, "*Ist es eine Erfahrung dass wir Denken?*") He does, to be sure, grant a good deal to anthropology of what he denies to psychology as a science. But if transcendental idealists have in fact often made contributions to empirical psychology, it is despite rather than because of the view that mind is above nature or is its creator.

In arguing for unconscious perception Leibniz does not fall into the modern vice of speaking of unconscious ideas. Perception is an internal state of a monad representing external things. Ideas belong to the field of "apperception" which is reflective knowledge of our perceptions.[5]

It may also help us to understand Leibniz's position if we remember that the term *perception* was used in an analogic or metaphoric sense by Bacon and others to denote the fact that even a physical object might show the effect that other objects make on it.

(i) *Unconscious Perceptions*. Leibniz advances the earliest and simplest argument for unconscious mind. To hear the noise of the sea, he assumes, "one must hear the parts which compose it, that is the noise of each wave." [6] But we are not conscious of the separate waves. Therefore we must have an unconscious perception.

This argument is aptly characterized by William James as an excellent example of the fallacy of division, i.e. of arguing that what is true only of a collection jointly must be true of each member of the collection distributively.[7]

A weight of pounds may move a lever or scale where a single ounce will not move it at all. Indeed, modern physics is full of instances of electricity, light, or chemical agents failing to produce any part of a given effect before they reach a certain intensity, frequency, or volume.

In any case even if each wave produced a physical effect on our nervous system, it would not follow that it must also produce a mental effect. It may well require many such physical effects before a conscious one is produced. It requires many days' rain before some reservoirs will overflow at all. As the errors of great thinkers are always instructive, it is well to reflect on this one from another angle.

The notion of infinitesimal perceptions was suggested to Leibniz by the insensible particles fashionable in the physics of his day and by the notion of infinitesimals which he introduced into the calculus. In all this accuracy was sacrificed to generality. The hankering after complete parallelism made him confuse physical with psychic analysis. But clearly, while extensive physical magnitudes are divisible into what Leibniz

---

[5] *Opera*, p. 715.
[6] *Opera*, p. 197.
[7] James, *Principles of Psychology*, Vol. I, p. 164.

called infinitesimals, purely mental magnitudes are not so divisible.[8] Minute physical particles like drops of water may have independent existence apart from the ocean but this is not the case with the mental elements which arise in perception. In recalling us to the fact that in perception certain wholes precede their parts, that we apprehend entire objects or situations before the parts can become clear, the Gestalt theory cuts the ground from under this atomistic conception of mental life. But the great lesson which we have to learn from Leibniz's error here is the difficulty of discriminating between the analysis of the physical object and the analysis of the perception of it.

If we keep this in mind we shall not be imposed upon by any of the numerous variants of Leibniz's argument for the existence of unconscious perceptions.

A scientifically impressive argument in favour of imperceptible mental phenomena is to be found in Stratton's interpretation of his experiments on differences in sensation. We cannot, for instance, clearly distinguish between the sensation of a weight of 100 grams and one of 102; nor between the latter and one of 104. But we can distinguish between the first and the third.[9]

Assuming that the sensation of weights of 100 and 102 grams cannot be really equal, Stratton, Ward, and others have argued that we have

---

[8] The fact that mental phenomena are not generally divisible, as are physical bodies in their spatial magnitude, is not a valid argument against the possibility of measuring the former.

In the first place, mental events have their temporal dimension. If, for instance, we mark the beginning and the end of an act of perception and note the time-interval, we have measured its duration as we measure that of any other event. With proper chronometers many interesting phases of mental life have thus been actually measured with some precision, e.g. the time it takes for one idea to evoke an associated idea, the time phase of recollection, the minimum duration for any distinct perception, the speed of learning, etc.

In the second place, many mental traits seem to be correlated, and we can measure the degree of correlation. Mental tests of "intelligence" are instances of this.

Finally, while psychologists generally do not share Fechner's view that the intensity of a sensation is a quantity made up of a number of increments added to a threshold sensation, it is still possible under proper conditions to measure the intensity of certain feelings, pains, etc.

In psychology, however, it is easier than elsewhere to rush off to measure something without considering what it is that we are measuring, or what the measurement means. In this respect some recent "measurements" are of the same logical type as Plato's determination that a just ruler is 729 times as happy as an unjust one.

[9] G. M. Stratton, in *Philosophische Studien*, Vol. XII, p. 531, and *Experimental Psychology*, p. 84. Cf. James Ward, *Psychologic Principles*.

here an obvious mental difference that is nevertheless imperceptible. But in the light of our foregoing analysis we need not grant this at all. The admitted facts are that the physical weights 100 and 102 are different but that our sensations in the given situation are indistinguishable. Under different conditions it is conceivable that greater sensitiveness might be developed but we should then have different sensations. If two sensations are indistinguishable as sensations, any distinction between their objects is physical, not mental. We must not in the interest of continuity ignore the actual fact that we do have indistinguishable sensations of physically distinguishable objects. Nor is there any reason for rejecting the fact that two sensations indistinguishable from a third are distinguishable from each other. Two individuals, each of whom is a cousin to a third, may not be so related to each other.

Another argument of Stratton's is based on Dunlap's experiments which show that shadows too faint to be perceived affect our judgment of distance almost to the extent which they would if they were visible. But the fact that imperceptible shadows influence our judgment does not prove that they were perceived. All sorts of physical facts determine our judgments but are not on that account mental.

When the doctrine of unconscious mentality is limited as in the examples just discussed to the perception of minimal or threshold differences, it does not seem to be capable of much mischief. But rigorous science requires that we keep our categories clean and the examples discussed do not justify calling into being mental elements that are neither bodily nor conscious. We may of course use the term *mental* for those bodily processes which condition our conscious life or which represent the existential basis of possible conscious events. But between bodies and consciousness or awareness there is no third intermediate entity, though historically we can well admit that consciousness grows out of life.

This brings us to the question of panpsychism.

(ii) *Panpsychism.* Evolutionists who accept the principle of continuity without qualification believe that since conscious life appears in the course of cosmic history it must have been latent in the cosmic dust or nebula from the beginning. If this latency is anything more than the bare fact that what happened was possible, this argument has force only to those who do not believe that real novelty or new combinations can appear in time—a view which goes back to the original eighteenth century form of evolution, as exfoliation of that which was originally involved or folded up. But the theory of epigenesis which replaced it in biology might replace it also in general philosophy. At any rate we need not go to speculative biology to see actual unicellular organisms develop into conscious individuals of the species *homo sapiens*. In this development it is hard to say when consciousness appears. But it is certain that before the development of the nervous system the foetus has not any consciousness in the sense in which we apply that term to the awareness which the adult has of the world about him. Why then should the consciousness be traced back to the cosmic dust? Those who do so are influenced by intellectual aesthetics. But there is no scientific value in the suggestion. It does not suggest any possible line of investigation. Thus it will always suggest itself to some minds and there can in the nature of the case be no refutation of it. But neither is any specific issue illuminated by it.

We can if we wish call the non-living a lower degree of life, and the non-conscious a lower degree of consciousness. But is anything gained by regarding our clothes or our household furniture as living and conscious beings?

(iii) *Unconscious Thoughts.* Popular and mythologic thought does not completely identify the soul with observable consciousness. The soul might have thoughts which are unknown to us. In popular psychoanalysis such unconscious thoughts, kept out of our consciousness by another occult power called the censor, largely determine our weal and woe.

Now it is well in meeting this contention to recognize that consciousness, like the light which illumines objects, is not all of the same degree of intensity. When I am absorbed or talking to some one in the street or in a crowded room, there are all sorts of sights and noises of which I am aware faintly, and a shift in my interest and attention may make them more vivid. As I look over my own body the visible part is not divided from the invisible part by a sharp line, but the former gradually shades into the latter just as the day shades into the night through a twilight zone. Thus the tip of my nose must often be in the field of my vision, but it is difficult to say how much of it is perceived. These twilight zones of dim consciousness are sometimes wider as in drowziness before falling asleep, and sometimes very narrow as in waking from sound dreamless sleep. But the difference between the conscious and the unconscious like that between day and night cannot thereby be denied.

We must also not overlook the fact that all sorts of experiences conscious or unconscious leave impressions in the bodily organism so that the recurrence of the experience is facilitated, even when the given experience is not consciously remembered.

With these two cautions we can meet all the arguments in favour of unconscious thought.

The distinction between conscious and unconscious life is not so sharp as that between the living and the non-living. Hence, as we have already noted in the chapter on biology, popular animistic biology or naïve rationalism speaks of "instinctive" or organic adaptations as if they were all due to conscious purposes. When reflection shows that we are not conscious all the time, our linguistic habits tend to make us resort to unconscious thoughts to fill in our unthinking intervals. A few examples will make this clear.

I am puzzled by a mathematical problem. I go to sleep, and when I awake (or after an interval in which I think of something quite different) the solution suddenly appears to me quite clear. The popular explanation is that my unconscious self solved it. But this, so far as our knowledge extends, is a quite meaningless remark, and certainly does not explain anything. Nothing has happened in this case that does not happen in all conscious activity as for instance in conversation or in writing. Some one asks me a question and I may answer at once before I think of what I say. Is it necessary to assume that an unconscious thinker within me has thought it all out and dictated my answer? This is only to duplicate the problem. Where does the unconscious thinker get his answers, wise or foolish as they may be? Or I sit down and hurriedly write a letter. Where do the ideas come from? Surely no explanation is offered by the notion of an unconscious thinker. The only light we can get is to study the history of the visible individual's experience in the phenomenal realm and the way he is conditioned by his inherited nervous system, his environment, training, previous occupation, etc.

Other popular examples bring out the same answer. I am hypnotized and told to go to a certain place at two o'clock the next day. At the appointed time I have an inexplicable desire to go to the given place. Does not this show the pressure on me all the time of the unconscious thought previously suggested? Or I go to sleep resolved to wake up an hour earlier than usual and do so. Does not this prove the presence of continuous unconscious thought? The answer is twofold:

(1) There is nothing in these cases differing in principle from our usual capacity to carry out resolutions to do things in the future. I decide to go home an hour earlier than usual. After the decision, I become absorbed in my daily affairs, and then all of a sudden, at the proper time, or some time before or later, the idea comes to me and I execute it. Why do I happen to think of my resolution or appointment at the time I do? Only a study of my particular habits and associations of ideas can suggest an answer.

(2) The idea of an unconscious thinker within me, separate from my own consciousness, does not explain anything. It is simply a reduplication of the fact that our organism is conscious and that we recognize certain acts of ours as fulfilling our intentions or desires.

There is no necessity of supposing that when I forget a name and then later recollect it the idea had a continuous existence in the subconscious. Ideas are not things, but events, and when an idea recurs it is the object meant or referred to that is the identical element in the two events. Otherwise they are two different events in the same organism.

It has been argued that even if an idea be viewed as an event like the glow of a burning object, it is natural to suppose that between the intervals, when it flares up, the process of burning continues though on a reduced basis. But this is quite unnecessary. A carbon may become incandescent when an electric current of a certain kind is passed through it, and become black again and have no incandescence at all when the current falls below a given strength.

Psychoanalysts, when pressed as to the logical difficulties of the concept of the unconscious mind, justify it on the pragmatic ground that it works, that it helps us to discover and remove the causes of human maladjustments. Undoubtedly the practitioners and many of their patients feel that way about it. But this is hardly scientific evidence. For many more people have been healed by Christian Science and by various other magical cults. Are the tenets of each of these faiths thus proved true? The ways in which people are seemingly cured are as complicated and baffling as the ways in which their distresses arise. Faith or confidence in some cause or person generally seems to be a potent influence. It is hard to say in what percentage of cases the patient has really been cured and what precisely has effected the cure.

Possibly the clinical skill and personal traits of the psychoanalyst may have more to do with his success than any of his dogmas about the unconscious mind. Certainly a scientifically trained psychoanalyst like the late Dr. Rivers does well to admit that the therapeutic success of psychoanalysts does not prove the truth of their theoretic assumptions. The psychoanalyst would also do well to admit more often the inherent difficulty of applying scientific method to his field. The individually complicated and private nature of the troubles, for instance, prevents verification of alleged facts, consultation with other experts, the publication of precise records, etc. This can be seen by comparing the whole course of the psychoanalyst's handling of a case with the procedure insisted on by modern hospitals for surgeons or physicians dealing with the more definitely recognizable and more easily reportable bodily ailments. Psychoanalysts must first of all develop more definite techniques for verifying the facts which they seek to explain.

The method of "free association" is not sufficiently determinate to prevent the psychoanalyst from reading some guess of his own into the case, and the patient who comes for relief is hardly in a position to correct the analyst.[10]

It is sometimes urged that the expression *unconscious thought* is just a suggestive metaphor, a fiction to denote the fact that our organism conducts itself as if it had conscious purposes of which we can find no trace in our consciousness.

Those who urge this regard the objectors to the term *unconscious thought* as myopic pedants. What harm can there be in a merely linguistic contradiction if the effect is to extend our vista in the field of human motives?

But this is too light-hearted an attitude to a fundamental corruption of our intellectual currency. The wealth of a country does not consist in

---

[10] That the analyst must perforce not only suggest things to the patient but more or less browbeat her or him into accepting such suggestions is obvious, and the testimony of Dr. Trigant Burrow to this effect is quite convincing. See the preface to his *The Social Basis of Consciousness.*

its paper checks or bank notes. Yet a country may be economically ruined by policies which prevent us from taking its notes at their face value.

(B) THE MENTAL AND THE PHYSICAL

Until recently scientific psychologists generally agreed that mental or conscious phenomena constituted the distinctive subject-matter of their study. These phenomena, e.g. perceptions, were assumed to be events that happened in time to certain organic bodies under ascertainable conditions.

The distinction between the mental and the physical latent in this view is inconsistent with both monistic mentalism and monistic materialism. The former denies that perceptions or thoughts are events in real or physical time, and the latter denies that there is anything other than the physical. Both of these denials, though in different ways, produce confusion and prevent a clear and full development of psychology as a natural science.

(1) MENTALISM. Those who from Bishop Berkeley to Bergson and Dr. J. S. Haldane implicitly or explicitly deny that human perceptions and thoughts are events in a physical universe, conditioned by bodily changes which are governed by physical (or physiologic) laws, are not actuated by any hostility to psychology. They wish rather, in the interests of their religious and moral beliefs, to destroy the mechanistic view of the world built up by Kepler, Galileo, Newton, Leibniz, and Kant, as something which "does not really set out to interpret reality, but only to discover and make use of certain limited practical purposes." [11] But by stretching the term *consciousness* so that it includes everything, they in fact deny it as a distinctive phenomenon; and by insisting that "there is no such thing as a physical world existing apart from consciousness" they prevent due consideration of the physical conditions of the latter.

The argument that things cannot exist apart from our perception of them (that their being is only our perception of them) rests upon a confusion between a truism and an absurdity. It is a truism to say that "reality is for us perceived reality" [12] if by *perceived* is meant *intel-*

[11] J. S. Haldane, *The Sciences and Philosophy.*
[12] Ibid., p. 231.

lectually perceived or thought of. We cannot intelligently mention or think of reality except by thinking of it. But it is absurd when we recall how seldom we think, to suppose that things do not exist except when we think of them. The whole meaning, the joys and tragedies, of life and of the cosmos are due to the influence of things that are undreamed of in our philosophy.

One of the most effective ways of meeting the dogma that all things exist only as ideas in the mind, which, as Hume honestly pointed out, no one can take seriously all the time—is to accept it consistently and thus eliminate the virus of nihilistic scepticism that it contains. Let us agree to call all possible objects "ideas of the mind." There still remains the difference between the idea of the loaf of bread which I have in my mind but cannot eat to remove my actual hunger, and the idea called bread which I do purchase, weigh, cut up into slices and use to satisfy the hunger of my children and myself. If the Berkeleyan argument is, in a sense, irrefutable, it is because it is a resolution to use the word *idea* for everything, not only for those things which are ordinarily called ideas but also for those things from which ideas are by their old meaning distinguished. The difference between day and night is not removed by applying the term *day* to the twenty-four-hour interval which includes the night. But much confusion results from playing fast and loose with our intellectual currency.

Of course, all things known are, in a sense, in the knowledge relation, but they are not, therefore, identical with the act of knowing them. The discovery of Australia or the death of Alexander the Great is not identical with my now knowing it. Indeed, the contrary is not only true, but is and must be explicitly admitted by subjective idealists. Thus Berkeley would not for a moment think of arguing that, because we know of God's existence, His existence is nothing but an idea in our mind. There surely cannot be anything like rational mind without ideas of the past and the future: yet no one can well maintain that past and future events are identical with the present act of knowing them. To assert that would be to make all time references meaningless. If, then, God and our own past and future ideas are not identical with the act of knowing them, why suppose that other human beings, our ancestors for instance, come into being only when we think of them or perceive them? And if sanity requires me to believe that other human beings exist apart from my ideas or impressions of them, and existed long before I was capable of knowing anything, why suppose that their physical bodies exist only in my perception? Indeed why doubt that my own physical body existed before the "I" and its power of perception came into existence?

> The baby new to earth and sky,
> What time his tender palm is prest
> Against the circle of the breast,
> He never thought that "this is I."

Once, however, we firmly recognize the difference between objects known and the act of knowing them, there is no longer any force in the assertion that all physical things known must be mental. For though the act of knowing is mental, the objects known are not identical with it.

One of the commonplaces of modern books on epistemology is a group of arguments to show that our senses break up and distort the impression which external objects make on them, so that when "the mind reconstructs these impressions into an object," the latter in no way resembles the object external to us. All these arguments, however, rest on the assumption that we know that there are objects, what they are, and how they act on the other physical objects that we call sense organs. If these assumptions are true we have true knowledge of the external (i.e. physical) world to begin with, as conditions of our knowledge of the facts of perception. If these assumptions are false, the whole argument about the nature of sense perception loses its basis and therefore its conclusiveness.

The original and still the most potent of all these arguments is Berkeley's argument about the perception of depth and the third dimension of space. Our eyes see only the surface of things, hence the third dimension is created or invented by the mind. Waiving the psychologic objections to this by Koffka [13] and others, we may yet insist that Berkeley's assumptions as to the nature of the retina, the object perceived, and the relation between them, assume that space is three-dimensional, before we perceive anything about it. Such an assumption cannot prove that the third dimension is nothing but a creation or invention of the perceiving mind. We know the tri-dimensionality of

[13] *Psychologies of 1930.*

things before we know the nature of retinal images. At best the "ideal-ist" argument would show that the physiologic conditions for our know-ing three-dimensional objects are different from those that condition our knowing two-dimensional ones. It in no way supports the conclu-sion that the third dimension does not exist before our perception of it. Knowing is not the same as creation. A thing does not exist before it is created, but it may exist before it is known, and it may be known as having existed before the date of the act of knowing it.[14]

(2) BEHAVIOURISM. Having dealt with those who would make a natural science of psychology impossible by regarding everything as mental, we now face those who deny that there are any mental phe-nomena or that they can be the object of scientific observation and study.

(i) *Materialistic Monism.* Logically, the case against those monists (whether they call themselves materialists, behaviourists, or what not) who categorically deny the existence of conscious phenomena is the same as against those who insist that everything is mental. They vainly try to wipe out a real difference by stretching a word to cover two entities or aspects that need to be distinguished.

The refusal to apply the term *mental* to a judgment about psychol-ogy, to a feeling of approval, to a melody or word that haunts us in an inarticulate form, indeed to any memory, regret, fear, or hope is just an arbitrary resolution to refrain from using words in their ordinary meaning. Its chief result is to confuse our categories by stretching some term like *material, physical, physiologic,* or *neurologic* to cover what it ordinarily does not mean. For there are verifiable differences between conscious and unconscious "behaviour," between losing consciousness in a faint or under the influence of an anaesthetic, and remaining conscious even while paralyzed. The empirical differences which enable us to

---

[14] If the reader wonders how it is that so many acute minds have accepted the Berkeleyan argument, let him remember how infrequently men say, "This argument would prove the truth of my contention but it is invalid all the same." Now the Berkeleyan argument has been used to defend not only the current religious view of the world, but also anarchic individualistic scepticism or such impressionism as that of Anatole France, which denies the possibility and necessity of seeking objective tests of truth. Historically, we may say, modern subjective "idealism" arose as a reaction to the Copernican revolution which displaced man as the centre of the universe. It admits that man is an infinitesimal item in the physical world, but it puts the latter entirely within man's consciousness. What can be more flattering to human vanity than to believe that we carry the whole physical universe in our heads?

tell when consciousness is lost and when it is regained will remain even if we do not use the term *conscious*. Furthermore, we must in the interest of scientific physics distinguish between the loaf of bread before me and the memory of the loaf that I had yesterday or my present idea of a new type of loaf which I should like to see baked. The latter idea cannot be called a physical object without upsetting all the fundamental tests of physical existence. We may well believe that all these mental events have their physiologic correlates—but two series that are correlated are two and cannot be identically one and the same. And to say that "for the behaviourist the mental series is but regarded as another neural series," [15] is irresponsible nonsense. To deny the existence of conscious phenomena because they are physically conditioned is like denying that the behaviourist's words mean anything because they depend for their expression on muscular movement and on sound or light waves.

The only theoretic argument against the existence of consciousness advanced by recent behaviourists is that of A. P. Weiss. Starting with the assumption that ultimately everything is constituted by electrons (and protons) in motion, he argues that consciousness must be reducible to the former. This naïve denial of facts because they do not fit in with a preconceived assumption is a clear case of the fallacy of universal reducibility. For granting that all things are combinations of electrons and protons in motion, it does not follow that combinations of these elements cannot contain forms not contained in the elements separately. We may see in the operation of this fallacy the typical nominalistic incapacity to recognize any existence except spatial and material terms, to the neglect of forms or patterns.

(ii) *The Insulation of Consciousness.* There are those who, without denying that they are themselves conscious, maintain that no one can know the consciousness of "the other one" and that therefore such consciousness cannot be the object of scientific study. This is supported by the argument that all we can know of our fellow men is their behaviour in time and space.

Watson indeed does identify meaning and outer action. Watch what a man does and "his action is his meaning." [16] But in this he clearly does

[15] A. P. Weiss, *Journal of Philosophy*, Vol. XVI (1919), p. 63.
[16] *British Journal of Psychology*, Vol. XI (1920), p. 103.

violence to the facts of silent thought or deliberation, which sometimes precedes action. Consider the simple case of one who wonders whether he should take a right or a left turn on a road. There may be all sorts of brain, muscular, and glandular activity in his body while he is hesitating; but such activity is not the meaning of the issue before him. In his subsequent action, one of the possibilities before him will be realized. The contemplated but unperformed act involved neural processes as well as the other. But only the latter also had exterior physical expressions. In rejecting the identification of the consciousness of others with their external conduct, William James had suggested as a *reductio ad absurdum*, the idea of a mechanical sweetheart. Professor Singer boldly takes up this example and by the use of the pragmatic test seems to turn the tables on the author of pragmatism. What difference could there be between a conscious sweetheart and a mechanical model that did all things that we expect of the human one? We may answer this challenge dialectically and empirically. Dialectically this argument proves nothing, for if there is a conscious element in the lady who has drawn our deepest affection, then according to the principle that every existing trait makes a difference, her conduct never will be completely identical with that of a mechanism, even if for a while we cannot tell the difference.

But empirically we not only have no difficulty in distinguishing a mechanical doll from a conscious sweetheart, but we can distinguish different degrees of this soulful quality among humans, and no tragedy is so common or so poignant as that of the lover who finds that the beloved does not sufficiently understand or feel his joy or his sorrow. Throughout our daily life we must constantly distinguish between conscious or intentional and unconscious or unintentional acts. We often make mistakes as to the presence of intention in others and even in ourselves. But the fact that we find out our mistakes proves that we are not without knowledge. Take a number of people engaged in the gentle game of "table rapping." They say that their hands just rest on the table and that they are not pressing. A dynamometer applied to their muscles proves that they are pressing with a force often as great as ten or fifteen pounds. So far as their muscles and glands are concerned they are pressing. But how about their report? All that we know of life indicates that most people are telling the truth when they say that they are not conscious of their pressure. We often perform acts which are generally conscious but which under certain circumstances are not conscious, and we recognize the given circumstances in this case. Additional knowledge may carry us further and make us conclude that in this case some are not telling us the truth, i.e. that they know that they are pressing. To recognize a deliberate lie is to recognize what goes on in the mind of another.

(iii) *The Attack on Introspection.* It is not necessary for the behaviourist to deny the existence or the knowledge of conscious experience. Sometimes he expressly admits that such an experience as aversion is mental. But he does insist that psychology can be scientific only if it substitutes entirely objective categories of behaviour such as muscular, glandular, or neural responses for mental categories, and he is especially insistent that introspection must be entirely eliminated.

Let us first consider the scientific value of interpretation; and to do this let us distinguish between the claims that introspection is inadequate in itself and that psychology can dispense with it altogether.

Though there are still those who, like Bergson, Croce, and others, assume that our intuition of the workings of our own mind is the most certain and reliable knowledge, nearly all psychologists recognize introspection to be a most difficult art leading frequently to results that are inconclusive because of their vagueness and great variability. In this they agree with the wise men of all ages who have realized how difficult is the maxim "Know thyself." When the poet says, "O wad some Power the giftie gie us, to see oursels as others see us!" he is generally understood to be expressing a craving for a kind of knowledge that is rare and difficult. Santayana well says, "Our notion of ourselves is of all notions the most biased and idealistic. If we attributed to other men only such sound judgment, just preferences, honest passions, and blameless errors as we discover in ourselves, we should take but an insipid and impractical view of mankind." [17]

Nevertheless there seems to be little reason to suppose that a science of human conduct can be built up without reference to properly controlled introspection. To describe human behaviour without any knowledge of our own conscious life may be possible if we limit ourselves to physical motions or to observable physiologic processes—about which

[17] Santayana, *Reason in Science*, p. 1.

more knowledge is highly desirable. But we cannot fully understand or conduct human life without some introspection.

One of the first to express his distrust of introspection was Auguste Comte, who argued that the mind, like the eye, can see objects but not itself. It is true, as he pointed out, that when we try to analyze an emotion, the emotion is no longer the same. But this knowledge itself is based on introspection. The fact is that mental states can be very complicated, and the most intense emotion can last long enough to be capable of observation in its less intense stages or in its memory traces.

All psychologists are anxious to add to our knowledge of man's actual conduct.[18] Let us take a concrete case of such conduct as the behaviourist sees it.[19] "On the psychological side we can describe a man's behaviour in selecting and marrying a wife. We can show how that event has influenced his whole life after marriage. In detail, how the increased responsibility stabilized certain emotional maladjustments, how the added financial burdens led him to work longer hours and to study the details of his profession so that his salary would be increased. It would not help us very much in the present state of science to be able to trace the molecular changes in cell constitution—they certainly exist, but are aside from our problem." Is it not fair to add that in the present state of science the knowledge that increased responsibility leads a man to study the details of his profession in the hope of increasing his salary has not been acquired through the study of muscular or neural or glandular responses? We should not be able to judge of human motives or purposes unless we knew something of our own. Thus we judge a man's intelligence by his success in attaining certain ends which we assume he desires. But we should not assume this if we did not have desires of our own. We do not ordinarily think of these implicit assumptions as introspective judgments. But there is in fact no sharp impassable line but only a difference in degree of development between a judgment as to a visible tree and a judgment as to my own act of seeing the tree. A judgment that I have a tooth is objective and a judgment that it pains is subjective and introspective. Does not the development of introspective judgments help to clarify our objective ones? Do we not

---

[18] Hunter, in *Psychologies of 1925*, p. 85.
[19] Watson, *Psychology from the Standpoint of a Behaviorist*, pp. 40-41.

as we become aware of our personal failings become more careful in trying to get objective results?

A simple example of reliance on introspection can be seen in medicine. The scientific physician or dentist does not hesitate to ask the patient whether he feels certain pains or whether he can estimate their intensity, duration, etc. One might argue that as medicine develops, the diagnostic value of felt pain is controlled by objective tests, just as the oculist uses instruments of objective observation to supplement the introspective judgment of what script the patient can see better. But such control is in fact built up partly on the basis of the introspective testimony of many patients and could not arise without such testimony.

Not only is the behaviourist over-critical in his rejection of introspection, but, like most revolutionists, he is too naïvely credulous in his expectations. Will the use of the terms *behaviour, stimulus,* and *response* solve all the inherently difficult problems of understanding human conduct?

Granting the most sweeping of its claims to novelty, behaviourism has been before the public at least for two decades and it is still a programme and a hope rather than a record of scientific achievement.

No one who realizes the desperate need of more scientific knowledge in human affairs would want to dampen any scientific worker's ardour, even if it were not logically difficult to refute a hope. But a critical analysis of the behaviourist programme shows that it is largely based on a too naïve conception of scientific method. The behaviourist talks as if all that is necessary for the building up of a science of human behaviour is to administer certain stimuli to the human organism and record the response. It does not require much reflection, however, to note that not everything called a stimulus will bring forth a recognizably determinate response—whatever these most vague terms may denote. I raise my hand in a certain manner, or emit certain sound waves before a human organism. What will be the response? If the stimulus is physically or physiologically determinate, can we say that the response will be determinate? Certainly not. For the response will vary according to an indefinitely large number of factors in the history of the individual who thus receives the stimulus. He may pay no attention to it because he is absorbed in other things, or because he does not see any significance in my performance, or chooses to treat me as beneath his notice, or he may understand me and return the salute gratefully and joyously or resent it bitterly, depending on all sorts of circumstances in his past history.

There is nothing new in the categories of stimulus and response which will guide us to the setting of problems that will admit of determinate solutions of the sort that throw light on the nature of human behaviour; and the behaviourists have not yet advanced any single hypothesis as to human behaviour which has proved a stimulus to fruitful investigation.

Taking the sum total of the positive scientific contribution of the behaviourist school (apart from their programme), judging it as generously as we may, it may fairly be characterized as either continuation of the work on conditioned reflexes begun by Pavlov and Bechterev, on the nervous co-ordination of the organism outlined by Sherrington, or studies in animal intelligence following the method of the epoch-making doctoral thesis of Thorndike (1898). The first three of these names belong more strictly to physiology and neurology, and Thorndike was trained by William James, and his experimental genius in limited problems seems independent of the behaviouristic ideology. Indeed when the behaviouristic army ceases to content itself with mere plans or calling fellow psychologists to arms and begins seriously to attack psychologic problems it is bound to break up into two disconnected divisions, one interested in physiologic researches, and the other concerned with the social situations in which men practically find themselves. Now it is difficult to state social situations without teleologic terms—witness the quotation from Watson above. The behaviourist must then choose his categories from common sense and applied social science, which are irretrievably mentalistic, or else he must confine himself to the causal analysis of physiology. If he abandons the belief in the categories of consciousness he can form no connection between physiology and social science.

Others [20] as well as behaviourists have tried to draw the distinction between psychology and physiology on the supposition that physiology deals with the activity of separate organs, while psychology deals with the behaviour of the organism as a whole. This is a very vague and ghostly distinction when we remember Aristotle's warning that the

[20] Woodworth, *Psychologies of 1930*, p. 328.

parts of the body are not physiologic organs except in relation to the whole body.

Is reproduction a physiologic or psychologic process? If we leave out the psychical element, we are dealing with physiology, and yet the whole organism is involved, certainly if the organism is unicellular.

The birth, growth, and death of the organism as a whole, its locomotion, its dependence on food and oxygen, on moisture, atmospheric pressure, temperature, etc., clearly belong to the province of physiology. Does the organism as a whole enter less into these processes than into the psychologic phenomena of a toothache or nausea? The truth is that the difference between psychology and physiology would disappear if we left out all reference to consciousness. We should then have physiology which looks back to physics and chemistry for its causal methods and guiding principles, and mechanistic sociology of the kind that might describe animal events but would not get close to the specific institutions of human life, which depend on science, religious beliefs, etc.

If we recognize psychology as a study of the conscious experience we need not bound it with sharp lines. We can recognize that it is impossible to exclude all reference to conscious experience in the study of human physiology. We can see this in works like Sherrington's *Integrative Action of the Nervous System* and the older studies of brain physiology such as those of Broca and Flechsig. For these involve some psychologic test as to whether the patient can or cannot see, hear, etc. On the other hand any psychology which is more than a dialectic development of introspective data must include physiologic knowledge. But psychology even of the behaviouristic kind uses social categories, which distinguish hand-shaking and hitting from muscular contractions.

We may, in summary, note that those who seek to eliminate the concept of consciousness entirely have more in common with those who reduce the whole world to forms of consciousness than is generally recognized. In general, any argument as to whether the world falls within one or the other of opposite categories must be largely verbal. But the common element of mentalism and behaviourism may be seen more precisely if we consider their relation to the primitive error of reifying the soul or mind, discussed in the first part of the present section. If, upon the assumption that the mind is a substantial thing, we look for its constituent elements, we find only such bilateral facts as seeing, desiring and wondering, in all of which "external reality" is implicated. Now either these relations are internal and constitutive of our original mind-substance, in which case the universe becomes an attribute of mind, or they are external, in which case there can be no empirical evidence for the existence of the mind itself. The former alternative is chosen by idealists, who (as, for instance, Bradley) call upon dialectical proofs that two independent substances cannot co-exist, to confirm their discovery that everything experienced is in experience. The second alternative is chosen by realists who have the same repugnance for a non-empirical reality "behind the veil" which leads idealists to reject the independent reality of experienced objects, but satisfy their empirical temper by eliminating "the mind" in favour of the concrete world of observation. In both cases the position assumed is explicable only if we conceive the mind as a substantial thing. With that assumption rejected, there is no difficulty in supposing that the bilateral facts of experience give us knowledge of two aspects of the universe.[21] The existence of these distinct aspects is the first postulate of a scientific psychology.

Monistic mentalists and monistic materialists are equally unable to discover a distinctive subject-matter for a natural science of psychology.

§ III. LAW AND CAUSALITY IN PSYCHOLOGY

(A) PURE PSYCHOLOGY

LET us consider an account of the world of consciousness such as we can find in any classical treatise on pure or introspective psychology from Hume's *Treatise on Human Nature* to Brentano's *Psychologie* or Stout's *Analytical Psychology*. Such a world is certainly not an anarchic chaos. Quite the contrary. It may seem too orderly as contrasted with the hurly-burly of outer nature with its cosmic weather. But allowing

[21] Our rejection of the notion of a substantial mind implies, of course, a rejection of that dualism which conceives of the mind and the "external" world as two independent substances.

The notion that there is a mind previous to any mental experience is, as we have indicated in another connection (pp. 137-140), rendered plausible only by a misleading metaphor; and the application of the term *external* (which literally applies only to spatial relationship) to denote the relation between things and experiences of them is a pernicious source of confusion.

for a certain vanity which makes us see more order in our conscious life than can be proved to a scientific mind (since different observers see somewhat different orders) we must still admit certain invariant relations which we can verify under proper experimental conditions. Various mental events are connected with each other as wholes and parts, as ends and means, and as associated in uniform co-existence or succession. Of this sort are the laws of memory, of the number of things we can perceive, of the order and mental condition of learning new ideas, and in general of the association of ideas. Certain ideas or emotions, fears, hopes, resentments, gratitudes, etc., are relevant to each other in definite ways.

Such a science at times may be rigorously logical as in its analysis of the various elements that constitute our perceptions, ideas, emotions, etc., or in the determination of what states, such as fears, hopes, resentments, and gratitude, are relevant to each other. Whatever we may decide to be the proper field of psychology, this clarification of our categories is an essential part of scientific work.

But this kind of a science has obvious limitations. Ordinary mathematics can have relatively little application to it because its measurable magnitudes (e.g. irritability, intelligence, etc.) are intensive and not extensive magnitudes so that we cannot apply addition theorems to them.

More serious, however, is the objection that such a science of pure psychology cannot give us the kind of causal relation that we have in physical science; for consciousness lacks continuity of existence, and thoughts or emotions which are related by laws of sequence may have temporal gaps between them which can be filled in only by the conditioning bodily processes. The conscious world is like a field of flowers. We cannot understand its causal relations unless we study plant economy. To explain why any particular mental phenomenon takes place in a given situation must always depend on factual physiologic knowledge. Important, therefore, as are the psychologic laws of apperception, of association of ideas, etc., they do not explain anything (any more than acts of the soul do) unless they are supported by knowledge of physiologic conditions. These considerations put us on guard against taking in a literal sense such metaphoric expressions as that the mind arranges its sensations or attends to them. The pure mind, apart from the body, is not a recognizable entity which does things in the way in which carpenters and other machinists do things. We can speak of our conscious life as a whole or of its general stream in relation to particular phenomena. But such relations are logical. They are not concerned with temporal sequences in the sort of extended continuum that characterizes the domain of physics.

(B) CAN THE MENTAL BE THE CAUSE OF THE PHYSICAL?

When in popular language we say that worry will bring about a certain bodily depression we use the word *worry* to denote a state of the entire person (mind and body) and *do not exclude the bodily processes which are necessarily involved.* So understood the statement is true enough. But when the term *worry* denotes the purely mental phase, abstracted from its bodily substrate, it is no longer true to say that it is *the* cause of the purely physiologic processes which constitute bodily depression. The whole science of physiology is based on the assumption that the loss of sleep, disturbed digestion, relatively increased katabolism, etc., which constitute the given depression, have their cause (or necessary and sufficient condition) in other bodily conditions (to wit, those accompanying the mental factor). These bodily conditions and processes are thus assumed to form a continuous series in accordance with chemico-physical laws such as that of the conservation of energy. The introduction of purely mental or non-physical terms into the series would destroy its homogeneity and would be inconsistent with the aim and method of physiology as a natural science. For while the physiologic system is continuous and antedates our conscious life, the latter is not only intermittent but there is a point of discontinuity between any purely conscious event and the purely physical change which immediately follows it. We may, and some often do, define the causal relation so as to include this kind of sequence. But such causality remains different in type from the kind we use in physiology and nothing but confusion can result from applying the same term to two different kinds of relationship in the same context.

The physiologist to be sure is as yet far from having explained all his phenomena or demonstrated that his assumed laws or continuous connections determine necessary and sufficient conditions. But his ideal offers the only definite programme of research dealing with possibly measurable and verifiable elements. It is certainly the only ideal of explanation along which steady progress has been made for centuries.

To attribute physical causation to purely mental elements is to open the flood gates on all sorts of superstitions. Can I, for instance, by prayers, bring a blight on my neighbour's pigs? A good many peasants believe this and pay in candles, etc., to bring it about. Psychical researchers and other believers in the open mind may argue that it is dogmatism to foreclose the issue, and that we ought to institute statistical inquiries in order to see how effective such prayers are in fact. But whatever the results of such inquiries, they would be indifferent to the pathologist. He would in any case, if he were faithful to the ideal of science, have to look for the cause of the blight in the physiological antecedents.

(c) CAN THE PHYSICAL CAUSE THE MENTAL?

Must we not by parity of reason reject the notion that bodily changes are the causes of mental phenomena? And yet can we not directly produce mental changes by purely physical means? Here again it is not the well-authenticated facts that are in dispute, but their accurate formulation. No one can well dispute that from the point of view of strict physiology, the drug that deadens our sensations, or stimulates our reveries is the cause only of those bodily conditions which are their invariable accompaniment.

If our fundamental laws of the conservation of energy be true, the energy of the physical cause is to be found entirely in the resulting physical effects and no part of it is lost to the physical world in producing a mental or non-physical effect. While this is largely a postulate there are no facts to cast serious doubt as to the possibility of ultimately carrying this through. This is not to doubt that if we bring about certain bodily states, we shall also have their mental accompaniments. And this is a most important fact of all human life, since all education and the influencing of our fellow men depends upon choosing the right physical expression or means to bring about the desired mental state. But the relation between the mental and the physical cannot be called causal in the sense in which we use that term in physics or physiology.

Why given mental and bodily states go together is not only unknown but, apparently, beyond the scope of the ordinary methods of physiology to explain. For the very possibility of physiologic explanation seems to depend on abstracting the purely physiologic aspect from the total or concrete reality.

It may well be urged that since mental events undoubtedly do take place in the natural world, they must have some natural cause or some sufficient reason to determine why consciousness takes place when it does rather than not. But not only is such reason as yet unknown, but if we found it it would have to be quite different from the causal relation that binds together physical changes according to the laws of mechanics, thermodynamics or electricity, since consciousness is qualitatively different from what these laws govern.

To assume that conscious phenomena have a continuous existence, not merely as bodily possibilities but as subconscious or other psychoid conditions, is to introduce new unverifiable hypothetical entities without in the least removing the mystery as to the origin and passing of consciousness in the individual organism, or in the history of nature.

This view has frequently been rejected by being branded as epiphenomenalism. An age that worships machines and the men who do things, recoils from the view that denies causal efficiency to consciousness itself. But this recoil is based on the failure to make certain distinctions. The sense of efficiency of which we are proud belongs not to mentality apart from the body but to the whole conscious organism, which most people regard as themselves. The efficiency of the conscious organism involves a combination of teleologic and causal relations. The teleologic relation is more complex and more variable than the physically causal one, so that some people are rightly said to be more efficient than others. But in inefficient as well as in efficient people, physiology supposes a definite causal relation between successive states of the organic energy.

These considerations will enable us to clarify another inherent difficulty or aporia in the conception of a natural science of man, viz. the question of free will and determinism.

(D) FREEDOM OF THE WILL

There is a widespread impression that on the subject of the freedom of the will it is futile and unwise to argue, that here reason has come to an impasse. You either believe freedom to be a fact of experience or something demanded by the dignity and responsibility of man, or else you hold the belief in it to be an outworn objection to science.

We need not, however, let a fatalistic impression prevent us from trying to arrive at clearer ideas on a subject which has absorbed so much human attention. The honorific association of the term *freedom* has caused it to be used in many senses, and many of those who technically hold to the word (e.g. Kant and Hegel) do not differ in substance from those who reject it (e.g. Spinoza).

In the first place we may dismiss as entirely inconclusive the argument that man is free because he is held responsible. What justice, it is asked, would there be in punishing any man for acts which he was not free to avoid? To which the obvious answer is that we cannot prove the existence of anything by the argument that if it did not exist our policies would not be justified.

But the argument might perhaps be put thus: Humanity would not hold any one responsible if it did not believe in freedom, and the generality of a belief raises some presumption in its favour. But it is not true, historically, that the religious and moral conscience of man has always held individuals responsible because it deemed them free. The most fatalistic peoples, Hindoos, Mohammedans, Calvinists, and others, have not hesitated to hold men responsible for their acts. Nor is sin in the primitive consciousness always an act of free will. The most fearful sins may be committed unintentionally, e.g. touching the Ark of the Covenant (I Sam.). Is not exclusion from heaven the worst of punishments? And do not orthodox Christians exclude all unbaptized children from that blessing?

Logically, the holding of men accountable is not only quite consistent with determinism, but involves it. What would be the use of rewarding or punishing people if we did not expect that certain desirable consequences will generally follow from our so doing? That is the only ground for discriminating in our punishments—for the same punishment will not produce the same results in the insane as in the sane, in infants as in adults, in those who are under certain external constraints as in those who are not, and so forth.

Turning to an examination of the issue itself, we must first distinguish between empirical freedom or the power to do what we will and freedom as the absence of determination in what we can will.

(1) EMPIRICAL FREEDOM. Freedom may be viewed negatively as freedom from restraint, or positively as power or freedom to do what we want in order to achieve our heart's desire. This is the freedom for which men care and for which they will often sacrifice life as worthless without it. In this sense, however, freedom is not something absolute or unqualified. It varies according to our native endowment and circumstances. Suppose that I want to move my arm or say something. I may be deprived of freedom by being paralyzed, by being tied up with ropes, or by the threat of serious harm. We generally think of non-freedom as due entirely to external restraints. Against these we feel we can generally maintain a certain degree of inner freedom. We can think or say to ourselves many things that we cannot do externally. Whatever external pressure the Inquisition might have applied, Galileo could not be deprived of his freedom to say under his breath or quite inaudibly, "Still the earth does move." There is nothing in this kind of freedom inconsistent with determinism—a theory that human conduct is governed by laws of its own. It is undoubtedly true that men feel that they exert energy when they actually will to do something. They like to believe that their wills are real powers in the world and that they themselves are not mere playthings in the hands of blind fate or foreign powers. From this point of view, it may be argued, men will never accept our analysis of the rôle of the conscious will in a physiologically determined organism. It will help to clarify the situation to consider this.

We may in passing note the contrary and equally true proposition that men do not object to being playthings of good fortune; indeed that they rather enjoy the feeling that no matter what they will do they are fated to succeed and to be happy, to be admired as clever, beautiful, honourable, etc. Fatalism does often liberate our energy in the task before us from numbing doubts, indecision, and the burden of having to deliberate, to decide, and to take responsibility. We do live in a society which often professes to believe that the highest moral value of life is effort itself, so that the complacent Jack Horner or superficially successful man of affairs is typical when he tries to attribute all his good fortune to his efforts and all his failures to hindering circumstances. However, our moral repugnances against fatalism, against regarding all effort as vain, do not apply to physiologic determinism or mechanism. Indeed, absolute fatalism in its popular form (that nothing we do or can do will have any effect) is extreme indeterminism. Determinism, on the contrary, means that if we had acted differently, the consequences would have been different.

The limits of our actual power are a matter of cumulative knowledge. To say to me: "No matter what you do you will die sooner or later," is to utter a truth grounded in experience. But experience also shows that there are things which previous thought or knowledge can prevent or help to bring about. Our previous analysis of psychologic causation shows that this efficacy of effort may be a fact if we look to the concrete personality which includes the organism and its conscious states.

As our will to do something is the expression of a certain organic "set" or impulse in a given direction, the bodily substrate of it may supply power to carry the act into effect.

Moralists often refer to a certain inner feeling of freedom, to a conviction that I might have chosen differently; and rigid determinists have dismissed this as an illusion. From the point of view of a determinism such as outlined above, there is more truth in the moralist's psychology than in that of the absolute determinist. If mental life is subject to laws it has a definite meaning to say I might have chosen differently, if other values or considerations had been present to my mind, and we can say this as truly as we can say that if a stone had not been in a given place the growth of a tree would have been different.

Here, however, the thoroughgoing adherent of free will may object that we have so far missed the whole point of the free will controversy, which is precisely the point as to whether our will still is determined by anything outside of it.

There are two forms of this contention: one of radical indetermination, the other that the will is self-determined. Radical indetermination is sometimes called the liberty of indifference.

(2) RADICAL INDETERMINISM. Empirically this doctrine means that whatever my actual choice I might have chosen differently even if all the circumstances had been the same.

This absolute initiation of an act of will can have no empirical evidence for it. The mere feeling that I might have chosen differently is not sufficient, for if I do not see why I chose as I did rather than any other way (which is not generally true of great and significant choices), I can never be certain that there was not something in my past that gave me a definite bent in the way I did choose. On the other hand we do observe definite uniformities in the choices which people make, and these uniformities constitute their character.

(3) SELF-DETERMINATION. In its transcendental form, the doctrine of self-determination contends: Our self may determine what we empirically shall want—so that if any one has wicked desires it is the sinful soul behind the scene that is responsible.

Against this view there are two objections: (1) We have no way of finding out the soul behind the concrete person as he or she appears on the earthly scene, and no reason for calling it the self. (2) All that we know indicates that our desires are partly determined by the particular organism that comes into the world at our birth or when we begin to choose, and partly determined by the character of our experience. As our experience is cumulative we may be aware at a given point that what we shall see or do depends upon what we have already seen or done. The doctrine of self-determination, in this empirical interpretation, is indistinguishable from determinism.

(4) SUMMARY. There is nothing in humanity's consciousness of freedom to throw doubt on the proposition that purposive human conduct exhibits discoverable uniformities. There is no illusion in the feeling that our efforts can be efficacious, nor in the feeling that we not only cannot do everything but cannot will everything, that our characters are more or less fixed by our equipment and by our cumulative experience. This by no means forecloses the metaphysical issue of absolute novelty in the world. If, as we indicated before, the world in time must show novelty, why may not that novelty appear in the psychic realm as well as in the biologic and the physical? There is every reason to suppose that there is such novelty. That in every moment the world of our feelings as well as the world of existence contains something new is as certain as that every movement is different from every other. But the meaning and dignity of human life does not lie in mere difference. In the sane life various conscious experiences must be related to each other. Such continuities as our lives reveal are the proper object of a natural science of psychology.

END OF BOOK II

Book III

REASON IN SOCIAL SCIENCE

Chapter One

THE SOCIAL AND THE NATURAL SCIENCES [1]

THE business of mapping out the proper domains of the various sciences was an integral part of the social philosophy of Auguste Comte, and a dominant interest of American sociology in its earlier years. The actual progress, however, of the various sciences since Comte's day shows that it is not only foolish but mischievous for the sociologist, philosopher, theologian, or moralist to lay down any law restricting the scientific work of the physicists, astronomers, chemists, or biologists. It is fortunate that Comte did not have the actual power to introduce his well-intentioned order into the business of scientific research. If he had, we should now be without our knowledge of the chemical composition of the stars and without the knowledge gained by studying chemistry and biology according to the methods of physics. Science is an exploration of the unknown; and it need not surprise us that prediction as to the outcome of few other human adventures is as hazardous as that concerning the direction which the future progress of any science will take. This reflection need not prevent us from trying to arrive at clear ideas as to the distinctive traits which characterize the social sciences. We may, however, be warned by it against the possibility of restraining the growth of science by the setting up of absolute boundaries on the basis of present incomplete knowledge. Our safest way seems to be to take account of the most influential views that have actually prevailed as to the relation between the natural and the social sciences, and to clarify the situation by submitting them to critical analysis.

The Aristotelian doctrine of the four causes suggests four points of comparison between the social and the natural sciences.

---

[1] The greater part of this chapter appeared as one of the essays in Ogburn and Goldenweiser, *The Social Sciences,* and is reprinted with permission of the publishers, Houghton Mifflin and Company.

§ I. THE SUBJECT-MATTER OF THE SOCIAL SCIENCES

THE distinctness of the *words* "social" and "natural" inevitably suggests that the subject-matters denoted by them must be mutually exclusive. Yet few seem willing to maintain such a sharp dichotomy between natural and social facts as to call the latter un- or non-natural. Social facts and the human beings between whom they take place are located in physical time and space. Deprive these social facts of their physical elements or dimensions and they lose their usual meaning and cease to have reference to anything existing.

These reflections suggest that we must not conceive the social and the natural sciences as mutually exclusive. Rather should we view them as dealing with parts of the same subject-matter from different standpoints. The social life of human beings is within the realm of natural events; but certain distinctive characteristics of the social life make it the object of a group of special studies which may be called the natural sciences of human society. In any case the empirical or historical fact before us is that many questions are clearly in both the physical and the social realm. We may, if we like, draw a sharp line between physical and social anthropology, between physical and economic geography, and perhaps even between individual psychology as a natural science and social psychology as a social science. But the distinction is in any case a thin and shifting one. When we come to the study of linguistics or of epidemics, or to the various branches of technology, we see the breakdown of all the sharp separations thus far suggested between the natural and the social sciences.

(A) ARE SOCIAL FACTS SIMPLY PHYSICAL?

Can we avoid all the foregoing difficulties by boldly declaring that all social phenomena are simply physical?

The affirmative is maintained by those monistic materialists who now call themselves behaviourists. In line with the modern positivistic traditions they speak the language of empiricism and induction; but clearly no mere accumulation of facts can adequately prove the absence of some factor in social phenomena other than those taken into account

334

in physics or biology. In point of fact, these monistic materialists like A. P. Weiss base their stand entirely on the a priori argument that, since social phenomena must manifest themselves in time and space, they must, like physical phenomena, be constituted by matter or electrons in motion. But the fact that social phenomena always involve physical elements fails to prove that they contain nothing else. The fact that X is a man does not disprove that he is a scientist. The behaviourists do not see that while physical elements are necessary they are not sufficient to determine the meaning of social phenomena.

What differentiates the group of facts we call social from other physical facts? Consider the simplest possible social event, e.g. the doffing of the hat to a lady. Physics describes the mechanism of the motion, and chemistry describes the energy transformation which makes the motion possible. But the social significance of the act is not thereby indicated. Social descriptions involve altogether different categories from purely physical ones. To describe men as showing deference or as seeking food, mates, etc., does not give us their physical co-ordinates. The behaviourist uses the categories of stimulus and response. But if the stimulus is purely physical and the response is equally so, how can we get anything as distinctive as the social categories? From the laws that are common to all physical phenomena you cannot deduce those that are peculiar to a given group. Something more and distinctive is needed for the description of social phenomena.

Moreover, the behaviourist himself must and does admit that there often are no determinate relations between the physical dimensions of stimuli and their social responses. The same physical stimulus, for example, the sight of pork, may be followed by the most diverse social responses on the part of Arabs and Russians. Also, all sorts of different physical stimuli, for example, a pistol shot or a light signal, may lead to the same social response, for example, the starting of a race.

The idea that if we take social facts on a large scale we shall find their determining physical conditions, seems to find support in certain statistical correlations between geographic (especially climatic) conditions and social behaviour. Crime, increased or decreased efficiency in economic production, and increase or decrease of marriage rates, seem to be correlated with definite conditions of temperature, atmospheric pressure, air currents, moisture, and so on. I say *seem to be* because I am not sure how many of these correlations will hold after careful analysis of the factors involved. Thus, greater crime during the summer months may not mean any direct effect of temperature on anti-social disposition, but merely greater social opportunity for the commission (and detection) of certain social acts called criminal.

But does not the assumption that summer conditions make for more open-air life and greater opportunity for certain crimes itself take for granted the causation of the social by the physical? Any answer here is confusing unless we clearly distinguish between necessary and sufficient conditions. Given social conditions otherwise the same, a longer day and a milder temperature will bring more people out of doors. But it is hard to say what social effects the mere fact of summer is adequate to produce by itself.

We can see all this more clearly in the usual illustrations of the geographic interpretation of history. The proximity of the sea, it is claimed, explains why the Greeks and Phoenicians were seamen: the presence of certain plants and animals, why certain tribes were agriculturists and hunters; the presence of certain minerals, why certain nations developed certain industries, and so on.[2] Now we may well grant that people cannot fish or sail boats if large bodies of water are inaccessible to them, and they cannot develop metallurgy if they have no ore. But history amply shows that the mere proximity of the sea will not develop mariners or fishermen, e.g. England before the Tudors, or certain parts of Ireland, and that the presence of clay will not make people use pottery. Nor will the presence of certain foods always be followed by their utilization. Thus, many peoples suffering from shortage of food do not utilize the milk of their domestic cattle, the eggs of their fowl, the fish in their rivers, the flesh of certain animals, etc. The reason for such failure is often attributed to irrational taboos, but often it is due to the more simple consideration that people have not learned to utilize resources which to us are obvious. Is it not true that if we take mankind as a whole throughout its history, the utilization

[2] "Since life in the desert develops courage . . . it should follow that the half-savage desert hordes are braver than other peoples. . . . The character of each nomadic tribe, however, varies with time. When these tribes settle in the fertile lands of the high plateaus and become accustomed to the abundance and well-being which these regions offer them, then their courage fades as do their brutality and the rudeness of manners acquired in the desert."—Ibn Khaldun (fourteenth century), *Prolégomènes historiques* (Paris, 1858-1868).

of natural resources is very limited and generally the result of a very slow *process of learning?* The mere presence of the resources is certainly not an adequate cause of men's learning to use them.

In general, those who wish to reduce social phenomena to nothing but physical elements are under the illusion that particular social facts can be derived from physical universals alone. They fail to note the distinction between the necessary and the sufficient conditions of social acts—a distinction which is emphatically explicit in the exact sciences.

(B) ARE SOCIAL FACTS SIMPLY BIOLOGIC?

No one, I take it, seriously questions that social phenomena are conditioned by the biologic processes of the human organism. Yet social phenomena are not merely biologic. The cry which is evolved by pain, the turning away of the head from the sight of something frightful, or the turning toward certain other objects, are all biologic facts. But they acquire social significance only when definitely related to the rules or ends of a community. The ceremonials of courtship, marriage, and so forth involve opposite sexes. But the specific character of these ceremonies as courtship or marriage is social and not merely biologic.

All this is very elementary. Yet an astounding number of widespread and influential errors result from overlooking it. Consider, for instance, the Spencer-Fiske theory that prolonged infancy is the cause of the family and of the growth of civilization. Obviously the helplessness of the infant will strengthen the family bond only when *both* parents already have a disposition to care for the infant. It will not affect parents whose relations are not already more or less permanent. The biologic fact of prolonged infancy cannot, therefore, by itself explain the permanent family.

The distinction between social facts and their biologic elements will likewise show the untenability of attempts to find the adequate cause of the family form in the biologic fact of sex. Sex impulse is fitful and variable, and social factors are necessary to explain the diverse rôles that sex plays in family life, in sacred prostitution, celibacy, and so forth.

The confusion between the biologic and the social point of view is increased by the fact that popular biologic thought has a large infusion of anthropomorphism in the form of Lamarckianism, in the Darwinian theory of sexual selection, and the like. Even natural selection is popularly conceived as if it were similar to conscious breeding, as if Nature literally selected certain forms because they are the fittest for Her preconceived purposes. This has led to a most deplorable confusion between biologic and moral considerations. In vain has Huxley clearly and eloquently pointed out (in his lecture on *Evolution and Ethics*) that the phrase "survival of the fittest" has no ethical connotation at all. It is only an analytic proposition to assert that those species survive who in a given environment and under particular conditions are able to breed at a rate sufficiently large to offset the death rate, and that hence if an unusual cold wave held sway over us those most likely to survive would be the Esquimaux, whereas a prolonged heat wave might leave none of us except some miserable Indians on the Amazon. Intellectual discrimination is not an easy virtue where strong moral prejudices can be defended as the Decalogue of (biologic) Science. But from the point of view of logic and scientific method we need have no hesitation in characterizing as downright fallacies most of the explanations of social institutions by the principle of natural selection. Will natural selection explain why the Macedonians and not the Egyptians had certain laws against incest? Is it true that those who have anti-social feelings leave less progeny than those who are devoted to the common good? Will those who are devoted and courageous be more likely to leave offspring than those who are selfishly and shrewdly accommodating? To ask these questions is to throw sufficient doubt on purely biologic selection as an explanation of social traits. For it simply is not true that every existing social trait has survived because it has helped the race to survive. Many social maladaptations, for example, prostitution or war, have persisted through the ages though they in no way make for an increased birth rate or a decreased death rate.

Similar to the confusion between biologic and social fitness or adaptation is the confusion between biologic and social heredity. It seemed beyond a shadow of doubt that sentiments, ideas, linguistic and artistic forms, manners, etc., spread and survive irrespective of whether the great originators of them—saints, prophets, philosophers, orators, artists, gallant ladies, and so on—left any offspring of their bodies. Yet most of the contemporary discussion of social traits seems to assume that social traits are carried along in the germ plasm.[8] This confusion is caused by ignoring the difference between biologic heredity through the germ plasm and social heredity through tradition, indoctrination, and imitation.

As the rudiments of these questions are precisely the ones that are most generally overlooked, it is well to insist upon them here.

There are two distinct facts of biologic heredity: (1) that offspring resemble their parents more than they resemble other individuals of the same species, and (2) that they continue to resemble other individuals of the same species much more than they ever resemble individuals of other species. As individuals of the same species are generally subject to the same environment, it is always difficult to determine to what extent their resemblance is due to the same environment and to what extent it is due to the continuity of germ plasm. Considering the almost infinitesimal amount of direct study of human heredity for an adequately large number of people and generations, and considering how few pure races we can find throughout history, it seems foolish to make any confident assertions about social traits belonging to any human group through biologic heredity. Foolish, in fact, have been nearly all the generalizations about the various races, for example, Gobineau's generalization that the black excel in art and that the Greeks therefore must have had a black strain in their blood, or Lord Acton's generalization that the Semites are monotheists and the Aryans pantheists or polytheists, or the popular generalizations about the social traits of the French, English, and German. The social composition of the latter peoples has changed relatively little in the last one hundred and fifty years, yet the popular view as to their distinctive cultural characteristics has undergone marked changes. Before the French Revolution great historians like Gibbon could look upon the English as turbulent in contrast with the orderly French. In the nineteenth century it was the French who were looked upon as fickle. The dreamy Germans, having the empire of the clouds at the beginning of the nineteenth century, appeared in a different guise at the beginning of the twentieth century. Yet the great fundamental changes of social life were in the main along the same direction in all three of these countries. Obviously, these social changes since the Industrial Revolution are not to be explained by fixed

[8] Cf. MacDougall, *The Group Mind.*

racial traits; and in general the spread of any social arrangements by imitation or learning—for example the use of rolling friction, metals, sewing machines, moving pictures, music, or even the same language—is independent of race.

(c) ARE SOCIAL FACTS SIMPLY PSYCHOLOGIC?

In recent years there has been a marked effort among economists and political scientists to give their work a basis in psychology, particularly of the behaviouristic variety. The uncritical acceptance by social scientists of the behaviouristic terminology with the consequent blurring of the distinction between physiological and purposive categories subjects their work to the criticisms already noted in our discussion of psychology. Psychology, we must remember, denotes three distinct enterprises: (1) the experimental science of psychophysics, or correlation between mental phenomena and their physical conditions; (2) analytic psychology, or description of mental phenomena in terms of their hypothetical elements, by introspective or speculative analytical methods; and (3) social psychology, which is often nothing but the description of diverse social phenomena in somewhat metaphorical and cloudy psychologic language. The first two of these studies clearly belong to the field of natural science, while only the third can appropriately be called a social science.

Social sciences such as economics, politics, and jurisprudence are not primarily concerned (as is psychology) with the individual's psychophysical responses nor with what will be revealed, by interpretation or analysis, as going on in the mind of an individual manufacturer, a political boss, or a judge listening to counsel. Their primary aim, rather, is to establish certain objective relations called economic, political, jural, etc. A description of the various systems of kinship, of the different ways of distributing land, paying rent, taxes, and so on, clearly belongs to social science, and not to psychology. In trying to understand the basis of these relations the contribution of psychology is doubtless of very great importance. But the establishment of economic facts is the affair not of the psychologist but of the economist. The psychologist can give us the psychologic explanation only if he first learns what the economic facts are. No one dreams of maintaining that human traits, habits, motives, hunger, fear, hope, and the like, can be eliminated from social science; but it is the economist, not the psychologist, who explains why cotton has to be shipped from Egypt to England to be sent back as cloth, or why it is cheaper for a merchant in Cincinnati to have his cloth made into garments in New York, rather than in his own city.

It is, of course, easy enough to define psychology as the science of all human and, therefore, of all social phenomena. But a unified science will not be produced by such arbitrary definition. The methods used to solve problems of cost-accounting or of legal procedure are not genuinely homogeneous with strictly psychologic methods used in determining the variations of hunger or why people generally start to run when others do. In the former problems we deal predominantly with abstract measurable relations, in the latter with physiologic processes or analysis of immediate personal feelings. Hence, if we wish to know the reason for the changes in women's clothes or why Tammany Hall wins so many elections, we are safer in consulting an expert "designer" or politician than a trained psychologist.

In general, the data from which psychology as a natural science proceeds are relatively simple present facts of direct observation and immediately personal reference. The data of the social sciences are more complex and refer to objective relations between different people.

(D) THE DISTINCTIVE SUBJECT-MATTER OF THE SOCIAL SCIENCES

The social sciences may be said to deal with the life of human beings in their group or associated life. But is there something that distinguishes this life from that of plant colonies or of social insects studied in biology or natural history?

Three answers have been offered: (1) that the social sciences deal with volitional conduct and judgments of value while the natural sciences deal with causal relations; (2) that the social sciences deal with concrete historical happenings while the natural sciences deal with abstract or repeatable aspects of natural events; and (3) that the social sciences deal with a peculiar subject-matter which I shall refer to as "tradition" or "culture."

(1) The Volitional or Teleologic Character of Social Facts

(i) *Individual Volition.* There can be no question that descriptions of social facts are largely in terms of purpose or final cause. It ought to be equally obvious that the causal relation cannot be eliminated from social considerations. There is obviously no antithesis between the two points of view if one remembers that when A is the cause of B, A is also a means of bringing about B as an end. The real issue is whether actual conscious purpose is always a direct and adequate explanation of social phenomena.

We must also draw a distinction between the microscopic view of human purpose, between the little drops of human volition, and the general social streams which result from them. Little does the respectable *paterfamilias* intend, when he begets and rears lusty children, to lay the basis of imperialistic wars or monastic institutions. The voyage of Columbus was undoubtedly one of the causes of the spread of English civilization to America. Yet Columbus no more intended to bring it about than the microscopic globigerinae could have planned the chalk cliffs of England which are the result of their life work.

That human volition by itself is, apart from favourable circumstances and mechanisms, inadequate to produce social results, is ancient wisdom. Yet after an event has happened we are prone to look upon the volition as the producing cause. A striking illustration of this fact is the way we explain inventions as due to the need of them. We are inclined to forget the great multitude of human needs that have gone unsatisfied through the ages. A visit to the cemetery of human hopes is needed for sobriety of judgment. Need or necessity may determine what invention will be generally developed; but inventiveness itself is the daughter of exuberant energy and favourable means. The history of science amply illustrates this.

In general, then, we must note that individual volition is not an adequate cause of large social changes, and that it does not suffice to distinguish the social sciences from the natural science of psychology.

(ii) *Social Teleology.* The exclusively teleological character of social science has been maintained not only from the psychological but also from the moral or jural point of view. Social phenomena are phenomena of conformity to regulation. All human conduct—our goings and comings, what we eat and what we wear, what we say and what we habitually do—is subject to social control. These actions are, therefore, judged not merely as illustrations of natural sequence in conformity with causal laws, but as conforming or failing to conform to social mores or norms. In general, we may say that the distinctively social point of view regards acts not as events in nature but as problems for us—how to choose the proper means, or to eliminate conflict in our aims. From this point of view the supreme unity of social science is to be sought not in the widest law of causal sequence but in such a conception of the ultimate social ends as will make possible a coherent science or system of judgments of human conduct.

The foregoing must be defended against the superficial positivism which on the basis of a misconception of the traditional logic restricts science to the study of existence and denies the possibility of normative science. We are thus faced with the insistence that economics must be restricted to the study of the causes of actual phenomena (Weber), that ethics as a science can deal only with the causes of certain actual practices (Lévy-Bruhl), and that jurisprudence can deal only with laws as customs which actually prevail (Rolin).

Such a programme is neither possible nor desirable. We cannot disregard all questions of what is socially desirable without missing the significance of many social facts; for since the relation of means to ends is a special form of that between parts and wholes, the contemplation of social ends enables us to see the relations of whole groups of facts to each other and to larger systems of which they are parts.

Those who boast that they are not, as social scientists, interested in what ought to be, generally assume (tacitly) that the hitherto prevailing order is the proper ideal of what ought to be.[4] This is seen in the writings of positivists like Comte, Gumplowicz, and Duguit, who heap scorn on the Utopists that are concerned with what ought to be. A theory of social values like a theory of metaphysics is none the better because it is held tacitly and is not, therefore, critically examined.

Lévy-Bruhl, who sees that normative considerations cannot be elimi-

---

[4] Even a geographical determinist like Ratzel can, after observing that "the whole life of the state has its roots in the soil," conclude: "A people should live on the land fate has given them; they should die there, submitting to the law."

nated from ethics, preserves the latter as a rational art. But is not all scientific reasoning a rational art?

If the prejudice against normative science were justified, it would render not only mathematics but all theoretic science impossible. For in developed sciences like physics, we are concerned with a theoretic development of the wider realm of possibility, and thus we must deal with ideal entities such as perfectly free, rigid, continuous, geometric bodies, with frictionless motion, etc. Only by such theoretic development can we fruitfully apply principles to actual sensible bodies, and see phenomena together in a new light. This is exactly what the study of social ideals does for the purely scientific understanding of social phenomena.[5]

Granted, however, that social institutions which have never existed are within the domain of the social sciences, we cannot suppose that concern with ideal entities suffices to distinguish the phenomena of the social sciences from those of other sciences. For an adequate account of the distinctive subject-matter of the social sciences we must take note of the element of tradition, of the ways whereby social conformity is brought about. To understand this we must take into account:

(2) THE HISTORICAL CHARACTER OF SOCIAL FACTS. It has been urged, notably by the followers of Windelband and Rickert, that while natural science deals with abstract aspects of phenomena that can be indefinitely repeated, social science deals with events which are unique. It is rather easy to refute this assertion by pointing to geology as both a historic and a natural science. Yet it would be vain to deny that the understanding of social phenomena requires a more extensive knowledge of the past than is generally the case in physical science. Questions of politics or economics, law or religion, generally refer to some particular historic state of human society.

Since social phenomena thus depend upon historic continuity, there can be no adequate knowledge of them without some reference to the past. It is, however, hasty and false to conclude that the full nature of social phenomena is to be found entirely in their history.

The extensive explanation of all sorts of queer social practices as

---

[5] If it be urged that the engineer or the applied physicist also views his field as a series of problems, the answer is that engineering involves problems precisely to the extent that it is concerned with human ends and therefore with social phenomena.

survivals of customs that originally had some use, is a form of inverted-or crypto-rationalism. There is nothing in history to indicate that our ancestors were more rational and more intent on doing things for definite purposes than we are. In any case the weakness of this type of historicism is that it tries to explain the present by the less-known past. It is largely motivated by dissatisfaction with the older rationalistic explanations of existing institutions, illustrated in Blackstone's fanciful "reasons" for the various laws of England. The study of actual history does undoubtedly show that the alleged reasons could not have been operative in originating these laws. Yet it is clearly fallacious to suppose that if we once learn from history how a given law or custom originated, we thereby fully explain its present existence. For obviously there are many old customs which no longer persist. If any law or custom, then, has survived, it must have something about it that has made it persist longer than other customs. The unusual liability of common carriers today may have originated in the fact that in Roman times the actual carrier was a slave, but the rule seems to have survived where others failed because of its serviceableness to modern needs. Good reasons as to social utility are, therefore, not eliminated by real or historical reasons as to origins.

There are, of course, social phenomena of long-time rhythm like the growth (or the decline) of the fundamental institutions of religion, language, social movements, etc., that are significantly contemporaneous only if the present includes a large chunk of what for other purposes is the past. But though history is thus a necessary condition for the extension of our knowledge, it will not enable us to dispense with the rational analysis of the present. Social science is this analysis or account of the abstract or logically repeatable aspects of social life.

We must conclude, therefore, that the fact that social material is less repeatable than that of natural science, creates greater difficulty in verifying social laws but it does not abrogate the common ideal of all science. Yet our discussion is not altogether negative. It leaves us with the conviction that in view of the historical character of social facts we ought to be especially on guard not to accept as law that which merely seems to hold true of present or past social phenomena. The process of distilling the essential law from social phenomena must be surrounded with many more cautions than the similar process applied to physical events.

(3) THE CULTURAL CONTEXT OF SOCIAL FACTS. If we keep in mind both the historic and the teleologic aspect of social life, we see an inter-action and a mutual dependence between the descriptive and the norma-tive, between the actual historic cause and the ideal demands of a given system. The subject-matter of social science, thus, differs from the subject-matter of the natural sciences, not only in introducing the prospective or teleologic point of view, which describes movements in terms of their goals, but in the more specific element of tradi-tion which sometimes takes the form of conscious teaching and learning. We may say that the distinctive subject-matter of the social sciences is culture in the sense defined by Tylor, viz. "the complex whole which includes knowledge, belief, art, morals, law, custom, and any other capabilities and habits *acquired by man as a member of society.*" I have stressed the last clause because that seems to be the sig-nificant clue to the distinction that we are seeking. The substance of culture, such as language, roads, tools, moral habits, and dispositions, are all modifications of the physical and the organic world. What makes them objects of the social sciences is that these modifications take place through social life and are handed on from one generation to another. It may be urged that there must have been a time when there was no human society and that therefore society cannot be an original factor and everything must ultimately be due to nature. It seems likely, how-ever, that man developed in some pre-human society, and in any case the absolute first moment of human society is not a phenomenon within the scope of social science. In the actual scope of the latter all the modi-fications of nature that we call culture take place under the influence of previous social life or culture. We thus have a distinct field or subject-matter for social science. The form of such science will depend upon the character of the connections between the various elements (of the phenomena of culture).

§ II. THE IDEAL GOAL (τέλος) OF THE SOCIAL AND THE NATURAL SCIENCES

So far as science means the rigorous weighing of all the evidence, including a full consideration of all possible theories (which is the true antidote for bias or prejudice), all sciences obviously have the same ideal. But in this sense the efforts of a critical historian like Thucydides are also scientific. A good deal of social and political science is thus scientific only to the extent that history is. When, however, we examine the nature of the general rules employed in evidence or explanation of particular facts, we notice that in the natural sciences these rules are the objects of explicit logical or mathematical development which makes it easier to verify them, while in the social sciences they are apt to be only implicit, and consequently unexamined and of uncertain truth-value. Consider, for example, the explanation of certain social phenomena as due to the gregarious or social instinct, or the explanation of other phenomena as anti-social manifestations of disruptive individualism. If these explanations have any point it is because we assume it to be a law, in the first case, that human beings fall in line and yield to the suggestion of others, and in the second case, that human beings are to some extent intractable, resent dictation, and so on. Obviously, both of these general propositions are to some extent true. But the question of how we shall measure the precise extent of these seemingly opposite laws of human nature has not yet received any treatment comparable in definiteness with that accorded to the laws of physics, chemistry, or physiology.

It is the aim of all natural science to rise above the historical stage and to become theoretic, that is, to attain the form of a theory or system in which all propositions are logically or mathematically connected by laws or principles. Loose words about science being practical, experimental, and inductive cannot permanently obscure this truth, made evident in the history of every branch of physics and biology. No science, for instance, can seem so hopelessly empirical and so immediately practical as chemistry, yet its whole growth through the Periodic and Moseley's Laws has been in the direction of a deductive system. Modern theories of heredity and variation are doing the same thing for biology. Clearly, some of the social sciences, for example, economics, aim at the same goal. Yet we can also survey the history of political science, as the scholarly and genial Professor Dunning did, and come to the melancholy conclusion that two thousand years of effort has brought us no farther than were the Greeks in the days of Aristotle. Indeed, it seems that professional pride rather than desire for scientific accuracy makes one deny the inferiority of the social to the natural sciences with regard to established and universally verifiable general laws. This is made manifest in the inferiority of our control over human nature when compared with our progress in the manipulation of physical nature.

The difference between the natural and the social sciences in this respect is not accidental and not readily removed by pious resolution. In the first place, the subject-matter of the social sciences is inherently more complicated in the sense that we have more variables to deal with than in physics or biology. In these sciences specimens are more easily obtained; we can experiment at will, varying the conditions one at a time, and thus more readily arrive at definite answers.

In the second place, there is the subjective difficulty of maintaining scientific detachment in the study of human affairs. Few human beings can calmly and with equal fairness consider both sides of a question such as socialism, free love, or birth-control. Opinions on these matters are not viewed with the ethical neutrality with which we view opinions as to the structure of protoplasm, the ether, the atom, etc. Emotional attachment to views which we habitually honour, and repugnance toward those views that good people are taught to despise, hinder free scientific inquiry. For the progress of science always depends upon our questioning the plausible, the respectably accepted, and the seemingly self-evident. This difficulty is present in a measure when social honour attaches to the holding of certain opinions on physical issues, e.g. the Ptolemaic astronomy, or the theory of special creation. With regard to physical questions, however, the fact that it requires elaborate training to follow them has taught the more intelligent part of the public some humility and the wisdom of suspending judgment. But how can one admit ignorance on a social question on which every one else has a confident opinion? One is tempted to say—to paraphrase a remark of Bertrand Russell—that the reason social scientists do not more often arrive at the truth is that they frequently do not want to. The desire to attain the truth is, after all, a late and relatively undeveloped human motive compared with the more vital and voluminous motives of social approval.

Because it is thus impossible to eliminate human bias in matters in which we are vitally interested, some sociologists (for example, the Deutsche Gesellschaft für Soziologie) have banished from their pro-gramme all questions of value and have sought to restrict themselves to the theory of social happenings. This effort to look upon human actions with the same ethical neutrality with which we view geometric figures is admirable. But the questions of human value are inescapable, and those who banish them at the front door admit them unavowedly and therefore uncritically at the back door. It is, then, better to aim directly at carrying the critical scientific spirit into the very study of moral values. Only critical reflection and a wider knowledge of the variety of human ideals can shake the naïve confidence in the absoluteness of our contemporary and local ideals. This does not mean that the variability of moral judgment disproves the possibility of a science of ethics any more than variability of vision disproves the possibility of a science of optics. If that which is deemed right at one time and place is deemed wrong at another time and place, it is because the judgment of right and wrong must include regard for the different circumstances. But the persistent confronting of the diversity of historic fact with critical judg-ments as to values enables us to overcome not only traditional absolut-ism, but also that narrow empiricism which vainly supposes that we can intelligently determine what is good in specific situations without regard for the ultimate issues involved. This is entirely parallel to the situation in the natural sciences where we must critically confront our experimental findings with general ideas in order to interpret the former and test the latter.

Social science can thus in the long run best attain its goal only when those who cultivate it care more for the scientific game itself and for the meticulous adherence to its rules of evidence than for any of the uses to which their discoveries can be put. This is not to deny that com-passion for human suffering and the desire to mitigate some of its horrors may actuate the social scientists. But the social reformer, like the physician, the engineer, and the scientific agriculturist, can improve the human lot only to the extent that he utilizes the labour of those who pursue science for its own sake regardless of its practical applications.

How, indeed, can we improve human affairs unless we know what actually *is*, and what is *better?* How can we attain certain knowledge except by a wholehearted respect for the rigorous rules of pure science? "He serves all who dares be true." But in the end we must remember that the knowledge of the truth, like the vision of beauty, is a good in itself.

To subordinate the pursuit of truth to practical considerations is to leave us helpless against bigoted partisans and fanatical propagandists who are more eager to make their policies prevail than to inquire whether or not they are right. The pursuit of pure science may not completely prevent our initial assumptions from being biased by practical vital preferences. But this is not to deny that the aloofness involved in the pursuit of pure science is the condition of that liberality which makes men civilized. If it be maintained, as it justly can be, that this ideal is an unattainable one, the only answer is that this is true also of the ideal of beauty, of holiness, and of everything else that is ultimately worth while and humanly ennobling.[6]

§ III. THE FORM OR METHOD OF RESEARCH IN THE SOCIAL AND THE NATURAL SCIENCES

THE great Poincaré once remarked that while physicists had a subject-matter, sociologists were engaged almost entirely in considering their methods. Allowing for the inevitable divergence between the sober facts and heightened Gallic wit, there is still in this remark a just rebuke (from one who had a right to deliver it) to those romantic souls who cherish the persistent illusion that by some new trick of method the social sciences can readily be put on a par with the physical sciences with regard to definiteness and universal demonstrability. The

[6] Actuated by the shallow activism of the "business man's" philosophy which conceives perpetual motion to be the blessed life, and by an illiberal or "Puritanic" contempt of pure play or enjoyment, American writers frequently try to ridicule pure science as an "indoor sport." But there is nothing ridiculous about a noble and liberal sport—certainly not as much as in the appeal to the queer compound of temporarily prevailing prejudices covered by the terms "red-blooded," "mechanistic," "democratic," and "Christian." (Cf. Wolfe's essay in Tugwell's *Trend of Economics*.)

maximum logical accuracy can be attained only by recognizing the exact degree of probability that our subject-matter will allow.

From the fact that social questions are inherently more complicated than those of physics or biology—since the social involves the latter but not vice versa—certain observations as to methodologic possibilities follow at once.

(A) THE COMPLEXITY AND VARIABILITY OF SOCIAL PHENOMENA

In the first place, agreement based on demonstration is less easy and actually less prevalent in the social than in the natural sciences, because the greater complexity of social facts makes it less easy to sharpen an issue to an isolable point and to settle it by direct observation of an indefinitely repeatable fact. The issue between the Copernican and the Ptolemaic astronomy in the days of Galileo was reduced to the question whether Venus does or does not show phases like the moon's, and this was settled by looking through a telescope. If Venus did not forever repeat her cycle, and if the difference between a full circle of light and one partly covered by a crescent shadow were not so readily perceived, the matter could not be so readily settled.

With the greater complexity of social facts are connected (1) their less repeatable character, (2) their less direct observability, (3) their greater variability and lesser uniformity, and (4) the greater difficulty of isolating one factor at a time. These phases are so dependent on one another that we shall not treat them separately.

The practical difficulties of repeating social facts for purposes of direct observation are too obvious to need detailed mention. What needs to be more often recognized is that social facts are essentially unrepeatable just to the extent that they are merely historical. The past fact cannot be directly observed. Its existence is established by reasoning upon assumed probabilities. In the case of physical history or geology our proof rests on definitely established and verified laws of natural science. In the case of human history the principles assumed are neither so definite nor so readily verifiable.[7]

---

[7] Thus it is difficult to refute the assertion that race differences are constant and not changed in the course of time by direct or selective influence of the environment. For if history fails to record any extensive intermarriage of a race like the Hebrew with other peoples and yet shows marked changes in physiognomy as members of the race

The greater variability of social facts may, if we wish, be viewed as another phase of their complexity. Any cubic centimetre of hydrogen will for most purposes of physics or biology be as good as another. But observation on one community will not generally be so applicable to another. Even purely biologic facts, e.g. the effects of diet, seem to be more variable in the human than in other species. Reasoning from examples in the social realm is intellectually a most hazardous venture. We seldom escape the fallacy of selection, of attributing to the whole class what is true only of our selected instances. To urge as some philosophers do that this is true only because physical knowledge is thinner and depends more upon the principle of indifference, is to urge an interpretation, not a denial, of the fact.

It is, of course, true that for certain social questions we can treat all individuals as alike. Thus, for vital statistics every birth or death counts the same, no matter who is involved. Likewise, in certain economic or juristic questions we ignore all individual differences. Yet there can be no doubt that the applicability of such rules in the social sciences is more limited and surrounded with greater difficulty than the application of the laws of the natural sciences to their wider material.

J. S. Mill in his *Logic* has raised the interesting question as to why it is that in certain inquiries one observation or experiment may be decisive while in other cases large numbers of observations bring no such certain results. In the main this difference holds between physical and social observation.

I venture to suggest a rather simple explanation of this fact—a fact that puzzled Mill because he did not fully grasp the logic of hypothesis. In any fairly uniform realm like that of physics, where we can vary one factor at a time, it is possible to have a crucial experiment, that is, it is possible to reduce an issue to a question of yes or no, so that the result refutes and eliminates one hypothesis and leaves the other in possession of the field. But where the number of possible causes is indefinitely large, and where we cannot always isolate a given factor, it is obviously difficult to eliminate an hypothesis; and the elimination of one hypothesis from a very large number does not produce the impression of progress in the establishment of a definite cause.

settle in different lands, it is still open to proponents of the theory to assert that they *must have* intermarried.

The last observation suggests that the greater complexity and variability of social fact also make its purely theoretical development more difficult. In general, social situations are networks in which one cannot change one factor without affecting a great many others. It is, therefore, difficult to determine the specific effects of any one factor. Moreover, social elements seldom admit of simple addition. The behaviour of the same individuals in a large group will not in general be the same as their behaviour in a smaller group. This makes it difficult to apply the mathematical methods which have proved so fruitful in the natural sciences. For these mathematical methods depend upon our ability to pass from a small number of instances to an indefinitely large number by the process of summation or integration.

Where the number of units is indefinitely large we can assume continuity in variation. But the application of continuous curves to very limited groups of figures to which our social observation is usually restricted produces pseudo-science, for example, the assertion that if our distribution is skewed we have a proof of teleology.

The relatively small number of observations that we generally have to deal with in the social sciences makes the application of the probability curve a source of grave errors. For all the mathematical theorems of probability refer only to infinite series (for which we substitute as a practical equivalent "the long run"). Where the number is small there is no assurance that we have eliminated the fallacy of selection. The mathematical error of applying a continuous curve to a discrete number of observations produces ludicrous results. It is vain to expect that the crudeness of our observation and the vagueness of our fundamental categories will be cured by manipulation of the paraphernalia of statistical methods. The mathematical theory of probability enables us to manipulate complex probabilities when we have some determinate ratio to begin with, such as that American pennies fall heads as often as tails. But social scientists seldom take the trouble to formulate the material assumptions of probability (or "indifference") which underlie their conclusions as to the probability of a given event, e.g. a rise in the price of gold. They thus endow their reasoning with the magical appearance of bringing forth material propositions out of the forms of pure mathematics. Actually the material assumptions of statistical workers are often purely aesthetic in origin, being dictated by an un-intelligent regard for smoothness and symmetry in graphs.

Physical categories have themselves been clarified by analysis. The dimensions of the different categories that we talk about—energy, action, force, momentum, etc.—are numerically determined in terms of m (mass), l (length), and t (time). In the social sciences the very categories that we use are hazy, subject to variable usage and to confusing suggestion. Does law determine the state, or the state make the law? How many thousands of learned men have discussed this and similar questions without fixing the precise meaning of the terms "state" and "law." [8]

It is a familiar observation that the difficulty of framing exact concepts in the social realm causes much confusion through ambiguity. To this it should be added that vague concepts make possible the constant appeal to vague propositions as self-evidently true. Open any book on social science at random and you will find the author trying to settle issues by appealing to what seems self-evident. Yet most of such self-evident propositions are vague, and when we ask for their precise meaning and for the evidence in their favour, our progress stops. In the natural sciences the questioning of what seems self-evident is relatively simple because when we have a simple proposition we can more readily formulate a true or an exclusive alternative. In social matters where difference of opinion is greater and demonstration more difficult, we cling all the more tenaciously to our primary assumption, so that our assumptions largely mould what we shall accept as facts.

Any one who naïvely believes that social facts come to us all finished and that our theories or assumptions must simply fit them, is bound to be shocked in a court of law or elsewhere to find how many facts persons honestly see because they expected them rather than because they objectively happen. That psychoanalysts, economists, sociologists, and moralists labour more or less in the same situation, the tremendous diversity of opinion among them amply indicates. Will a classical anthropologist admit that some Indians had a patriarchal form of kinship before adopt-

[8] Before Fourier definitely established the exact "dimensions" of the various physical categories, physicists could dispute (as the Cartesians and the Leibnizians did) as to the proper measure of "living" forces. Social science likewise needs a system of categories the exact dimensions of which are so clear as to make impossible the many confusions of which the example in the text is only one illustration.

ing the matriarchal type? Is it a *fact* that the suppression of certain desires, deliberately or as a result of imitation, necessarily produces pathologic states of mind? One has but to scrutinize the statement to see how much must be assumed before it can be shown that a fact is involved here.

Is corporal punishment in schools, or free divorce, an evil or not? Under the influence of general opinions one can readily maintain it as a fact that all the consequences of such practices are evil. But one who refuses to admit that these practices are evils can be equally consistent.

Is this true in the natural sciences? Certainly not to the same extent. Because theories do not to the same extent influence what we shall regard as physical or biologic fact, false theorems have never been such serious obstacles to the progress of natural science. The statements in popular histories that the Ptolemaic, the phlogiston, or the caloric hypothesis stopped the progress of science have no foundation. On the contrary these and other false theories in physics were useful in suggesting new lines of research. It is this fact that led Darwin to remark that false observations (on which others rely) are much more dangerous to the progress of science than false theories. Now in the social sciences we certainly do not have the elaborate safeguards against false observation that the natural sciences with their simpler material and many instruments of precision find it necessary to cultivate. The very circumstance that social facts are apt to be more familiar makes it easier to be misled as to the amount of accurate knowledge that we have about them.

From another point of view we may express this by saying that in the social sciences we are more at the mercy of our authorities with regard to what are the facts. The social worker or field anthropologist has less opportunity to preserve his specimens than the naturalist or the laboratory worker. If a later social worker or field anthropologist finds the fact to be different from what was reported by his predecessor, there is the possibility not only that they have observed different things but also that the social facts have changed.

In this connection it is well to note that the invention of a technical term often creates facts for social science. Certain individuals become *introverts* when the term is invented, just as many people begin to suffer from a disease the moment they read about it. Psychiatry is full of such technical terms; and if a criminal is rich enough he generally finds experts to qualify his state of mind with a sufficient number of technical terms to overawe those not used to scrutinizing authorities. The technical terms of natural science are useful precisely because they carry no aroma of approval or disapproval with them.

(B) ARE THERE ANY SOCIAL LAWS?

In view of the paucity of generally recognized laws in social science it is well to ask categorically if the search for them is fully justified. The existence of similarities in different societies at different times and places has been used as a proof of the existence of "a uniform law in the psychic and social development of mankind at all times and under all circumstances." [9] But *similarities* of customs and beliefs, even if they are not superficial or due to the prepossession of the observer, are not laws. As human beings resemble one another in their physical, biologic, and psychologic traits, we naturally expect that their social expressions will have points of resemblance, especially when the outer material is similar. If the number of human traits were known and within manageable compass, the principle of limited possibilities (enunciated by Dr. Goldenweiser) might be a clue to the laws of social life. But even a finite or limited number of facts may be too large for our manipulation.

Physical laws are in fact all expressed in relatively simple analytic functions containing a small number of variables. If the number of these variables should become very large, or the functions too complicated, physical laws would cease to be readily manipulated or applicable. The science of physics would then be practically impossible. If, then, social phenomena depend upon more factors than we can readily manipulate, even the doctrine of universal determinism will not guarantee an attainable expression of laws governing the specific phenomena of social life. Social phenomena, though determined, might not to a finite mind in limited time display any laws at all.

Let us take a concrete example. A man says to a woman, "My dear!" The physical stimulus is here a very definite set of sound waves, and we have reason to believe that the physical effect of these waves is always determinate. But what the lady will in all cases say or do in response

[9] Lester Ward, *Pure Sociology*, pp. 53, 54.

356

depends upon so many factors that only an astonishing complacency about our limited knowledge of human affairs would prompt a confident answer.

The a priori argument that there must be laws is based on the assumption that there are a finite number of elements or forms which must thus repeat themselves in an endless temporal series. But why may not the repeatable forms and elements be only those which enter our physical laws? What guarantee is there that in the limited time open to us there must be a complete repetition of social patterns as well?

In any case, those who think that social science has been as successful as physical science in discovering and establishing laws may be invited to compile a list of such laws and to compare the list in respect to number, definiteness, and universal demonstrability, with a collection such as Northrup's *Laws of Physical Science*.

We may approach this issue more positively by considering three types of laws:

(1) Every general fact that can be authenticated can be regarded as a law. Thus, that gold is yellow is the assertion of a law, i.e. whenever you find a substance having a certain atomic weight, etc., it will also be yellow in colour. We do not generally call statements of this sort laws, because a long list of such statements would hardly constitute what is distinctive of advanced *science*. In the latter such statements are connected with others by logical principles. Nevertheless such laws or facts are basic to science, and in the social realm they do not seem so numerous or so readily authenticated. Is it a fact, for instance, that the negro race is not ambitious?

(2) The second type of law is that of empirical or statistical sequences, e.g. "Much sugar in a diet will be followed by decayed teeth." In natural science we regard these also as but the starting points of the scientific search for the third type of law. Why do such sequences hold? We answer this question if we find some more general connecting link, e.g. some chemical process connecting sugar with the tissue of the teeth.

The situation in the social sciences is logically similar. Empirical sequences are not scientifically satisfactory laws. We are more apt, in the social realm, to find correlations that confirm our opinions and to neglect items that do not. If the graduates of a college get into *Who's Who* the college compiles such lists and claims credit. If they get into jail the college does not hold itself responsible.

(3) The third type of law is the statement of a universal abstract relation which can be connected systematically with other laws in the same field. Such laws we may indeed find in the social sciences, e.g. the laws governing the exchange of goods under free competition as worked out mathematically by Walras, Pareto, H. Schultz,[9a] et al., but always these laws are on a plane of abstraction from which translation to actual experience is difficult and dangerous. The ideal entities represented by "free competition" and "economic man" are, to be sure, no more abstract than the ideal entities of physics, but nevertheless economics is a poorer basis for predicting actual events than is physics. For in physics the transition from laws governing rigid bodies to those applying to (say) lead bullets is made on the basis of new laws of compressibility or elasticity, while in economics the transition from our laws of supply and demand in an ideal market to those determining (say) New York Stock Exchange transactions is still largely a matter of guess-work.

On a similar plane is the law of social inertia, parallel in motive to Newton's first law of motion. If, following the suggestion of the law of inertia, we assume that all social phenomena persist unless something is brought into play to change them, we have a useful principle of methodic procedure. Similarly we may assume the law of social heredity—all social institutions will be transmitted by parents to children, or people will believe and act as did their fathers before them except in so far as certain factors produce changes in our social arrangements and in our ideas and sentiments. Similar remarks may be made about the law of imitation or the law of differentiation in the division of labour, which certainly help to explain certain elementary aspects of most social phenomena.

The law of imitation has of course to take account of the impulse to do things differently in order to attract attention, and the law of social differentiation has to note the factors which make for uniformity, just as the biologic law of differentiation must be checked by noting the phenomena of convergent evolution. But until opposite tendencies or forces can be more precisely measured, these laws will not go very far to explain actual happenings.

[9a] See *The Meaning of Statistical Demand Curves*.

All laws are abstract. They state what would be true of a given factor if all other things remained indifferent. In physics other things do remain in a measure indifferent. But in the social field the variation of one factor produces all sorts of disturbances in others.

Physical phenomena, in other words, do show certain abstract uniformities of repetition which enable us to predict what will happen with greater certainty than in the social realms.

These observations are reinforced by considering the precise meaning of social causation and social forces—terms borrowed by the social from the natural sciences, generally without regard for their precise meaning or their applicability to the questions of social science.

(c) SOCIAL AND NATURAL CAUSATION

The notion of cause originates in the field of legal procedure. A cause ($\alpha i \tau i \alpha$) is a case or ground for an action. The Stoics, basing themselves on certain notions of Heraclitus, brought the notion of law into the conception of natural happenings. Law to them meant not mere uniformity that just happens to exist, but something decreed by the World-Reason or $\lambda \acute{o} \gamma o \varsigma$. Violations of it are possible but reprehensible. This is still the popular view, which speaks of certain acts as unnatural and of nature punishing all violations of her laws. The notion of a law of nature as a non-purposeful but absolute uniformity, so that a single exception would deny its validity, arises from the modern application of mathematics to physics. The proposition that all x's are y's is simply false if one x is not a y.

Modern physics seeks to attain such laws by the process of abstraction. Thus, the proposition that all bodies fall to the earth suggests itself as such a law. But if we remember the behaviour of smoke, of birds or balloons, some modification of this statement is necessary if universality is to be attained. This is achieved in the statement that all bodies attract one another. For in the case of bodies which do not fall we can show the presence of some force which counteracts the attraction of the earth so that the latter force is thus recognized. If the counteracting forces did not themselves operate according to a known law, the law of gravitation would be useless. We can predict phenomena only because the gravitational and the counteracting forces are independently measurable. Unless similarly social forces are measurable and there is some common unit or correlation of social forces, the whole notion of law as employed in the physical sciences may be unapplicable. When religious and economic interests pull individuals in different directions, which force will prevail? Such a question can certainly not be answered on any scientific basis. We do not know how many units of one social force will counteract another. All we can say is that in some cases religious motives prevail over economic ones, in some cases the reverse is true, and in most cases we cannot separate the motives at all.

The difference between social and natural causation is confused by the doctrine that social "forces" are psychic, and that at least one of them, desire, acts like a physical force—indeed, that it obeys the Newtonian laws of motion.[10] Obviously if social phenomena are not merely physical, the term "social force" can at best be only a metaphor and we should be careful to note its real difference from physical force.

This difference is ignored when a popular sociologist speaks of social motion as following the line of least resistance even more closely than does nature herself.[11] In natural science we know what a straight line is before considering any given physical process. But what is denoted by the metaphor, the "straight line" or the "line of least resistance" in any given social process, is something that we arbitrarily tell only after the event. Psychic forces are not physical forces.

If purely psychic forces operate at all, it is in the way of desire as an actually felt state of mind. But as we have seen before in our discussion of psychic causality, desire can be said to bring about results only if there happens to be also some adequate physical or physiologic mechanism at hand, so that the relation between the desire and what follows does not replace the physical causation but is an entirely different type of relation.

Similarly, social forces are not merely psychological. What is called social causation may be regarded as a teleologic relation. But the fact that in social relations we deal with large groups enables us to depart from individual psychology. We can thus say with greater certainty that an economic opportunity will be utilized, or that the religion of their fathers will be followed by a large group of persons, than that it

[10] Lester F. Ward, *Psychic Factors in Civilization*, pp. 94, 123.
[11] Ross, E. A., *Foundations of Sociology*, p. 43.

will be utilized or followed by a single individual whose specific disposition we do not know. We cannot tell what a given individual will think of the next war, but we can be fairly certain that every nation will, like Rome, manage to be convinced that its side is the just one.

Social causation, then, need not be like that of individual purposes. The overcrowding in cities does not intentionally bring about certain social diseases, any more than the invention of the cotton-gin was intended to bring about the economic changes which led to the fall of the older southern aristocracy and the political changes which led to the Civil War. It is of greater importance to recognize that social science is for the most part concerned not like physics with laws expressing the invariant repetition of elements, nor with laws of individual psychic events, but with laws about the relation of very complex patterns to one another.

Consider a number of examples of social causations. It is surely significant to inquire as to the effects of density of population. Is feminism a cause or an effect of the greater economic opportunity open to women? Is poverty the cause or the effect of a higher birth rate? In all such cases a causal relation means some connection not between individual events, or mere sums of such events, but between diverse patterns of distribution, sometimes of the same group of events. If social institutions as specific groups of events are themselves called events, we must distinguish the different levels of the term *events*. It will, however, prevent confusion if we remember that a social institution is a mode of viewing or grouping a number of events and is, therefore, strictly speaking, not a datable event, although the constituent events may occur between two dates.

Thus it is that in social causation the cause does not disappear when it produces an effect, but can be said to continue and to be modified by its effects. A system of education may affect the commerce of a people and that in turn may modify the system of education. That is possible because "system of education" is not a single temporal event but a pattern of events actually coeval with the pattern of events called "the commerce of a people." The causal relation or the interaction between them is predominantly a matter of logical analysis of groups of phenomena.

The purely scientific interest is thus best served by isolating some one aspect of social phenomena—e.g. the economic, the political, the religious—and tracing the effect of changes in that aspect. Even the historian must select and restrict himself to certain phases of social events. But the practical interest in social outcome is not immediately satisfied by uniform sequences nor by the merely necessary conditions of social happenings which are too numerous to be very interesting. It needs, rather, a knowledge of the quantitative adjustments of *all* the factors necessary to produce a desired effect. This is seldom attainable. We can under certain conditions tell, for example, that a reduction of price or certain forms of advertising will increase sales. But the variation of any one factor due to local conditions is very large, and our concrete practical knowledge always involves guesswork. Hence, it can never guarantee us against fatal errors.

(D) TENDENCIES AS LAWS

One of the most usual ways of generalizing from insufficient instances and ignoring or lightly disposing of contrary facts is to call our generalization a tendency. There is an apparent analogy to this situation in the popular formulation of the laws of physics, e.g. when the law of gravitation is stated as a tendency of two bodies to move toward each other. The word *tendency*, however, can always be eliminated from physics if we remember the law of composition of forces. As the force of gravity and the resistance of a table to a falling body can both be independently measured, there is no logical difficulty in saying that the law or force of gravity is operative even when the body is brought to rest on the table.

But in the social sciences, where single factors cannot be easily isolated and independently measured, an essential indeterminateness in discussion is inevitably produced by reliance on the notion of "tendency." For conflicting schools or parties can begin with the assertion of opposite tendencies and never really join in a definite issue. Thus one party may base its political theory on the assertion that all men love or tend to love liberty, and dismiss contrary facts as sacrifices which men make in the interest of peace, etc. The opposing party can base itself on the opposite assertion that men inherently tend to love order and to fear or hate the burden of responsibility so much that they prefer to obey even insane tyrants; and the facts which cannot be so described can be attributed to exceptionally unbearable conditions, etc. Obviously if the strength of these opposing tendencies were measurable and determined, the seemingly opposite assertions might be seen to be theoretically equivalent, that is to say, they would lead to the same predictions as to whether people will or will not obey under given conditions. Of course the emotional bias in favour of one set of words over the other will, so long as words have emotional associations, make people of different experience or temperament cling to different formulae. But the outstanding fact of methodologic importance is the indetermination involved in the use of the notion of tendency, and the vain and interminable disputes that it makes possible.

It is of course scientifically useful to resist the suggestion of proposed plausible generalization by discovering contrary "tendencies." Also the existence of opposite "tendencies" must be considered before we can proceed to measure them. But the temptation to set up tendencies as laws makes social science essentially indeterminate in the sense that diverse schools set up diverse principles all with the same show of truth. Thus when the Durkheim school of anthropology says all religion is totemic in origin, or that all magic is simply an illegitimate use of the supernatural, many phenomena are aptly described, but others are ignored. Rival schools start with these other facts and deny Durkheim's theory in toto. Both sides are right if they admit that they are describing some facts, and both are wrong if they pretend to describe all the facts of religion or magic. The way out of such typical sloughs of social science is to recognize that while full description of some of the facts may be needed in the beginning of social science, the ideal end is to attain universal statements about partial aspects of all the phenomena in a given class.[12]

---

[12] To call a correlation a tendency is to admit that we have not yet purified the concepts correlated of the foreign ingredients which block the regular manifestation of some causal relationship.

If what the grocer sells as sugar is in fact not uniformly sweet, we may say that grains of the stuff *have a tendency to be* sweet. It will require the careful work of a chemist to distinguish the $C_{12} H_{22} O_{11}$ that *is* sweet from the other substances that *are not* sweet. It is just this work of analysis and refining, a hundred times more difficult with social institutions than with sugar, that a complacent reliance upon "tendencies" as the substance of social science discourages. Like fictions, metaphors, and false theories, "tendencies" are valuable stimulants for scientific thought, but they will not take the place of food.

(E) TYPE ANALYSIS

Various philosophers, Dilthey, Spranger, Rickert, Troeltsch, and other opponents of the interpretation of history on the basis of natural laws, have urged that it is individual concepts like humanity, Christianity, the Renaissance, etc., that enable us to organize history and the cultural sciences. It does seem, at first blush, that concepts of this kind do enable us to build up science, in the sense of a significant synthesis of facts; and yet they seem to have nothing to do with laws of repetition but denote unique and essentially unrepeatable objects. But critical reflection will, I think, show that here as elsewhere there is no extension of definite systematic knowledge except in terms of abstract universals which deal with repeatable elements. Historical concepts like *the Renaissance* are convenient symbols to sum up a group of facts. But to understand the meaning of these facts and what determined their interconnection we must resort to assumptions of psychology, economics, politics, and other social sciences. If you ask why is one who has caught the spirit of the Renaissance able to appreciate facts which others would not notice or understand, the answer is not found in the magical potency of the individual concept, but in the fact that familiarity with any field of human experience develops many definite habits of expectation, i.e. implicit assumptions as to laws which bind phenomena together. This is also true in other fields, e.g. physiography and field biology. Such knowledge becomes more and more reliable to the extent that these implicit assumptions are made explicit and critically tested.

The theory of type analysis does, however, call our attention to a significant phase of social inquiry. To the extent that social science is descriptive or follows the method of natural history, it must begin with actual wholes or complexes rather than with clearly defined elements. If we begin some social study, e.g. of panics, slavery, totemism, or the different moral codes in different classes of a community, we necessarily begin with vague terms such as "forced labour," "the average man," "the prosperous classes," etc. These are not finished geometric diagrams, but hints of meaning pointing in a general direction. It is the task of the inquiry itself to make these vague concepts more definite by applying them to the concrete material. The greater complexity, however, of social facts makes it more difficult to apply our concepts directly to any given social fact or situation. We therefore take many facts or situations and make a sort of composite photograph in which individual variations fade out. We let memory select the significant features. There can be little doubt that this ability to see the type in phenomena is characteristic of social insight. This is particularly true in historical and anthropological studies, which lose all scientific value when they sink into anecdotal or merely factual description.

However, in social science as in natural sciences like mineralogy, chemistry, and biology, the study of types, if it leads anywhere, only makes explicit fundamental laws.

A type is an ideal configuration of distinguishable but not always separable features. It has a scientific value in that it enables us to look for and identify certain characters after we have recognized others that co-exist with them in a given type. (When the type includes rhythms or patterns of successive events we obtain laws of succession of events analogous to those of natural science.)

The advantage of operating with types is that while laws deal with highly abstract elements which become more and more insufficient as our phenomena become more complicated, types are more definite combinations or patterns which become familiar to us through continual reference. They are, in Santayana's phrase, concretions in discourse and thought. The procedure which begins with a study of types thus follows the order of learning in daily life, where we do not begin with sensations or other hypothetical elements but with complexes involving both sense-perception and conceptual thought, which are progressively refined in the growth of knowledge.

It is well also in view of the complexity of social inquiry always to keep in mind the picture of the totality denoted by the type. For in isolating abstract elements we are apt to forget how they function in the actual totality from which they are abstracted. This is illustrated in legal theories of free contract that ignore the fact that in actual life the labourer is not free to bargain as to terms, or in economic theories which ignore the fact that the desire to buy in the cheapest market may not operate in actual instances owing to sentiment, prejudice. unintelligence, etc.

In general we may say that it is not science which initiates the general belief in universal laws. Science finds such beliefs and tries to test and refine them. In practical life we all seek to justify our beliefs or acts by very wide general propositions or premises. It is only when the *result* of such premises appears distasteful that people resist their plausibility, and then the very contrary propositions are likely to be asserted, though the latter also lead to difficulties. The intellectually timid who refuse to give justifying reasons may frequently escape the embarrassment of inconsistency and acquire a reputation for sobriety of judgment. But if rational scientific knowledge has greater potency in preparing us to meet a changing world (not immediately but in the long run), it is precisely because it depends upon our isolating or abstracting the relevant and recurrent elements and studying their temporal succession. An intelligent use of type analysis therefore depends upon this very ability of neglecting in the phenomena before us all that is irrelevant and non-typical. The essence of scientific genius (whether in the natural or the social sciences) is just this ability to discover points of view from which new arrangements of facts are visible and under which order and system can be introduced into what has hitherto appeared as hopeless chaos. The weakness of the ordinary account of induction is that it minimizes this inventive genius. It talks of the facts suggesting their explanation as if all one has to do is to look at the facts, whereupon the true explanation is automatically precipitated from their sensible traits. A more adequate view of the learning process, which, by allowing for its a priori element, can explain how this scientific insight is both possible and rare, is held today by the Gestalt psychologists. We see the world not in terms of pure sensa—yellowness, bigness, roundness, etc.—but in terms of organized configurations, e.g. bananas, baseball games, beautiful women, and bunches of grapes. How such configurations are organized depends not only upon the objective character of our environment but as well upon the observer's intellectual and emotional make-up. Such a view can account for the fact that Wordsworth, looking at a daffodil, sees more than Peter Bell, and that Roentgen, looking at a spoiled camera plate, sees something that no other observer ever saw.

On this analysis, the fundamental unity of common sense, social science, and natural science is apparent. But it will not do to dismiss the differences in certainty, accuracy, universality, and coherency that actually distinguish these fields as "mere matters of degree." The word "mere" is itself not only a term of degree but the one quantitative word in the English language that has a superlative without a comparative (i.e. without any ordering relation which could give its degrees a meaning). Thus it is a dangerous invitation to slipshod thinking. Not only are differences of degree as important as any other differences, but the similarity of different things which leads the superficial to renounce further thought is the necessary basis for significant distinctions. The differences in approximation to the ideals of science between common opinion and the professional thought of men like Boas, Pound, and Maitland, or between such thought and that of Einstein, Mendeléyev, and Loeb, demand our appreciation and reflection. What we have called type analysis, involving as it does a certain lack of explicitness in its assumptions, of definiteness in its terms, of coherency and accuracy in its conclusions, may not be the ultimate ideal of social science, but it is all we can expect in the foreseeable future. Though it is proper to distinguish this level of thought from the plane attained in the natural sciences we should not deny it the adjective *scientific* unless we are ready to eliminate such names as Galen, Leeuwenhoek, Pasteur, and Darwin from the rosters of Science.

§ IV. THE FACULTY OR TYPE OF MENTALITY INVOLVED

THE foregoing considerations suggest the element of truth in the Aristotelian view that while physical science depends on theoretical reason ($\nu o \tilde{\upsilon} \varsigma$), practical social science involves more sound judgment ($\phi\rho\acute{o}\nu\eta\sigma\iota\varsigma$). Sound judgment means ability to guess (or intuit) what is relevant and decisive, and to make a rapid estimate of the sum of a large number of factors that have not been accurately determined. In practice the statesman, the business man, and even the physician may often find the suggestive remark of a novelist like Balzac of greater help than long chapters from the most scientific psychology, since the latter deals with elements, whereas in conduct we deal with whole situations. This frequently gives rise to a philistine anti-rationalism. What is the use of speculating about the ultimate good? Why not, rather, use our intelligence to increase the sum of justice and happiness in actual cases? But *can* decision be intelligent if inquiries as to the ultimate meaning of justice and happiness are prohibited? How will the restricted use of intelligence in that case be different from the uncritical acceptance of traditional judgments as to what is good and what is bad in specific cases?

The efforts of the human intellect may be viewed as a tension between two poles—one to do justice to the fullness of the concrete case before us, the other to grasp an underlying abstract universal principle that controls much more than the one case before us.

None of our works shows these forces in perfectly stable equilibrium. The problems of engineering, medication, administration, and statesmanship generally depend more upon not overlooking any of the relevant factors. But in pure science as in personal religion and poetry intense concentration on one phase rather than justice to many is the dominant trait. To the extent that the social sciences aim at the adjustment of human difficulties, they involve more judgment and circumspection. To the extent that they aim at insight or θεωρία, they are at one with pure science and with religion and poetry.

Chapter Two

HISTORY VERSUS VALUE[1]

Two principles are generally relied on as axiomatic in the popular philosophy of the day, viz. (1) that nothing is explicable except in terms of its history, and (2) that the value of anything is independent of its history. In the popular mind these two principles dwell side by side in millennial peace. But the mission of philosophy is to bring a sword as well as peace. It must not only reconcile conflicting considerations, but it must also find oppositions where none would otherwise be suspected. Nor is it difficult to see in the two principles before us, in spite of intermixture of blood, the representatives of the two warring houses of empiricism and rationalism.

Unbounded faith in the omnipotence of the historical method is typical of nineteenth-century thought, just as faith in naïve rationalism is held to be typical of eighteenth-century thought. Such distinct representatives of popular thought as Carlyle and St. Beuve express it in aphorisms such as: "History is not only the fittest study, but the only study. It is the true epic poem and the universal divine scripture"—and "History, that general taste and aptitude of our age, falls heir in effect to all the other branches of human culture." Though this wave of historicism was in its inception closely related to the romantic movement, and was supported by the authority of Schelling and partly by that of Hegel, it became in its conscious maturity predominantly positivistic,[2] realistic, and empirical, intimately related to the apotheosis of induction and distrust of reason which came in vogue after the first third of the century. In reaction to the boldly a priori Hegelian method of writing history and the vagaries of the *Naturphilosophie* of Schelling's followers, there followed a general ideophobia which reached its height

---

[1] Read before the American Philosophical Association, December, 1913, and here reprinted, with slight modifications, from the *Journal of Philosophy*, Vol. XI (1914), p. 701.

[2] Cf. Comte, *Cours de Philosophie Positive* (Martineau's trans.), VI, esp. Chs. 1-6.

when papers by Helmholtz and Clausius on the correlation of forces were rejected by orthodox physical journals as too metaphysical.

The empirical and realistic temper of this historical movement shows itself clearly in the writings of its acknowledged leader, Ranke, according to whom the essence of historical method is not in passing judgment, but in gathering all the available data and finding the facts or events as they really happened.[3] "First of all we must understand the world, and then desire the good."[4] To attain this understanding we must distrust the employment of abstract principles, for "the spirit which manifests itself in the world is not of so conceptual a nature" (*begriffsmässig*).[5] The characteristic ontology of this mode of thought finds expression in Ranke's repeated assertion that "the genus appears only in the species."[6]

The triumphs of this point of view have led to the belief that all questions can be settled by appeal to history. Hence more or less lengthy historical introductions to all sorts of axiologic discussions are quite the fashion. As a rule it will be found that the historical introduction is very much like the chaplain's prayer which opens a legislative session or political convention: very little of the subsequent proceedings are decided by reference to it. But there have not been wanting brave souls who have taken the historical faith quite seriously and have actually attempted to make the historical point of view replace or supersede all independent valuation.

We might, in passing, note the significant fact that the more developed a science is the less use it makes of history. Thus history has no applications in mathematical investigations, and next to none in physical researches. With the recent growth of experimental and scientific methods in biology and the realization of the inadequacy of the supposed law of parallelism between ontogeny and phylogeny, the historical point of view has been losing the importance it once had in the study of life

---

[3] See the preface to his first book, *Geschichte der romanischen und germanischen Völker* (1824), and the appendix to the same, *Zur Kritik neuerer Geschichtsschreiber.* Cf. his *Sämmtliche Werke*, Vol. XXXIV, p. vii. For Ranke's attack on Hegel see his *Weltgeschichte* (*Sämmtliche Werke*, Vol. IX, Pt. II, p. xi).

[4] *Weltgeschichte*, Vol. IX, p. 236.

[5] Ibid., p. xi.

[6] For useful collections of Ranke's theoretical views, see Nabaldian, *Rankes Bildungsjahre und Geschichtsauffassung*, and O. Lorenz, *Ranke, die Generationslehre und die Geschichtsunterricht.*

phenomena. However, we shall here concern ourselves only with the values which are the study of the *Geisteswissenschaften.*

§ I. THE RELEVANCE OF HISTORY TO SOCIAL VALUES

(A) ECONOMICS

IN THE economic field the question whether economic history can, of itself, give us a theory of value has been the object of a long controversy lasting over sixty years. In the middle of the nineteenth century, when the classical school of economics seemed to be suffering from intellectual anemia, the historical school arose first in Germany and then in England, in protest against the whole abstract or deductive procedure in this essentially human field. I am acquainted with no argument for humanism in philosophy which was not in effect applied by the historical school to the field of economics. Thus it was pointed out that the economic man is a mere abstraction, having no exact counterpart in time and space, that actual men are not selfish calculating machines, that economic action is a response to a total situation in which diverse uneconomic factors enter, and that economic systems are not static, but constantly changing, etc.[7] In the light of those considerations, it was boldly asserted by men like Knies and Hildebrand, Cliff Leslie, Ingram, and others that the abstract deductive method of valuation must be abandoned, and that only by historical methods can we get at the essence of these phenomena.[8] We need not here examine the counter-attack by the Austrian school, though Menger's *Untersuchungen über die Methode der Sozialwissenschaften* deserves to be better known among philosophers as a keen analysis of scientific method, and particularly noteworthy for its demonstration of the indispensable character of abstraction or isolation and deduction in all scientific procedure. For

[7] Comte, *Cours de Phil. Positive*, VI, Chs. 2-3. List, *Das nationale System der polit. Ökonomie* (2d ed.), pp. i, li. Roscher, *Grundriss* (1843), preface; also *Grundlagen* (1877), I, pp. 26, 31 ff. Hildebrand, *Nationalökon. der Gegenwart*, p. 209; also *Jahrb. für Nationalökonomie, etc.* (1863), pp. 5 ff., 137 ff. Knies, *Polit. Ökonomie vom geschicht. Standpuncte*, III, § 3, p. 237.

[8] List, op. cit., pp. lix-lx, p. 17; Knies, op. cit., p. 35; Cliff Leslie, *Essays in Political and Moral Philosophy*, p. 189; Ingram, *History of Political Economy*, pp. 237 ff.; Ashley, *English Economic History and Theory*, preface.

our present purpose it is sufficient to point out that when the leaders of the historical school came to such topics as the nature and function of capital, money, and credit, they invariably resorted to deductive or mathematical methods.[9] This was not, let it be noted, a mere personal failing or relapse into old or accustomed habits, but, as their followers now admit, a retreat from an untenable position.

Schmoller, the acknowledged leader of the newer historical school in economics, who was at first inclined to subordinate economic science to economic history,[10] now admits the indispensable character of the deductive method.—Nay, he goes so far as to admit that history has done less to extend the theory of economics than have its practical applications.[11] Similarly, the most creative mind of the newer school, Karl Bücher, whose investigations have opened up new fields in the relation of economics to psychology and anthropology, says, "The only method of investigation which will enable us to approach the complex causes of commercial phenomena is that of abstract isolation and logical deduction. The sole inductive process that can likewise be considered, namely, the statistical, is not sufficiently exact and penetrating." [12]

There is thus today an acknowledged consensus among economists that the attempt to make history supersede abstract or deductive theory of value has hopelessly failed.[13] Nor is the reason for this failure far to seek. The historical school was misled by the crude inductive theory of science according to which a collection of facts can of itself establish a theory. As even a chronologically ordered series of facts cannot of itself

[9] Knies, *Geld und Credit*, Pt. II, Ch. XII, § 2.

[10] *Zur Litteraturgeschichte der Staats- und Sozialwissenschaften*, p. 279.

[11] *Grundriss, etc.*, I, pp. 122 ff., and § 14 of his important article in Conrad's *Wörterbuch* (3d ed.), Vol. VIII, p. 458. Cf. Toynbee, *Ricardo and the Old Political Economy*, p. 10.

[12] *Industrial Evolution* (tr. Wickett), p. 148.

[13] It may be pointed out that even the more moderate hope of the newer historical school, that a vast collection of monograph studies in economic history will supply a wider basis for economic theory, has proved vain; and the foremost students of economic history admit that their work is only of secondary or indirect help to the student of economics. See Conrad's address in Vol. II of *International Congress of Arts and Sciences*, p. 211. Veblen, reviewing Schmoller's *Grundriss*, says: "There seems no reason to regard this failure [of the historical school] as less than definitive." *Quarterly Journal of Economics*, Vol. XVI, p. 7. Cf. A. Voigt, in *Zeitschrift für Sozialwissenschaft*, N. F., 3 (1912), pp. 241, 311, 383. Below, *Vierteljahrschrift für Sozial- und Wirtschaftsgeschichte*, 5 (1907), pp. 482 ff. Hasbach, *Archiv für Sozialwissenschaft*, 24 (1907), p. 29. Tönnies, *Archiv für systematische Philosophie*, I (1895), pp. 227 ff. Pierson, *Principles of Economics*, I, pp. 33-36.

establish causal relations,[14] economic history cannot be written except by one already trained in economics, just as geologic history cannot be written except on the basis of established physical and biologic theories. By extending the sphere of known facts, economic history, doubtless, supplies us with an improved check or control over our economic theories. We are thus able to say that certain generalizations based on present-day conditions are not of absolute validity. But in itself history does not suffice, either for the settlement of controversial questions of economic policy, or for the establishment of an adequate scientific theory of value.[15]

(B) JURISPRUDENCE

The contrast between eighteenth-century rationalism and nineteenth-century historicism was first and most sharply drawn in the field of jurisprudence. In opposition to all eighteenth-century attempts to change actual legal institutions in accordance with the rights of man (deduced from rational principles), the historical school of jurisprudence founded by Eichhorn and Savigny maintained the supreme or exclusive importance of historical study. Law, Savigny maintained, is always the expression of a deterministic development of a national spirit (*Volksgeist*). Hence history is not merely a collection of examples, but "the only way to attain a true knowledge of our own condition." [16] Hence, also, all legislation, like the Napoleonic code, not based on a complete knowledge of the history of law can be only worse than useless.[17]

A close examination shows that the pillars of this faith are four characteristic dogmas, viz. determinism, organicism, evolutionism, and

[14] For the possibility of causal laws in this connection, see Marshall, *Principles of Economics*, I, Ch. IV, § 3, and Wagner, *Grundlegung d. polit. Ökonomie*, I, § 83. As to the possibility or impossibility of "laws" in history, see Xenopol, *La Théorie de l'Histoire*, Ch. IX, and K. Menger, *Untersuchungen*, pp. 146 ff. Cf. *Bulletin de la Société française de Phil.*, July, 1906, and July, 1907.

[15] Gide et Rist, *Histoire des Doctrines Économiques*, IV, Ch. I; Lifschitz, *Die historische Schule der Wirtschaftswissenschaft*, pp. 140-198, 254-288; Max Weber, *Archiv für Sozialwissenschaft*, Vol. XIX (1904), pp. 33 ff., and *Schmollers Jahrbuch für Gesetzgebung, etc.*, 1905, pp. 1324 ff. (reprinted in *Gesammelte Aufsätze zur Wissenschaftslehre*).

[16] See Savigny's introduction to his *Zeitschrift für geschicht. Rechtswissenschaft*, Vol. I (1815), p. 4.

[17] Savigny, *On the Vocation of Our Age for Legislation and Jurisprudence*, §§ 6 ff.

relativism. (1) Since the past completely determines the present, "the idea that each generation can make its legal world for good or ill according to its power and insight is the essence of the unhistorical view." [18] (2) Law is not a separate affair, but is, like language, the expression of the organic national spirit. Hence there can be no free borrowing or adaptation of the law of one people by another nation. (3) As each national spirit develops, it must pass through certain stages, and (4) what is created in one stage cannot be adapted to another. Hence legal institutions must be studied, not with reference to general or abstract principles, but with reference to the particular time and place under which they arose and functioned.

Much has been and is still to be said about these doctrines; but it is certain that though the historical school has been in the ascendency for nearly one hundred years, it has never succeeded in harmonizing them so as to present a consistent doctrine. If a determinism in which conscious human effort plays no part is taken seriously, how can we attach any practical importance to the *historical knowledge* of jurisprudence? If we accept the doctrine of organic connection of all social institutions in the national spirit, how can we explain the fact that peoples *have* successfully borrowed each other's laws? For our present purpose it is, perhaps, sufficient to point out that history itself does not bear out this faith in the exclusive importance of the historical approach to jurisprudence. No one can dispute that under the influence of eighteenth-century theories of natural rights, the constitutional law, the criminal law,[19] and a good deal of the civil law of the world was radically transformed and improved. The Napoleonic code, framed by men who, as Savigny clearly showed, were grossly deficient in legal history, has successfully spread and has become the basis of the law of most of the European countries, various African communities, all of Latin America, Quebec, and Louisiana, and has exercised influence even on the German Civil Code,[20] while all the labours of the historical school, excellent though they be in point of thorough historical scholarship, have little to show that is at all comparable. The crude, unhistorical rationalism of Bentham stirred into life reformative forces in all branches of the

[18] Savigny, *Zeitschrift, etc.*, p. 3.
[19] F. von Liszt, *Das deutsche Strafrecht*, I, § 7.
[20] E.g. in the doctrine of possession, § 932.

common law, but the Anglo-American historical school (founded by Maine) has not a single reform or constructive piece of legislation of any magnitude to its credit.[21] Indeed, the historical school has been a positive hindrance to any improvement or enlargement of the law—precisely because those who think of new problems exclusively in terms of historical analogies get tangled up in their own traces and think that what has been must remain forever.[22]

How can history help us to evaluate the laws of today or proposed changes? How, for instance, are we to be guided in determining proposed penal legislation? The answer of the historical school is: that is sound which is in harmony with the general European or American tendency as revealed by history. But this test taken seriously either bars all real changes, or else leads nowhere in particular. All real changes must be contrary to what has hitherto prevailed.[23] A historical study of the Roman law, or of our common law, may reveal to us exactly what Roman jurisconsults or English judges said and meant. But unless we are to suppose these worthies were endowed with omniscience, how could they have foreseen and solved all the perplexing and complicated problems which modern life presents? The actual efforts of the historical school to govern modern conditions with ancient texts has resulted, as Jhering and Pound have pointed out, in a series of pious juggling of irrelevant texts and old decisions made with reference to bygone conditions, or, more frequently, in an ultra-rationalistic shuffling of concepts—*Begriffsjurisprudenz*—which is none the better because it is unconsciously metaphysical.[24]

The historical school has thus not succeeded in eliminating the abstract methods of evaluation of the old natural law. By setting up the system [25] of the Roman or the common law as the embodiment of

---

[21] Jurists sometimes draw a sharp distinction between law and legislation. But no one can really understand law apart from law in the making.

[22] It is one thing to understand how the complicated rules of evidence grew up, but quite another to answer the question whether the whole body of such rules might not advantageously be wiped out today.

[23] Somlò, *Archiv für Rechts- und Wirtschafts-philosophie*, Vol. III, pp. 510 ff.; Kantorowicz, *Monatschrift für Kriminalpsychologie*, Vol. IV, pp. 79, 92 ff.

[24] Jhering, *Scherz und Ernst*, especially essay entitled "Im juristischen Begriffshimmel"; Pound, *Harvard Law Review*, "The Scope and Purpose of Sociological Jurisprudence," Vol. XXV (1911), pp. 598-604; Ehrlich, *Grundlegung der Soziologie des Rechts*, pp. 295 ff.

[25] The "system" of the Roman law is a rationalistic construction due to Donellus in the sixteenth century.

absolute principles valid for all times, it has simply substituted a conservative natural law for the old revolutionary or reformative one, presupposing the values of conservation instead of the values of creation or change.[26]

(c) POLITICS

The claims of history as the only basis of enlightened politics have been put forth so vigorously by historians like Freeman and Droysen,[27] that it has almost become an accepted commonplace. But the fact that history has always readily supplied weapons to all parties, democratic, monarchic, etc., has induced most modern historians to discard the hope of organizing the lessons of history into a systematic science of politics and to content themselves with aiming simply at discovering the truth as to past events.[28]

In practice, also, the knowledge of history is of comparatively little direct use to the statesman.[29] It is only a hopelessly amateurish spirit that would guide the policy of the United States by parallels drawn from the history of the Roman republic, just as the doctrinaire leaders of the Russian Revolution of 1905 expected Russia to go through exactly the same stages as France did after 1789.

Sometimes, indeed, we find a question of policy, like the veto power of our courts over legislation, argued almost entirely on the basis of

[26] As to the failure of the historical school to avoid a "natural law" of its own, see Bergbohm, *Jurisprudenz und Rechtsphilosophie*, pp. 280 ff.; Stammler, *Die Methode der geschichtlichen Rechtsschule*, pp. 4 ff.; Stammler, *Lehre vom richtigen Recht*, pp. 118, 135-136.

That the historical school has not really succeeded in refuting all standpoints of "natural law," can be seen in the present revival of "natural law" theories among jurists of the most diverse schools. See Ch. II of Cosentini's *La Réforme de la Législation Civile*; Jung, *Das Problem des natürlichen Rechts*; Saleilles, in *Revue Trimestrielle du Droit Civil*, Vol. I (1902), pp. 80-98, and Charmont, *La Renaissance du Droit Naturel*.

[27] Freeman: "Historical study does more than anything else to lead the mind to definite political creed." *History of Federal Government* (ed. Bury), pp. xiv-xv. Droysen: "Especially is historical study the basis for political improvement and culture." *Principles of History* (tr. Andrews), p. 56.

[28] Fustel de Coulanges, *Questions Historiques*, p. 11. F. York Powell, in Langlois and Seignobos, *Introduction to the Study of History*, p. xi. "Science has no other object than the truth, and the truth for its own sake, without regard to the consequences, good or bad, regrettable or fortunate, which that truth may have in practice." Gaston Paris, quoted by Masci in the *Rendiconti della Reale Accademia dei Lincei*, Vol. XXII (1913), p. 376.

[29] Morley, *Notes on Politics and History*, p. 103.

history, viz. as to what were the actual intentions of the Fathers of 1789. But this argument, it need hardly be pointed out, derives all its force from the political maxim that it is well to do only what our fathers intended. It is doubtless a theoretical gain when the study of history destroys the naïve illusion that we can always wipe out all the institutions of the past and start out anew on a rational basis. To suppose, however, that what has been must always remain is equally vain. What is needed, and what history alone cannot supply, is a quantitative social science which will deal not with absolute flexibilities or immobilities, but will enable us to compute the strength of social inertia and that of the forces available for change.

(D) ETHICS

In the attempt to make history the basis of ethics, we may distinguish two stages, the theologic and the biologic.

The attempt to derive theologico-ethical values from history begins with Augustine and his disciple Orosius, and continues to the middle of the nineteenth century in such works as Baron Bunsen's *God in History*. Without doing injustice to the powerful intellect of St. Augustine, we may safely say that the attempt to make the facts of history prove the truth or validity of Christian ethics is convincing only to those who are determined to be convinced beforehand. To the devout Christian it may be difficult to see in the decline of the Roman Empire anything but the effect of the moral corruption of the ancient world, but the anti-Christian makes out just as strong a case for the contention that the decline of Rome was due to the introduction of Christianity and the spread of monasticism.[30]

Though the theologic colouring has now definitely disappeared from our histories and it is no longer in good form to use such phrases as, "And thus we see the hand (or finger) of God in history," the essence of the method persists with the slight change that biologic terms have replaced theologic ones. Instead of the City of God or "the far-off divine event to which all creation moves," we have the goal of progress, and instead of Providence we have the struggle for existence. But whether we use the old terms or the new, history remains a branch of

[30] Cf. E. L. White, *Why Rome Fell.*

apologetics, an attempt to justify the powers that have been victorious. The essence of the matter is the dictum of Schiller, *Die Weltgeschichte ist das Weltgericht*, i.e. the belief that if the facts of history are allowed to tell their own tale, they will, like the poetic justice of the old-fashioned drama, always show the suicidal character of injustice and the ultimate triumph of the worthier types of civilization. This belief seems to me to rest on a peculiar dullness to the pathetic and tragic elements in history, such, for instance, as the crushing of several types of civilization in western Asia and eastern Europe by the brutal power of Genghis Khan, the loss of Bohemian independence. There is something inexpressibly brutal in the dogma of necessary universal progress,[31] which is simply the old dogma that this is the best of all possible worlds in a temporal form, to wit, that every change in the world is a change for the better. Like other forms of brutality, this glorification of the historically actual is due to a lack of sympathy or imagination which prevents us from seeing all the finer possibilities, hopes and aspirations, at the expense of which the triumph of the actual is frequently purchased. The doctrine that right always triumphs is but an insidious form of the immoral doctrine that what triumphs (i.e. might) is always right.

In terms of cold logic, my point is that all attempts to derive ethical values from history really presuppose or assume the very values to be derived. Suppose history capable of showing that certain courses of conduct lead to national extinction. That of itself cannot give us an ethical rule except on the assumption that national existence should always be desired. As a matter of fact, a great deal of the seeming success of evolutionary or biologic-historical ethics in suggesting solutions of moral problems is due to the unconscious assumption which underlies all these attempts, that mere life (i.e. biologic duration), or else the type of life which is the mode today, is the highest or most valuable end. It is sometimes said that history, the story of human success and failure, is the great laboratory of the ethics student. But unless we are in possession of some standard as to what we should con-

[31] I speak here only of the idea of *necessary* progress. As for the claim that the facts of history show that on the whole humanity *has* made actual progress or improvement, I can only say that our knowledge of the past is too fragmentary and our social sciences not sufficiently advanced in quantitative determination, to enable us to add the diverse gains and losses with any degree of justifiable confidence. Our control over nature has, doubtless, increased, but that the value of life has thereby been always enhanced is extremely doubtful.

sider success and failure, the experiments in our laboratory can have no meaning to us.

We may conclude, then, that nothing has yet been advanced that refutes the argument of Sidgwick, that the history of ethical opinion or practice cannot be the decisive factor in determining its validity.[32]

(E) RELIGION [33]

In religion, the historical method has frequently been regarded as primary by orthodox and heterodox. Take such controversies as that over the historical existence of Jesus of Nazareth, or the question of a historical succession of the Bishops of Rome from St. Peter to Leo the Great. They seem to have far-reaching religious bearings, but can any one pretend that the religious values of these questions would have arisen in the mind of an impartial student, for example, a Buddhist? As a matter of fact, the dogmas in question arose prior to all questions of historical research.

Religious liberals frequently claim that history is fighting their battle; and, doubtless, so far as orthodox religious teachings assume certain historical dogmas, modern historical research puts difficulties in their way. The history of the Old and New Testament, based on the methods of the higher criticism, certainly removed extraneous artificial difficulties in the way of accepting their religious teachings. But can any one maintain that the higher criticism tends to make converts for Judaism or Christianity?

Problems of religious value cannot be determined exclusively by history because the latter is dependent on psychologic and metaphysical consideration in determining what is held valuable in religion.[34]

[32] *Methods of Ethics*, III, Ch. I, § 4. *Philosophy, its Scope and Relations* (especially the lectures entitled "Philosophy and Sociology").

[33] The genetic fallacy in the supposition that the history of art can supply the answer to the questions of aesthetic valuation or critical appreciation, seems to me so clear that I shall pass over it. I may refer to K. Lange's *Das Wesen der Kunst*, pp. 13 ff., and Babbitt's *Masters of Modern French Criticism*.

[34] Troeltsch, "Religionsphilosophie," in *Festschrift für Kuno Fischer*, p. 142.

(F) PHILOSOPHY

Finally, we come to the history of philosophy. Since Hegel's attempt to present the history of philosophy as a rational system, the belief in the philosophic value of the history of philosophy has never lacked adherents. Indeed, in Germany there has been a marked tendency to sink all philosophy into its history, and in many of its systematic treatises the systematic part is a sort of appendix to the historical portion. Without denying to the history of philosophy a high value as a part of the history of culture or civilization, we may flatly deny that the truth of philosophic doctrines is dependent on their chronologic order.[35] Is it any argument for or against the truth of their teaching that Epicurus came after Plato, or that Sextus Empiricus came after Aristotle? I fail to see an argument in Epicurus that Plato has not met, or attempted to meet. Philosophic doctrines, in truth, have no necessarily continuous existence, and it can easily be shown that few of the great philosophers were acquainted with the writings of all their important predecessors. The attempt to present the history of philosophy as an independent continuous stream following out an inner necessary dialectic has many aesthetic charms, so that it will always be attempted, but it has no claim as genuine history or significant philosophy. Whatever philosophy be, it is not merely a branch of archaeology.

§ II. HISTORICAL METHOD

IN SPEAKING of history up to this point, we have been assuming that there is such a thing as historical truth which is to be found in accordance with definite methods of historical search—that, for instance,

[35] Those who believe that the history of philosophic doctrines can determine the question of their truth have seldom faced courageously the problem as to the nature of truth to which their position leads. If every philosophic system is an advance on its predecessors, we seem to be driven to the dilemma that either the historian's own point of view is the absolute truth and all previous systems but partial embodiments of it, or else that all views (including that of the historian) are true only relatively to their time or epoch. Hegel alone seems to have had the courage to accept the first alternative and view his own philosophy as the final revelation of the Absolute, so that henceforth no more history of philosophy would logically be necessary. The difficulties of the second horn of the dilemma are the familiar ones inherent in all theories which assume the relativity of knowledge.

the existence of certain laws, economic practices, or ritual observances in the past is to be determined by definite evidence, and the fact that they are revolting or shocking is irrelevant to the consideration of their historical existence.

Such an ideal of history, however, is, as a matter of fact, difficult to maintain, for history is a fine art (a branch of imaginative literature) as well as a science. The actual data of history consists of contemporary facts in the form of remains and documents. Historical science consists in criticizing this material, that is, in applying the laws of probability to it. The result of this process is to fix a number of points through which the historical curve is to be made to pass. The invention of such curves must be the result of creative imagination closely akin to the dramatic imagination. The historical material, as it issues from the fire of scientific criticism, never of itself presents a complete picture. It either offers too little (as, for instance, in the early history of Russia), or it offers too much (e.g. modern Prussia). In the former case the historian has to supplement the facts before him with hypothetical ones—in which process he is obviously dependent on his general philosophy of life or schema of relative values; and in the second case, he must select from the great mass of facts those which he considers most important, which again involves a process of valuation—since importance is distinctly a category of value. Hence we can understand the fact that no great historian has actually succeeded in making the objective or the scientific motive eliminate altogether the tendency to edification; and all historians, consciously or unconsciously, make their histories preach the gospel of the particular party or epoch to which the historian belongs.

Now, there are two attitudes which may be taken to this personal or subjective element in history. We may try, as the scientific school of historians is doing, to eliminate or minimize it by definite rules, or we may glorify it as a principle, as Droysen, Treitschke, and patriotic historians generally have actually done.[36] The attempt to do both frequently passes today as the evolutionary or genetic method in the social sciences. At this point I may barely indicate somewhat dogmatically

---

[36] "That bloodless objectivity which does not say on which side is the narrator's heart is the exact opposite of the true historical sense." Treitschke, quoted in Gooch, *History and Historians*, p. 150. The requirement in public schools that history should be taught so as to foster patriotism leads as a matter of fact to the subordination of history to national apologetics.

that the genetic or evolutionary method in the social sciences represents an unstable mixture of incompatible elements of rationalism and empiricism.[37] Popular Hegelian dialectics, fortified by the analogy of biologic principles that are fast being discarded by those engaged in actual biologic work, gives a general formula of progress with distinct stages through which all social institutions must necessarily pass. Thus the family must everywhere have passed through the stages of promiscuity, group marriage, matriarchal and patriarchal clan, etc. Industry must everywhere have passed through the hunting, nomadic, agricultural stages, etc. All this is bewitchingly simple, but the student who has been brought up on the mathematical and natural sciences finds in the mass of desiccated anthropological anecdotes that fill our treatises on social evolution nothing that can be called scientific evidence for the actual or necessary existence of these stages. Social evolution through necessary stages is a mythology, not as picturesque as the old theologic mythologies, but equally effective in quenching the thirst for genuine knowledge with the Lethean waters of the aesthetic imagination.

Two concessions to historicism appear from the above survey: (a) that it has certain pedagogic value, and (b) that it may effectively negate values set up by absolutistic and unhistorical systems. Both concessions, however, ought to have a *caveat* attached to them.

(A) PEDAGOGIC VALUE

That the historical or genetic method has alluring pedagogic value in such fields as economics, ethics, or philosophy cannot be denied. Instead of analyzing a subject and dealing with its abstract elements (which always requires intellectual concentration), we clothe them with historical existence and present them as the necessary stages in a temporal process. The gain thereby is so great as frequently to justify some loss of accuracy and distortion of facts; and we say that the letter killeth, but the spirit reviveth. A certain amount of conventionalization seems absolutely necessary in all teaching; but the danger of the genetic

---

[37] For criticism of evolutionism in the social sciences see Stammler, *Wirtschaft und Recht*, pp. 662 ff.; Vierkandt, in *Zeitschrift für Philosophie und Philosophische Kritik*, Vol. CXXVII, pp. 168 ff.; Diehl, in *Jahrbücher für Nationalökonomie*, Vol. LXXXIII, pp. 823 ff.; Tönnies, in *Archiv für Sozialwissenschaft*, Vol. XIX, pp. 88 ff.; Sidgwick, *Philosophy, its Scope and Relations*, Lectures 6-9.

method, like that of legal fiction, is that the teachers themselves may grow to believe it true. Thus the myth about the stages of industry, hunting, nomadic, agricultural, etc., was taken so seriously that an attempt was made to build a system of education on it (I refer to the culture epoch theory). Again, it is its pedagogic attractiveness that causes so many people to believe the baseless dogma that all history is an evolution from the simple to the complex. The slightest familiarity with the facts in the history of language or law shows that if any absolute generalization must be made, it should rather be that we are dealing here with growths in the direction of simplification.

(B) HISTORY AND ETHICAL ABSOLUTISM

Historical arguments frequently seem most effective against absolutistic theories of value. Thus if it is claimed that an aristocracy alone can give us good government, it seems relevant to point out the egregious selfishness and inefficiency of the English, Polish, and Venetian aristocracies. But the adherent of aristocracy is not thereby silenced, since the follies of former aristocracies may be ascribed to any one of the numerous circumstances under which the aristocracies of former days functioned, but which no longer exist. There is doubtless a strong probability that any one who, through history, becomes acquainted with beliefs and practices other than his own, will no longer affirm with such unquestioned assurance that his own beliefs and practices are the only ones possible, or even the best possible for every one at all times. In this respect history, like human geography, widens the social and intellectual horizon. But no one seriously questions the value of *history as a genuine method of extending the span of our experience.* Doubtless, also, the wider experience will enable one, as a rule, to judge more wisely in questions of value. But to expect that on any controversial question of today the teachings of history can be decisive, seems vain. The contention of some leading teachers of history, such as Professors Robinson and Seignobos,[88] that history favours the values of change or reform by curing people of the morbid dread of change, cannot be accepted as universal. History seems impartial and readily supplies aid

[88] Robinson, *The New History*, pp. 252 ff.; Langlois and Seignobos, *Introduction to the Study of History*, pp. 320-21.

and comfort to both sides. Any one with sufficient enthusiasm for half-truths (which characterizes most controversy in this field) can readily give the appearance of finality to the contention that rationalism is revolutionary, and historicism, with its tendency to glorify the actual, is the refuge of the conservative. In politics ardent reformers or revolutionists are almost always firm believers in principles, while the conservative is always drawing lessons from history that these things have never been and are not, therefore, practicable. In jurisprudence, both German and American historical schools find their reason for existence in their opposition to the revolutionary codifiers. Professor Robinson cites the use of history by the socialists, but this instance is rather instructive the other way. Socialism as a concrete human movement reflects, of course, the mixed and complicated motives which characterize actual human conduct. Political Marxism, however, is a fixed philosophy capable of definite analysis. Now the Marxian programme was, in his mind and in that of Engels and of all their orthodox followers, sharply opposed to the older revolutionary socialism of St. Simon, Fourier, etc.[39] Marx and Engels laugh at these rationalists who would establish all things on principles of reason and thus rest the world on its head. They insist, over and over again, that the past cannot be wiped out and that only through history can we see the future. The consequence of this was that under the influence of the Marxian political programme socialism ceased to be really revolutionary. As a political doctrine it no longer asks its adherents to do anything to bring about the social revolution, but only to keep the faith and wait for the catastrophic day of judgment —a political quietism like the Lutheran or Calvinistic distrust of good works, and an abounding faith in the omnipotence of the economic deity. Indeed, Marxism became dominant in the socialistic movement only after the failure of the revolution of 1848 and of the Paris Commune, when events showed the inferiority (in open conflict) of disorganized though enthusiastic revolutionists to a disciplined soldiery. In countries like Spain and Russia where, for various reasons, the revolutionary embers continued to smoulder, Marxism never received the same ascendency which it did in the more peaceful countries. Revisionism and syndicalism today indicate that socialists are beginning to be

---

[39] See especially Engels, *The Development of Socialism from Utopia to Science.*

dissatisfied with a religious peace purchased at the price of practical political disenfranchisement.[40]

The foregoing survey has perhaps indicated enough to suggest the conclusion that historicism, like its sister materialism, while professing empiricism, is really the offspring of vicious rationalism. Both are obsessed with the dogma that only the factual can have true being. The attempt to banish real possibilities from the world results in the common dogma of determinism. But if everything which is today is completely determined by its past, there can be really nothing new today. And if there be nothing new today, neither was there anything new yesterday, and history is lost in rational mechanics.

The denial, on the part of historicism, that there may be any order of values independent of historical sequences, is ultimately based on the nominalistic dogma that only particular entities in time and space are real. But values, like mathematical relations, may involve characteristics independent of the time order. This independence does not, of course, deny their intimate union in our common life, but it warns us against straining the principle of parsimony by trying to sew the vesture of the universe out of a single piece of cloth, or trying to weave that cloth without having the threads cross each other. Value and historical existence are independent of each other in the same sense that the two blades which form a pair of scissors are independent of each other. Both are necessary and intimately connected, but neither can absorb or, by a process of sublimation (*Aufhebung*), transcend the other.

[40] I leave this paragraph as it was written in 1913 when the prestige of the German party was at its height in the socialist world, and the leading Marxists in France, Russia, Austria, Belgium, Italy, and the United States, men like Jaurès, Plekhanoff, Victor Adler, Van der Velde, Labriola, and Hillquit, followed Kautsky's interpretation of the Marxian programme. Since then the leaders of the majority of the German socialists like Scheidemann have consistently discountenanced revolution, while the revolutionary turn to Marxism in Russia and in the Third International has been brought about not only by practical exigencies, but also by the entry into the Communist Party of revolutionists who formerly looked to men like Blanqui and Bakunin. Nor is this strange if we remember that apart from his theoretic writings Marx was a revolutionist in 1848 and later co-operated with Bakunin, Blanqui, and other violent revolutionists in organizing the International Working Men's Association.

Chapter Three

COMMUNAL GHOSTS IN POLITICAL THEORY[1]

§ I. THE REALITY OF GROUPS

A CERTAIN awe for the word *social* is one of the outstanding phenomena of current intellectual life. The triumphant elation and solace with which the social nature of man is announced and individualism denounced seems to presuppose the belief that previous generations were not aware of the fact that men live together. But long before the word *social* received its present vogue men reflected profoundly on the nature of family, economic, political, and religious association. Plato's *Republic* and Aristotle's *Politics* bear testimony as to the vitality not only of their own but also of previous Greek thought in this field. But, though Plato draws a significant analogy between the individual and the body politic, he does not speak of a communal mind distinct from the minds of the individual philosophers. Nor is Aristotle responsible for the famous dictum, "Man is a social animal." He asserted, indeed, that man is a political animal, but he expressly maintained that man's highest achievements are those rare moments of real insight which are also moments of divine isolation. Nor will any one acquainted with the long history of Hebrew and Christian thought as to the nature of Church and State and the relation of the individual soul to God, be inclined to view the current glib contrast between the social and the individual as the first and final revelation of the truth in the matter. The recent rise of the term *social psychology* may have lent some colour to a general impression that now at last we have discovered a real social mind distinct from the individual minds of men and women. But surely no scientific psychologist who studies the behaviour of men in groups makes any such claim.[2]

[1] The following chapter, except for minor changes, constitutes §§ 2-5 of "Communal Ghosts and Other Perils in Social Philosophy," *Journal of Philosophy*, Vol. XVI (1919), p. 673.

[2] Wundt is sometimes referred to as an exponent of this view (Gierke, *Wesen der*

The doctrine of a real communal soul in the form of a Folk Ghost [3] (*Volksgeist*) seems first to have received prominence in the romantic reaction against the French Revolution and the doctrines of the Enlightenment as to the rights and powers of reasonable man. Against the doctrine that we can make laws on the basis of reason or a priori principles, Savigny and his disciples urged that the laws of any community are and should be the historic product of the national ghost of its people. But while Savigny and his Romanist disciples attributed a real ghost only to the State, the Germanist Beseler and his disciple Gierke extended it to other associations—though not, be it noted, to all business associations. Gierke's theory has been introduced into Anglo-American thought mainly by the brilliant work of Maitland and Figgis and is now represented by Mr. Laski.[4]

It would take us far afield to attempt here an adequate account of the enormous literature that has grown up around the question as to whether the legal personality of associations denotes something real or fictional.[5] As the controversy has for the most part been carried on by jurists and historians and not by philosophers it is full of arguments as to the practical consequences of different theories, but naturally rather deficient in clear analysis of the philosophical principles involved. We may, indeed, eliminate most of the legal considerations by observing that legal personality is quite distinct from natural personality. There are natural persons who for some reason or other do not possess legal personality at all, e.g. slaves. That does not mean that the law denies the fact that these natural persons have organs, dimensions, feelings, etc. To paraphrase the words of a famous beadle, if the law did that it would indeed be an ass. On the contrary most legal systems that allow slavery

---

menschlichen Verbände, p. 11) but he in fact maintains that no actual *Gesamtgeist* exists apart from and independent of individual minds—*System der Philosophie* (1889), pp. 592 ff. Durkheim and his disciples, also, while insisting on the tremendous importance of group life in the constitution of the individual, still maintain that society exists only in and through individual minds. *Elementary Forms of Religious Life*, pp. 17, 221, 346.

[3] I am aware of the fact that *spirit* rather than *ghost* is the usual translation of *Geist*. But I think the notion of a substantial spirit which is also a person is best represented by the word *ghost*.

[4] Maitland, *Introduction to Gierke's Medieval Political Theories*; also *Collected Papers*, Vol. III. Figgis, *Churches in the Modern State*. Laski, "The Personality of Associations," in *Harvard Law Review*, Vol. XXIX (1916), p. 404.

[5] See Saleilles, *Personalité Juridique*, p. 1; also Enneccerus, *Lehrbuch d. bürgerliches Recht*, § 96.

recognize the natural personality of the slaves and may even protect it to some extent by diverse rules and regulations, while denying them legal personality or the right to sue in their own names. Perhaps the distinction between legal and natural personality may be seen even more clearly when we observe that some natural persons like infants and women are legal persons for certain purposes and not so for other purposes, while legal personality may be bestowed to certain funds (the fisc) and foundations to which no one has yet attributed real personality. Whether, therefore, certain groups should be regarded as legal persons is a practical question as to whether they should be made collectively the subject of certain rights and duties, and whether their liability should be limited to the extent of the corporate or collective funds. But the fact that our legal system draws a sharp distinction between the property of the corporation and that of the individual members or owners of it, does not determine the question of the real personality of the corporation, any more than the fact that certain proceedings are brought against the ship and not its owners determines the question as to whether a ship is a person.

Let us then examine the question as to the personality of groups as a question of fact. When we take a unified nation like France or an established church like the Roman Catholic, or a society like the Jesuit Order, there seems a clear prima facie case for saying that not only are there Frenchmen, etc., but over and above these there is the spirit or ghost of France, of the Roman Church, or of the Society of Jesus, which endures while individual men come and go. Omitting the supernatural claims of the Catholic Church and viewing the matter from the naturalistic point of view it seems quite clear that this contention for real group personality may be regarded as either true or false according to the meaning we attach to the word *personality*. If we mean to assert that every group has distinctive group marks and that there is something uniting the different individuals so that they act differently from what they would if they were not so interdependent, no one can well deny such reality, whether you call it personality or give it any other name. But if it is asserted that the French nation and the Roman Church literally have all the characteristics of those we ordinarily call persons—that the state is masculine and the church feminine, according to Bluntschli—we are dealing with the kind of statement which is believed because it is absurd. Groups are not begot through the union of father and mother, they do not suck their mother's milk, do not play children's games, do not spend weary hours in school, do not work for wages, strike for shorter hours, and do not suffer the trials and joys of anxious parenthood. Having no sense organs, they cannot in any strict sense of the word be said to have sensations or feelings, and it is not literally true to say that they feel praise or blame, hope or disappointment, love, hunger, colds, toothaches, ennui, the creaking of old age, or the perplexities of a world that to the honest mind must always contain unsolved and perhaps insoluble problems.

The defenders of the real personality of groups, like Gierke and Laski, distinguish, of course, between the personality of groups and the personality of natural persons. The two kinds of personality, they admit, are different and are called by the same name only because there are real analogies between them. By stretching the term *personality* beyond what it ordinarily denotes, they really change its meaning or connotation, precisely as the mathematician has stretched the term *number* by applying it to surds or "real numbers" which are not, in the original sense of the term, numbers at all. This tempts us to conclude that the quarrel between those who believe in the reality of corporate personality and those who believe it is fictional is a quarrel over words. For the most distinguished adherent of the fiction theory, Jhering, has pointed out [6] that this use of the language of identity for two different things that are in some way analogous is precisely what constitutes the nature of fiction. But though it is true that a good deal of the controversy would be eliminated if each side defined accurately the meaning it attached to the term *personality*, it would be a mistake to conclude that the issue is merely verbal and of no real significance. In the first place no question of this sort can be *merely* verbal, because words are most potent influences in determining thought as well as action. Theoretically we may be free to decide to use a word like *personality* in any sense we choose, but practically we must recognize that intellectual resolutions cannot rob words of their old flavour or of the penumbra of meanings which they carry along with them in ordinary intercourse. The attempt therefore to use old popular words in new senses is always productive of intellectual confusion. Thus when we personalize a group we are

[6] *Geist d. römisches Recht,* § 68.

apt to forget that "its" action may be simply the action of certain individuals in authority—the others, though they may be also responsible, being in fact passive or even ignorant of what has taken place. This confusion seems to me to show itself in Mr. Laski's contention that a corporation (as a mind distinct from that of its officers or members) can have the feeling of gratitude (or perhaps even the capacity to eat dinners).[7]

Apart, however, from the practical question of stretching words to include unusual meaning and thus confusing our intellectual currency, there is between the adherents of corporate personality and their opponents a fundamental philosophic issue: the extent to which the principle of unity should be hypostatized or reified (I wish the use of the word *thingified* were more common, since that which it denotes, the tendency to think of relations and operations as *things*, is one of the most common sources of philosophic error). All are agreed that groups are characterized by some kind of unity, and the fundamental issue is whether this unity shall be viewed as an entity additional to the entities unified and of the same kind, or whether it shall be viewed for what it is, as just the unifying relation. The tendency to personify groups, ships, storms, debates, and everything else is as old as human thought and is in some measure unavoidable. For we must always depend on analogies, and personal analogies give our language a vividness without which our hearers may be entirely unmoved. But modern mathematical logic has taught us to avoid the old form of the issue between nominalism and (the older) realism by recognizing the relational character of unity. When any one oracularly informs us that the whole is more than the sum of its parts, we reply that that depends upon the meaning of the word *sum*. Of the things that can in any definable sense be added the whole *is* just the *sum* of its parts and nothing else. There are, however, at least three recognizable types of unity. There is the physical or synthetic unity of a house or ship in which the constituting parts which existed before the whole are still recognizable. There are chemical unions in which the pre-existing parts lose their identity in the whole, but may be restored to their original state. Lastly, we have the organism or biologic unity, which we cannot freely create out of pre-existing parts nor break up into parts such that the whole can be reconstituted. Now

[7] *Harvard Law Review*, Vol. XXIX (1916), p. 483.

diverse human associations are characterized by all these types of unity in diverse ways. To the extent that our membership in certain racial, religious, national, or language groups, is not a voluntary act, these groups have something of organic unity. But to the extent that increasing civilization increases the freedom of associations, men can and do choose their language, country, religion and the intimate associations that give social importance to race. The most intimate union in human life is that of husband and wife. By that union the character of the constituent parts is profoundly modified, but they maintain their separate identities. The union may be dissolved and in certain legal respects the parties may return to the position in which they were before forming their union, though in other respects they can never be the same and possibly can never reconstitute the same happy family. Gierke, Figgis, and other protagonists of corporate personality are, however, too much in reaction against social contract theories to think highly of voluntary and possibly dissolvable unions. They think more highly of states and churches into which individuals are born, and in which they necessarily inhere as qualities inhere in a substance. The state or the church is the permanent reality of which individuals are the phenomenal appearances. Gierke, who has become a sort of patron saint of political pluralists, goes to the greatest extremes in this hypostatizing of the principle of unity.[8] But the history of philosophy from Aristotle to Bradley has fully shown the vicious infinite regress which follows when our substance becomes an additional quality, or when our unifying reality becomes an additional thing. When two persons are united in the marriage relation the unity is not in itself an additional person, though such unity makes possible many things which could not otherwise happen.

The reaction against social contract theories has led to absurd denial of the voluntary element which plays a part in all associations, even in that of the state. History, United States history especially, shows many examples of voluntary formations of states; and recent events show that such unions may also break up and new ones be reconstituted. The unity of France or of the Catholic Church rests in the mode of thought and action which millions of Frenchmen and Catholics habitually follow. If by an impossible event they should all simultaneously lose all memory and habitual manner of responding, the French nation and the

<hr>

[8] See his *Genossenschaftsrecht*, Vol. III.

Catholic Church would cease to exist. Every group involves some definite mode of interaction between its members. The more permanent the grouping the more permanent are these modes of action. When we become conscious of these ancient modes we call them traditions. But these traditions, though embodied in many material things, books, works of art, clothes, buildings, machines, etc., cannot maintain their significant character apart from a continuous current of individual minds.

Professor Dicey [9] seems to have put his finger on the chief difficulty which, in the absence of the relational formula for which I have been contending, meets those who ask: What more does a corporation involve than individual members? He says: "Whenever men act in concert for a common purpose, they tend to create a body which, from no fiction of law but from the very nature of things, differs from the individuals of whom it is composed." But when two oxen are yoked together they not merely tend to but do create a body, to wit, a team, which "from no fiction of the law but from the very nature of things differs from the individuals of whom it is composed," for a team of oxen can really do things which two oxen separately cannot. But that does not prove that a new ox is thereby created. Similarly when Jones and his two brothers form the Equitable Button Co., Incorporated, they do not create an additional soul or mind. If the Equitable Button Co. prospers we speak of "its" reputation, "its" assets, liabilities, etc. But that does not mean that there is "the red blood of living personality" in the corporation apart from the human individuals who are its owners. The same is true when people unite to form a debating club, a dining club, a church, a railway company, a bank, or an incorporated town.

§ II. GROUP RESPONSIBILITY

THE question of fact as to corporate personality is independent of the legal or ethical question of corporate responsibility. But as the discussion of personality is frequently confused by consideration as to responsibility we must consider the latter topic also.

If the impecunious agent of a corporation does a wrong, justice may demand that the stockholders on whose behalf it was done or who gen-

[9] *Law and Public Opinion, etc.*, p. 165.

erally profit by such acts, should be compelled to pay for the wrong out of corporate funds. This is in line with the general principle of making the master liable for the torts of the servant; but it does not prove that the corporation is a real mind separate from the minds of the individual officers and stockholders. But the question of corporate responsibility becomes more complicated and in itself more significant when we come to the responsibility of nations or states.

Who, for instance, is rightly responsible for the damage done to Belgium by Germany? Not the Kaiser alone, nor his immediate advisers, nor the members of the Reichstag who voted supplies, nor even all the citizens who supported the war. Germany as a whole is held responsible and that means that those who opposed the war as well as generations of Germans yet unborn must be made to pay. This certainly does not agree with the prevailing theory that no one should be punished except for some fault of his own. But most people believe both in individual and in collective responsibility—certainly German publicists are in no position to question the latter, since at the time of the Serbian invasion they justified the cruel sufferings imposed on innocent individual Serbians on the ground that the Serbian people must atone for the crime of the Karageorgevich dynasty.

In the presence of the obvious conflict between the principle of individual responsibility and that of collective responsibility, the philosopher is tempted to decide for one or the other of these principles. But humanity continues to profess both and to disregard both whenever necessary. Thus many tens of thousands of people are killed every year by what are called accidents in our mines, railways, factories, etc., and no one feels responsible. Most of these accidents could certainly have been prevented if people were willing to pay the cost of such prevention. If I tell my neighbour that the coal he uses is soaked with the blood of miners and brakemen killed in the mines and in the transportation service, he may see the truth of my contention, but he would resent my statement that by using coal he is participating in these killings and that the blood of these men is upon his head. In any case he will go on using coal; and in this respect I think the children of the world are wiser than (some of) the children of (reflective) light. For more harm may result by giving up the use of coal, railways, and factory products than now results from their use. King David refused to drink the water brought to him by his heroes from the well of Bethlehem at the price of blood. But many of us live in cities where the entire water supply is tainted with the blood of the toilers killed in building the tunnels and aqueducts. Does any morality require us to refrain from drinking it? Are not the portals of our houses sprinkled with the blood of our sons who bled to death that we may be safe? We call it a sacrifice on our part when we remember the ties which bound the dead to us. But when we ignore the ties which bind members of a community together, we are quite certain that we have no right to order people to be killed in order to prolong *our* lives.

These reflections suggest that in the face of the complicated situation before us we cannot unqualifiedly accept either the principle of individual or of collective responsibility, nor absolutely deny either. In our ethics the principle of individual responsibility, that each man shall be rewarded or punished according to his own deed, has been unquestioned. But in practice it is often disregarded, because inapplicable. It is impossible to isolate, in a complicated system of interaction between countless individuals, past and present, the part of the result due to any individual deed. The principle of individual responsibility postulates a world in which each individual can be the sole producer of definite results, a world where each individual can be the sole master of his acts and fate. This, I submit in all seriousness, is not the world in which we find ourselves. We find ourselves in a world where, not to speak of our involuntary physical heredity and early training, we are all in different measures benefited or harmed by the acts of others, and where no man can act or be punished without affecting untold others in diverse ways.

But while the principle of individual responsibility has remarkably little to commend it as a primary principle, it is none the less useful as a secondary one. In a world where individual fears, hopes, and ambitions are real sources of action, general carefulness and increased productivity can certainly not be promoted by disregarding entirely these individual emotions. Some rationalized system of individual rewards and punishments is, therefore, necessary to weight the natural consequences of action in such a way as to bring about more desirable results. Nor is it difficult to resolve any collective responsibility into a complex of personal responsibility. The responsibility of the community for an undue number of railway accidents is a complex of the responsibilities of railway commissioners, governors or presidents who appoint them, voters and politicians who elect these officials, railway managers, their directors, shareholders, bankers, etc. The national debt of Great Britain is not the debt of his Majesty (though the treasury, the army, and the navy are his), nor of the Cabinet, nor of the members of Parliament, nor even of the total present population of Great Britain. It is not the debt of a National Spirit or Ghost, but rather a complex of obligations on the part of certain officers to pay money out of certain funds to be obtained in diverse ways from a now indefinite number of Britishers present and future. Nor is it shocking to the general sense of mankind that future generations shall pay for our mistakes, or that they shall, without any struggle on their part, benefit by our efforts or good fortune. An absolutely strict debit and credit account between the members of a general community is neither possible nor desirable.

If collective responsibility is thus viewed not as a rigidly binding principle, but as a social necessity, we can see why our elementary sense of justice is not shocked when it is claimed that a country should pay the debt which a despotic ruler contracted, and the proceeds of which he squandered. As between the members of his country and those who stand in the place of the lenders, there may be many reasons for apportioning the loss on the former. But as we are dealing with a general maxim rather than with a rigid principle difficult cases are sure to arise. Thus I think there is a great deal of justice in the refusal of the Russian Revolutionary government of 1918 to pay the debt contracted by the late Czar in 1906 in his effort to suppress the opposition which arose because he revoked the people's constitutional rights—especially as the revolutionists at the time warned the European financiers. But while the leaders of Revolutionary Russia may be within their rights in refusing to pay such a debt, they may thus wrong the Russian people by cutting off their credit and, in consequence, necessary means of sustenance. Thus must principles lose their rigidity in the actual storms of experience.

§ III. POLITICAL PLURALISM

I DO not wish to leave the theory of communal minds or ghosts without paying a tribute of respect for the recent impressive movement of political pluralism represented by guild socialism, the ecclesiasticism of Mr. Figgis, the syndicalism of M. Benoist or M. Duguit, and the plural sovereignty theory of Mr. Laski. These theories have shaken political philosophy out of its torpid or somnambulent worship of the omnipotent State as the god on earth. They are peculiarly timely in so far as they attack the theory of an omnicompetent state at a time when the state has actually shown itself to be the strongest power on earth, much stronger in its power to dispose of life and substance than church, economic union, or the ties of language and race. The newer political philosophy has already rendered a great service in pressing the need for decentralizing our vast modern states, many of which have populations much larger than that of the Roman Empire at its height. Nothing can be more inimical to the human sense of power than for the individual voter to feel that after all he can accomplish very little politically since it is necessary to move millions before the action of the state can be modified. Large unified states undoubtedly tend to produce an oppressive uniformity that is profoundly inimical to the development of distinctive individuality. The spiritual need of local loyalties to offset this danger has been expressed by no one better than by Josiah Royce, whose later philosophy might be called a spiritual reflex of American federalism.

Nevertheless it seems clear that political pluralism is open to serious practical and theoretical objections. The partisans of pluralistic sovereignty ignore or minimize two dangers which human experience has shown to be very grave.

The first danger is that small groups or communities may be far more oppressive to the individual than larger ones. Men are in many ways freer in large cities than in small villages. Indeed it is precisely because of the intolerable oppression by local and guild sovereignties in mediaeval society that the modern national state was able to replace it. It is because the kings' courts were able to deal out what was on the whole better justice that they were gradually able to replace the local and vocational courts. The fact that our trade unions or southern states do not have absolute sovereignty in their own realms and that there is a possible appeal from their acts to the law of the land, certainly prevents them from oppressing some of their members more than they do. At any rate, the distinctive note of modern social and political philosophy (before the romantic and Hegelian reaction) is to be found in the long struggle to free the individual by means of natural rights from the claims of groups; and while it is doubtless true that individualistic, natural-rights theories have overestimated the powers and opportunities of the individual detached from some group, it would be hazardous to claim that the whole work of modern philosophy was unnecessary.

The second danger is that if the state gives up its sovereignty over any group there will be nothing to prevent that group from oppressing the rest of the community. Many who think that we must give up the notion of popular sovereignty in the same way as we have given up the notion of the sovereignty of kings reject the logical consequences of this position in the face of a strike by policemen. Policemen like other individuals are entitled to just treatment by the employing state, but no community can allow policemen or any other group to paralyze its whole life. We may try to set a line dividing the internal affairs of a church or trade union from those of its activities which affect the public at large, and contend vigorously that under no circumstances should the state as the organ of the larger community meddle in the internal affairs of the smaller society. But apart from the practical impossibility of drawing in advance any such line between the actions which do and those which do not affect the public at large,[10] this attempt really breaks down the whole theory of plural sovereignty, since in the last analysis some one will have the last word as to where that line is to be drawn, and it is logically impossible where groups conflict that each shall draw the line. To prevent the inconvenience of interminable conflicts, the power to terminate them by a deciding word is given to the state as the organ of the general community. The power to have the last word in any dispute is just what sovereignty is. The wisdom of large measures of home rule or autonomy to be accorded to various local, vocational, and religious organizations, need not be questioned. But we must recognize that the community cannot irrevocably part with its power to revise

[10] Every rule affecting a member of a union also affects a citizen.

such grants and that it is impossible for all the parties to a dispute to have the last word. Mr. Figgis, for instance, sets up the right of the church in matters of conscience as absolute against the state. Taken literally, as applied to individuals, the absolute right of free conscience would make all human organization impossible, since past experience has shown that there is no social institution, from property and marriage to the wearing of shoes and buttons, or the cooking of one's food, against which some individual conscience has not rebelled. While the greatest freedom in this respect is desirable, the state cannot give up its reserve rights to limit any form of conscience which it deems a nuisance. Nor is the matter much improved if, instead of individual conscience, we substitute the organized conscience of established churches. The churches in the South believed in slavery, but those in the North believed it to be iniquitous. The Mormons believe in polygamy as a divinely ordained institution, while others believe it to be adulterous. The Catholic Church believes in the use of images or icons, and another sect believes in the duty of breaking such images. If all of these are to live in the same community, somebody's right of conscience must necessarily yield. The matter is still more clear if, as in Mr. Laski's theory, we should attempt to bestow absolute sovereignty not only on churches, but also on trade unions and other groups. The evils of an absolute state are not cured by the multiplication of absolutes.

§ IV. CONCLUSION

THESE fragmentary considerations do not pretend to throw much light on the nature of the community and the problems of political science. But at least they may help to make our discussion more cautious by calling attention to the danger of reifying social relations as if they were additional substances. This fallacy of reification, responsible not only for the Folk-ghosts of political science but as well for the psychoids of biology, the subconscious mind of psychology, and a host of other intellectual bastards, is committed by nominalists and realists alike. Old-fashioned realists tended to regard universals, relations, or abstractions as new objects demanding in turn their own predicates and individuality. Nominalists attack this position (rightly) but since they deny the possibility of non-individual reality, such abstract aspects of the world as they cannot ignore are made into new individuals. The remedy for this confusion is the recognition that individuality and universality are polar categories.

In this connection, the principle of polarity also calls our attention to two modes of argument which are particularly vicious when used in social philosophy. These are the too facile antithesis of first principles and the too facile reconciliation of incompatible alternatives.

The first mode is illustrated when we argue that political democracy, idealism, individualism, or monism has broken down, and hence we must believe in industrial democracy, realism, collectivism, or pluralism. The facts of social life are clearly too complicated to allow such broad simple principles to be directly proved. Nor can either set of principles be categorically refuted. Difficulties *ad libitum* may be raised on both sides. In this connection I should like to call attention to the admirable procedure exemplified in Dean Pound's treatment of the Interests of Personality.[11] The individual interests worked out by the individualistic philosophy of natural rights are all restated in terms of social interests, but there is no pretended refutation of the older philosophy. Indeed, though Dean Pound's method has distinct technical advantages over the older method, it does not preclude the possibility of any one working out a complete theory of public and social interests on the basis of the individual rights or interests of personality. We can draw more than one true picture of the social world, provided we do not claim that our picture is *the* true one.

The second mode of argumentation against which I wish to raise a warning voice has vitiated our metaphysics and, as under the name of the organic point of view it still holds sway, we must be on our guard against it. Thus to dismiss the conflict between mechanism and purposive action, as a recent writer does,[12] on the ground that both are false abstractions, seems to me an arrogant shirking of a real problem, which may be all the more tempting and more dangerous in social philosophy. Social problems are generally difficulties which arise because we do not know how to attain what we want without also having something which we do not want. We want, for example, complete freedom of the press,

[11] *Harvard Law Review*, Vol. XXVIII (1915), pp. 343, 445.
[12] Dean Inge, *Plotinus*, p. 3.

but we do not like to see wicked people poisoning the sources of public information. The solution is obviously not some banality like liberty without license or other cheap evasion of a real difficulty. The social interests in freedom and in truth are not logically contradictory, but they are in fact incompatible in a world where many things are subjects of opinion. And this incompatibility is not to be removed by dialectic manipulation of principles, but by some specific invention similar to the invention of boats, which solved the problem how to get across the river without getting wet. In the infancy of science there may have been some excuse for philosophy to be associated with the search for magical formulae and panaceas; but now it seems time for philosophy to accept the division of labour and learn the vanity of trying to solve everybody else's problems.

Another writer, zealous for social philosophy, and for the gratuitous assumption that the philosopher is called upon to be the leader of the community in questions of statesmanship, speaks contemptuously of "epistemologic chess." [13] I am far from condoning the grievous sins of epistemology, but I think the implied condemnation of the play instinct in philosophy a much more grievous error. The history of philosophy and pure science will show, I think, that there never was a man who made a great discovery in the realm of ideas who did not keenly enjoy the play of ideas for its own sake. But in intellectual as in other play, we must follow the rules, and one of the primary rules of the intellectual game is that ideas must submit to the most rigorous criticism and to the test of fact. Therefore, to rush into social generalization without making sure of the consistency of our ideas or their adequacy to meet the ocean of complicated fact is much worse than epistemologic chess. The least that the community can expect of philosophers is that its toil and suffering shall not be made the subject of pompous frivolity.

[13] *Journal of Philosophy*, Vol. XVI (1919), p. 576.

Chapter Four

NATURAL RIGHTS AND POSITIVE LAW [1]

§ I. A PRIORI OBJECTIONS TO THE DOCTRINE OF NATURAL LAW

To DEFEND a doctrine of natural rights today, requires either insensibility to the world's progress or else considerable courage in the face of it. Whether all doctrines of natural rights of man died with the French Revolution or were killed by the historical learning of the nineteenth century, every one who enjoys the consciousness of being enlightened knows that they are, and by right ought to be, dead.[2] The attempt to defend a doctrine of natural rights before historians and political scientists would be treated very much like an attempt to defend the belief in witchcraft. It would be regarded as emanating only from the intellectual underworld. And yet, while in this country only old judges and hopelessly antiquated text-book writers still cling to this supposedly eighteenth-century doctrine, on the Continent the doctrine of natural law has been revived by advanced jurists of diverse schools, in France, Germany, Belgium, and Italy, and stands forth unabashed and in militant attire.[3]

There are, of course, important differences between the new and the old brands of natural law, which show that the attack on the old natural

[1] The greater part of this chapter has been printed in two separate articles, "Jus Naturale Redivivum," *Philosophical Review*, Vol. XXV (1916), p. 761, and "Positivism and the Limits of Idealism in the Law," *Columbia Law Review*, Vol. XXVII (1927), p. 238.

[2] Thus in an address before the American Historical Association, Dr. James Sullivan referred to popular discussion of inalienable rights as only serving to "illustrate the wide gulf which separates the scholarly world from the general public. The world of learning has long abandoned the state of nature theory." (Report of the American Historical Association for 1902, pp. 67-68.) The assumption, however, that with the fall of the "state of nature" theory all questions of inalienable rights are eliminated is quite gratuitous and in no way borne out by Dr. Sullivan's own evidence.

[3] One of the first to point out that the historical school of jurisprudence had not really succeeded in refuting the standpoint of natural law, was Stammler, in his *Ueber die Methode der geschichtlichen Rechtstheorie* (1888), pp. 4, 28-48. Since then he has pressed his conception of "natural law with a changing content" in all his important works. See his *Wirtschaft und Recht* (2d ed.), pp. 165, 176, 181, 456, and *Lehre vom*

law was not without some justification. Yet the name "natural law" is not inappropriately applied to the new doctrines, which are, in essence, a reassertion of the old in a form more in harmony with modern thought. That this reassertion is scientifically possible I shall try to show by a critical examination of the four usual arguments against the theory of natural law, namely, the historical, the psychologic, the legal, and the metaphysical.

(A) HISTORICAL ARGUMENT

The first and most popular argument is the historical one. This argument assumes that the old doctrine of natural law rested on a belief in the actual existence of human beings in a state of nature prior to organized society; and as history has not shown that such a state ever existed, natural law falls to the ground. To this very simple argument the reply is that the old doctrines of natural law rested on no such foundation. Even Rousseau disclaims it in his maturer work, as is well known to those who take the unusual course of actually reading his *Contrat Social*. When Grotius, Hobbes, and their followers speak of a state of nature they do not as a rule mean to refer to a past event. The "state of nature" is a term of logical or psychologic analysis, denoting that which would or does exist apart from civil authority. It is logically, not chronologically, prior to the "civil state." Similarly the "social contract" is not a past event, but a concept of a continuous social transformation.[4]

richtigen Rechte, pp. 93 ff., 196 ff., also L. v. Savigny, *Das Problem des Naturrechts*, in *Schmollers Jahrbuch*, 1901. Similarly Del Vecchio, in the three works now translated under the title, *The Formal Basis of Law*, Rensi, *Il fondemento filosofico de diritto* (1912), Platon, *Pour le droit naturel* (1911), and Charmont, *La Renaissance du droit naturel* (1910). On the positivistic side, Ardigò, *Sociologia* (1886), pp. 50 ff., and *La Morale dei positivisti* (1901), I, Pt. II, Ch. I. Jung, *Problem des natürlichen Rechts* (1912), and Cosentini, *La Réforme de la Législation Civile* (1913), I, Ch. II. I omit Herbert Spencer, for his arguments as to the nature of "absolute justice" are substantially of the eighteenth century type.

Influenced even more by purely legal than by philosophical considerations, are Hennebic, *Philosophie du droit et droit naturel* (1897); Picard, *Le droit pur* (1899); and Saleilles, *École historique et droit naturel*, *Revue trimestrielle du droit civil*, Vol. I (1902).

In England Pollock has been prominent in appreciating the importance of natural law doctrines; see his *Expansion of the Common Law*, Lecture 4, *Continuité du droit naturel*, *Annales internationales d'histoire* (1900), Sec. 2, and *Journal of the Soc. for Comparative Legislation* (December, 1900).

[4] This comes out most clearly in Kant, who discusses the whole matter on purely

There is, doubtless, a good deal of a priori history to be found in seventeenth and eighteenth century thought. But consider the general and now almost classical belief that human progress passes through certain necessary stages, and that by the proper handling of our scant and crude information about certain savage or primitive races, we can reconstruct the universal history of mankind. Is not this likewise a priori history? Yet, would it be fair to reject entirely a legal philosophy such as Kohler's for no other reason than that it assumes a necessary succession of matriarchal and patriarchal stages which, from the point of view of scientific history, are entirely mythical? The essence of the old natural law was an appeal from the actual or merely existing to an ideal of what is desirable, or ought to be, and historical considerations alone will not settle the matter.

It would be absurd, of course, to deny all value to historical study as an aid in the correction of the aberrations of the old natural law theories. Historical study has helped to break down what might be called either the absolutism or the provincialism of the old natural law, under the aegis of which people assumed their own local ideals to be valid for all times, places, and conditions. But historical study has been only one of the elements which have brought about our cosmopolitan thought. The widening of human geography by purely physical means, the increased ease of communication between different peoples, the more intimate acquaintance with oriental and other types of social life, are other and in some respects even more important elements. In the writings of Hegel, Karl Marx, and the German historical school of jurisprudence, the real nature of historicism as an inverted or romantic form of rationalism becomes apparent. The absolute, the system of production, and the *Volksgeist* simply take the place of ordinary human reason. They function in an entirely a priori rationalistic way. Instead of refuting the normative standpoint of the old natural law, these writers substitute an unconscious natural law of their own. Instead of the revolutionary dogma of the complete plasticity of social institutions, they substitute

ethical postulates. That Hobbes, also, kept free from historical assumptions is clearly brought out by Dunning, *Political Theories*, Vol. 2. There are two or three passages in Locke and one, at least, in Kant—not to mention Rousseau's immature discourse—in which the "state of nature" is spoken of in the past tense. But these lapses into the common way of speaking cannot be shown to have had any influence on the general ideas of Locke or Kant.

the equally absurd conservative dogma of the futility of human effort. Even the English historical school of jurisprudence has been shown by Professor Pound to be guilty of the same offence of setting up its own idealization of the prevailing system as necessarily valid for all times.[5] The great enemy in our field is not rationalism, but the identification of the definitive universal goal with something merely historical, as the supposed condition of the Hebrews before the monarchy, the system of equity in Lord Eldon's day, or the Prussian state of the time of Hegel. Even more vicious and extravagant is the dogma that every actual state is the best for its time.

(B) PSYCHOLOGICAL ARGUMENT

The second type of argument, the psychologic, is based on the assumption that all theories of natural law must be intellectualistic and individualistic. The eighteenth century was undoubtedly intellectualistic, in the sense that it attributed entirely too much to conscious experience, deliberate invention, or consensual contract,[6] and too little to slow unconscious growth; and we can but smile at the astounding naïveté of such views as that religion is an invention of the priests. But while a shallow mechanical intellectualism did colour all the speculation of the Enlightenment, there is no necessary connection between it and the theories of natural law. Certainly the jural views of Grotius, Hobbes, and Spinoza, or even those of Locke and Rousseau cannot be so easily condemned. Moreover, a great deal may be said for the view that would prefer the shallow intellectualism of the Enlightenment to the romantic distrust of human reason, which denies (as do Hegel, Karl Marx, and, in part, Savigny) that reflective thought can aid in the transformation of jural and political institutions.[7]

[5] "The Scope and Purpose of Sociological Jurisprudence," *Harvard Law Review*, Vol. XXIV (1911), pp. 600-604.

[6] The classical theory of natural law embodied in the Canon Law or in the writings of St. Thomas is entirely free from this tendency to reduce all obligations to contractual ones. Anglo-American historical jurisprudence, however, influenced by Maine's maxim, "Legal progress is from status to contract," has gone far in reading fictional consent or contract into the law.

[7] It may seem strange that a panlogist like Hegel should be such a contemner of reflective reason. But that he was so influenced by the Romantic reaction against the Enlightenment and the Revolution, his *Rechtsphilosophie* proves beyond doubt.

Similar considerations hold in regard to the supposed individualism of all natural law theories. Eighteenth and even seventeenth century speculation did undoubtedly err in attributing a self-sufficiency to the abstract or isolated individual which modern psychology holds he could not have in the absence of organized society. Government and laws, we now see, are not mere external checks over affairs which might prosper without them, but necessary conditions of organized social life. There might be physical objects, but there can be no property, industry, or family continuity without property and family law adequately enforced.[7a] But while laws and government protection create legal rights, the effectiveness of this process depends on the recognition of previously existing fundamental psychic or social interests. To the extent that these interests exist and demand protection even prior to the specific law which meets their demand, they are the raw material of natural rights. There is no property in ideas or published works before the existence of patent and copyright laws; but interests and claims do exist prior to and not as creatures of these laws which they call into being. The latter must justify themselves by the services they render to these and other interests.

(C) LEGALISTIC ARGUMENT

The third or purely legal argument has received its definitive form in Bergbohm's *Jurisprudenz und Rechtsphilosophie*. It is not unfair to represent his attitude to natural law as parallel to that of the pious Mohammedan to learning outside that of the Koran: Natural law either repeats the rules of positive law, and is futile, or it contradicts them and is illegal and not law at all. Conditioning this is, of course, the belief in the all-sufficiency of positive law as a closed legal system to regulate all possible cases. Unfortunately, however, the distinction between positive and natural law is not as well defined as that between the Koran and all other books. If positive law means law actually enacted by some human agency devoid of supernatural omniscience, it is clear that it cannot foresee and regulate all possible contingencies. The domains of life thus not provided for in the positive law are regulated by the customary rules of what people think fair, which thus constitute a natural or non-positive law. Where such rules, though non-legal, are

---

[7a] See "Property and Sovereignty," *Cornell Law Quarterly*, Vol. XIII (1927), p. 8.

fairly well established, judges will be bound by them (except in cases where their own sense of fairness asserts itself). Those who believe in the closed or all-comprehensive character of the positive law have tried to save their theory from these facts by saying that the positive law is only formally, not materially, closed. But this, like the fiction that whatever the sovereign has not prohibited happens by his command, gives us very little insight into the life of the law. It certainly ought not to hide the fact that ethical views as to what is fair and just are, and always have been, streaming into the law through all the human agencies that are connected with it, judges and jurists as well as legislature and public opinion. Indeed, the body of the law could not long maintain itself if it did not conform in large measure to the prevailing sense of justice.

Reviewing Lorimer's *Institutes of Law,* Pollock says of natural law that it "either does not exist or does not concern lawyers more than any one else," and "I do not see that a jurist is bound to be a moral philosopher more than other men." But twenty years later he says: [8] "Some English writers half a century behind their time still maintain the absolute Benthamite aversion to its name (natural law). Meanwhile, our courts have to go on making a great deal of law, which is really natural law, whether they know it or not, for they must find a solution for every question that comes before them, and general considerations of justice and convenience must be relied on in default of positive authority."

But if the sense of justice must necessarily exercise an influence in any growing law,[9] it becomes of utmost importance to the jurist that the

[8] Pollock, *Essays in Ethics and Jurisprudence,* pp. 19, 23, and the *Expansion of the Common Law,* Ch. IV.

[9] I have developed this point at length in "The Process of Judicial Legislation," *American Law Review,* Vol. XLVIII (1914), p. 161, and need not repeat the arguments there made. I may, however, mention two arguments that have been advanced against the position there indicated. (1) Judges, we are told, have no license to legislate at will in the interests of justice and mortality. Certainly not. But neither have they authority to decide any cases that are not presented to them. When, however, cases *are* presented, they must decide; and when issues come up, as they certainly do nowadays, which require the weighing of considerations of public policy, social welfare or justice, judges must legislate. (2) It is also urged that in new cases, judges must depend on the analogy of established principles. But it is a poor lawyer who cannot meet an analogy against him with another one in his favour, and upright judges, in choosing or weighing the force of different or competing analogies, must inevitably rely on their sense of justice.

principles of justice or natural law should receive the careful and critical treatment which we call scientific method. Hence, the Continental jurists who are giving up the view that legal interpretation is a mechanical process of extracting from the words of a statute a peculiar and magical essence called the will of the legislator, and who recognize that jurisprudence must necessarily be growing and creative, are also beginning to recognize the need of a systematic science of justice or natural law.

As foreign ideas, however, may seem undesirable immigrants in the field of American jurisprudence, we can press the last point in the field that is peculiarly native to us, that is, our constitutional law. The bills of rights of our Federal and State Constitutions embody certain popular principles of justice, and in spite of over a century of judicature, such phrases as "due process of law," "equal protection of rights," etc., are still essentially more or less vague moral maxims—the effort to transform them into legal principles of fixed meaning being thwarted by the imperative need of making an eighteenth century document fit the needs of twentieth century life. Hence, the problem of justice remains an inescapable one in the field of constitutional law.

Those who would denude the phrases of our bills of rights of their moral connotation urge that it is not well for courts of law to become courts of morals. Without wishing at this point to discuss the advisability of a system of government whereby a very small number of non-elective judges must, on the basis of a few hours' argument, say the deciding word on grave public questions, such as railroad rates, industrial combinations, etc., and without wishing to pass judgment on the actual results of our courts' efforts to enforce the body of moral principles (some at least of which they hold to be independent or anterior to all written constitutions), we may still urge that the fear of our courts becoming censors of morals is not well taken. The objection is not well taken because it fails to distinguish between individual morals, which must take into account personal motives, and questions of right and wrong in external and enforcible relations. The moral principles of our bills of rights are entirely of the latter kind. It is for that reason that I think it advisable to keep the old distinction between the science of natural rights or justice and the science of personal morals or ethics. The principles of justice applicable and enforcible in public relations may be regarded as part of social ethics, but they form a distinct group of problems relatively as independent of the other problems as the questions of economics are of the other questions of sociology. If the work of our courts in applying maxims of natural law has proved unsatisfactory, it does not follow that principles of justice cannot or ought not be worked into the law. Legal history shows that they always have been the life of the law.[10] So far as the use of the moral maxims of our bills of rights has actually proved unsatisfactory, the causes are to be sought in the specific conditions under which our courts have done their work. Of these conditions not the least harmful is the belief that jurists need no special training in the science of justice (either because law has nothing to do with justice, or else because what constitutes justice under any given condition is something which any one can readily determine by asking a magical arbiter called conscience). It may well be that such phrases as "due process of law," "cruel and unusual punishment," "republican form of government," and "direct tax," are too hopelessly vague to serve as definite legal rules. But these phrases will not prevent courts giving the stamp of constitutionality to legislation the justice of which can be shown to them.

The essence of all doctrines of natural law is the appeal from positive law to justice, from the law that is to the law which ought to be; [11] and unless we are ready to assert that the concept of a law that ought to be is for some reason an inadmissible one, the roots of natural law remain untouched. Now, it is true that the issue has seldom been so sharply put, for to do so is to espouse an amount of dualism between the *is* and

[10] There is no adequate direct history of the interaction between positive and natural law; but material will be found in the writings of M. Voigt, in Landsberg's *Geschichte*, F. von Liszt's *Das deutsche Strafrecht*, and Gierke's *Genossenschaftsrecht*, Vol. IV.

[11] For this reason I must reject Professor Fite's attempt, in his *Individualism*, to reduce natural rights to a question of intellect-power, or intelligent self-assertion. It seems to me a subtle way of reducing questions of right to a species of might. Basing his theory on an analysis of consciousness, Professor Fite consistently arrives at the position that the unintelligent have no rights. If that were so, we would have to say that infants, before the age of self-consciousness, and the senile or demented have no rights whatsoever, and any one who takes advantage of his superior intelligence in dealing with them is exercising his rights. In one case, at least, Professor Fite does not hesitate to follow his theory to such a conclusion. A nation, he tells us, which allows valuable public lands to pass into private hands through lack of interest and intelligence should not complain of being robbed. If Professor Fite were consistent, he would have to say, not only that the public has no right to complain of being robbed because of its ignorance, but that the robber is perfectly justified, so long as the public does not know a way of recovering it. This, indeed, would be reducing questions of right to questions of might, but it would really make the predicate *right* devoid of all meaning.

the *ought* which is shocking to the philosophically respectable. The respectable dread to admit the existence of real conflicts in our intellectual household; they would rather conceal them by ambiguous terms such as *natural* or *normal*. This is most apparent in the most philistine of all philosophic schools, the Stoics, whose tremendous influence in jurisprudence has brought about much intellectual confusion. There have not, of course, been wanting intellectual radicals who, in the interests of a strident monism have clearly and conscientiously attempted to eliminate the chasm between the *ought* and the *is*, either by denying the former, or by trying to reduce it to a species of the latter. Thrasymachus's definition of justice as the interest of the stronger, finds its modern form in the definition of right as the will of the sovereign, of the people, or of the dominant group. But few of these radical positivists have had the courage of their convictions; they smuggle in some normative principle, such as harmony with the tendency of evolution, social solidarity, etc., as *the* valid ideal. Marx may have boasted that he never made use of the word *justice* in his writings; but his followers would dwindle into insignificance if they could not appeal against the crying injustices of the present "system." The most courageous of all such positivists, Hobbes and Nietzsche, have not escaped the necessity of admitting, in a more or less thinly disguised form, a moral imperative contrary to the actually established forces. Our analytical school of jurisprudence, pretending to study only the law that is, has been repeatedly shown to be permeated with an anonymous natural law.

The boldest attempt in history to do away with the antithesis between what is and what ought to be is, of course, the Hegelian philosophy, with its violent assertion of the complete identity of the real and the rational. And it is one of the instructive ironies of fate that this most monistic utterance of man should have led to the widest rift that ever separated the adherents of a philosophic school. To the orthodox or conservative right this meant the glorification or deification of the actual Prussian state. To the revolutionary left, of the type of Karl Marx, it meant the denial of the right of existence to the irrational actual state. Nor need this surprise us, as the Hegelian philosophy is at least as fluid as its object, of which, indeed, it professes to be the outcome. By its own dialectic it sets up its own opposite, so that its assertion of the identity of the real and the rational gives way to the insight that

409

in the necessary opposition between these two we have the clue to the process or life of civilization. If the jurist objects that this is indeed fishing in muddy rather than deep waters, and that the science of law has, fortunately, nothing to do with all this, our answer is that this is precisely the muddy condition in which the legal theory of this country finds itself today. In the prevalent legal theory we find the conflicting assertions that the law is (and ought to be) the will of the people, and that it is (and ought to be) the expression of immutable justice; and an unwillingness to recognize the inconsistency which this involves. Professional philosophers, it seems, are not the only ones to take refuge in a twilight zone when their eyes are not strong enough to face sun-clear distinctions. The intellectual motives which lead to this disinclination to admit a sharp distinction between what is and what ought to be, come out perhaps clearest in the noble efforts of physicians engaged in teaching sex hygiene and furthering sex morality. They are afraid to characterize certain practices as immoral. They think it is more scientific to use such terms as unnatural or abnormal, knowing full well that these practices are natural in the sense that they are due to what are called natural causes, and normal in the sense that they are, alas, quite usual and widespread.

One of the roots of this error, which is also the basis of all empiricism, is the assumption that science necessarily deals only with the actual. It would take us far afield to point out that this is based on the inadequate analysis of scientific procedure which is embodied in the Aristotelian or scholastic logic with its underlying assumption that all propositions are of the substance-attribute type. If instead of the classificatory zoology, which is the science that Aristotle had in mind, we look at the sciences which use the hypothetic-deductive method, we get a different perspective. The objects of two contrary hypotheses cannot both exist; yet in every branch of any developed science progress depends upon such rival hypotheses receiving equally careful scientific elaboration before either can be rejected. Indeed, every branch of science aims to assume the form of rational mechanics or geometry, in which we do not directly deal with the realm of existence, but rather with the realm of validity or the valid consequences of given hypotheses or axioms. And not only scientific progress, but all practical activities, such as those of statesmanship, depend upon reasoning of the form, "What *would happen* if this engine were perfect or frictionless?" even though we know such perfection to be impossible. Intelligent action demands that we know what will happen if we turn to the right and what will happen if we turn to the left, though it is certain we cannot do both.

(D) METAPHYSICAL ARGUMENT [12]

The metaphysical objection to the possibility of a theory of natural law or justice runs thus [12a]—"Questions of justice are relative to time, place, and the changing conditions of life. Hence there cannot be such a thing as a definite science of these matters." The widespread prevalence of this view, even in high places, shows how wofully unfamiliar is the logic of science. The objection ignores the difference between a substantive code and a science of principles, a distinction which ought to be as clear as that between the directions of the engineer to the builder and the science of mechanics. The temperature or the time of sunrise of different places undoubtedly varies, yet that does not prove the absence of a rule or formula for computing it. Similarly substantive rules such as those of property cannot be well drawn without taking into account specific agricultural or industrial conditions. But this does not deny—on the contrary, it presupposes—the existence of a general rule or method for the determination of how far any property rule justly meets the demands of its time.

Moreover, there are indications that the variability of social judgments, such as those with regard to justice, has as a matter of fact been greatly misunderstood. The first impression of savage life as gathered from the reports of scientifically untrained travellers and others interested in noting striking differences, together with the intellectual intoxication produced by the frenzied acceptance of the principle of universal evolution, have combined to produce an over-emphasis on the diversities of human culture. As soon, however, as we get over the disposition to run wild with the concept of evolution, and examine the

---

[12] The reader may note a close parallel between these four arguments against the possibility of natural law and the four general arguments against rational method treated in the first chapter of this volume. The "legalistic argument" of this chapter is obviously empiricistic, while the "metaphysical argument" is an assertion of the relativism which takes temporal form in Chapter One as the "argument of kineticism."

[12a] See Lévy-Bruhl, *La Morale et la Science des Moeurs*, pp. 257, 260, 279.

matter somewhat soberly, the fundamental resemblances of all human races and modes of life will be seen not to have lost significance. Historians as radical and free from metaphysical preferences as Robinson find it necessary to emphasize the fundamental unity of human history, as opposed to the differences which separate us from the Greeks or Assyrians, and critical ethnologists like Boas are pointing out that the unscientific, uncritical reports of untrained observers as to so-called primitive life have produced false impressions of radical moral differences, and that the actual variations of moral opinion are largely explicable by the variation of social conditions.[13] In ordinary affairs and in public discussion we all do undoubtedly assume a large amount of agreement as to what constitutes justice. And while such agreement is not conclusive, it offers a sufficiently definite starting point for a critical science, which, according to the Platonic method, consists in positing ideals (or hypotheses) and criticizing or testing them in the light of ascertained social fact.

Militating against this programme are the prevalent views (1) that questions of justice are all matters of opinion, and (2) that all things are in a flux but that there is no *logos* (reason or formula) to determine the fact that things are changing, and no definite measure according to which they do so. Against the first, or Sophistic position, it is sufficient to point out that no one in practice disbelieves that one opinion may be better founded than another. Against the blind worship of the dogma of universal and absolute change, it ought to be sufficient to point out that change and constancy are strictly co-relative terms. The world of experience certainly does not show us anything constant except in reference to that which is changing, nor any change except by reference to something constant. We may generalize change as much as we like, saying that even the most general laws of nature that we now know, such as the laws of mechanics, are slowly changing, but this change can be established and have meaning only by means of or in reference to some logical constant. The belief that the world consists of all change and no constancy is no better than the belief that all vessels have insides but no outsides.

---

[13] Robinson, "The Unity of History," in *International Congress of Art and Sciences* (1904), Vol. II; Boas, *The Mind of Primitive Man.*

§ II. DIFFICULTIES IN THE PATH OF NATURAL LAW

APPROACHING the subject from the point of view of the requirements of a scientific theory, let us ask what is the character of the principles of legal justice or natural law, and how are they to be established? The traditional answer from the Stoics down to our own day is that they are axioms whose self-evidence is revealed to us by the light of natural reason. This belief is implied in the way in which these principles are appealed to in popular discussion of natural rights. In a Catholic manual of socialism, we have a long list of such eternal first principles, which are put in the same class with such axioms as "The whole is greater than any part," "The cause must be equal to the effect," and the like. As the model for this view is presented by the Euclidean geometry it is suggestive to apply to these self-evident axioms the criticism which modern mathematics has applied to the Euclidean system. The discovery of non-Euclidean geometry and the whole trend of modern mathematical thought has led us to discard as unreliable the self-evident character of axioms or principles. Such principles as that two magnitudes equal to the same are equal to each other, or that a straight line is the shortest distance between two points, are seen to be simply definitions, while others are either hypotheses or assumptions or else rules of procedure or postulates, whose contraries may not only be just as conceivable but even preferable in certain systems of mechanics. If now we apply the same criticism to our assumed principles of natural law, such as "All men are equal before the law," or "All men have the right to life, to the product of their labour, etc.," it becomes evident that it will not do to rely on their apparent self-evidence, and that the only way to defend them against those who would deny them is to show that like other scientific principles, e.g. the Copernican hypothesis in astronomy, they yield a body or system of propositions which is preferable to that which can possibly be established on the basis of their denial.

Like other scientific hypotheses they are to be tested by their certainty, accuracy, universality, and coherency. No science of justice can be built up by an intellectual *coup de main;* patient analysis of the multitudes of fact as well as the proper use of principles is required. No mere postulating of principles nor unimaginative abandonment to the infinity of details will enable us to make the necessary progress. We must control the work of philosophy with the wealth of the facts revealed by legal science, and analyze these facts in the light of the best available philosophy.

In particular, we must observe the limitations which are imposed upon the ideal of justice, as upon any ethical norm, by the positive factors in our problem, factors which, by varying in different environments, justify the diversity and reveal the unity of natural or ideal law. This much, indeed, we may grant to the critics of the classical natural law theories: that the proponents of such doctrines, either because of limited historical and anthropological learning or because of a cosmopolitan emphasis upon the similarity of all races and all ages, have generally minimized the rôle played by contingent, empirical facts in the materialization of our formal principles of justice.

To a certain extent the problem is common to all normative science. If our ultimate standards be formulated in terms of desire, we must ask what, as a matter of brute fact, the people of a given time and place do desire, if in terms of happiness, what will actually make present-day Americans (say) happy, and so on. Even the narrowest ethical commandments, e.g. "Respect private property," or "Cure the sick," demand non-ethical investigation of extreme complexity and difficulty if they are to achieve a practical concreteness. In part, however, the problem of interpreting natural law for a given social milieu is specifically one of legal science. What, we must ask, are the limitations which law, as an instrument of social control, imposes upon the ideal that it serves? And finally we must frankly recognize the dependence of natural law upon ultimate ethical principles. No doctrine of natural law can claim a greater degree of certainty and completeness than attaches to the basic ethical principles which it presupposes.

Our rejection, therefore, of the various a priori arguments against the possibility of natural law leaves us with a problem in which three difficulties must be faced, namely (a) the indeterminateness of our jural ideals, (b) the intractability of the human materials with which law works, and (c) the inherent limitations of general rules.

It is a common illusion to suppose that for all questions that can possibly arise our ideal of justice determines a specific answer. Our ideals are in fact much hazier than we ever care to admit.

We all begin by accepting the judgments prevailing in our community. Popular judgments and proverbial wisdom, however, though they may contain a kernel of enduring truth, seldom express it with a high degree of accuracy or consistency. It is because of this vagueness and inconsistency in popular judgment that science is needed as a corrective. The greater simplicity of their subject-matter, the more definitely elaborated technique, and a certain amount of ethical neutrality as to their result, enable the natural sciences to depart from popular opinion to a far greater extent than is possible in the social sciences. This, together with the natural difficulty of seeing any justice in doctrines or sectaries abhorrent to us, makes the elaboration of a consistent ideal of justice a task of the utmost difficulty. That this difficulty has not been fully overcome in treatises on ethics, natural law, or legal philosophy, becomes obvious when we ask: What precisely is the content of justice and on what evidence are its disputed claims based?

The classical legal definition of justice makes it consist in rendering every one his own: *suum cuique tribuere*.[14] But what is rightly anybody's own is precisely the problem which the law must determine according to some principle of justice. Otherwise the just becomes simply the legal and there is no possibility of unjust law.

If we go to books on ethics we are told that justice accords to every one that which he has earned, or that a just law protects every one in the enjoyment of the product of his labour. But what in a co-operative social state is the product of any one man's labour? In practice this is settled somewhat arbitrarily, by appeal to the law which happens to exist. Moreover even if we could determine on principle what is the product of any one's labour it would be very doubtful morality if one could keep it all when others, the sick, the infant, or the very aged, were to perish because of the exercise of this right.

That legal justice in some way demands the principle of equality seems certain. But what exactly do we mean by equality before the law?

[14] *Dig.* I. 1. 10.

In the end nothing more than fidelity to the classification that the law has already laid down. If all creditors are to share alike in a bankrupt's estate, equality before the law means that the judge must not favour any one party. Obviously, this does not enlighten us as to whether the classification that the law has created is itself just. Should not the law give certain creditors precedence, for example, those who are creditors of a railroad by virtue of having suffered bodily injury? Legal justice is formal. It is important that the law, once created, should be justly, i.e. faithfully, administered. But it is also important that the law should be just in content to begin with and purely formal considerations are insufficient to determine this.

The Kantian and neo-Kantian efforts to derive conclusions as to specific questions of justice from purely formal principles ignore the logical fact, made clear by modern logic, that from pure universals no particular existential propositions can properly be deduced. Pure universals are hypotheses and you cannot prove a fact by piling up nothing but suppositions. This can be seen in Kant's own efforts to derive the rules of perfect and imperfect obligations from purely formal considerations. These rules follow only if we accept certain empirical ends of life and assume certain conditions of life to remain permanent. For our present purpose we can see this best in Stammler's legal philosophy.[15] Stammler's ideal of a community of free-willing men appeals to us for various historic reasons, chiefly because of linguistic associations of the word *freedom*. Critical reflection, however, shows that his ideal itself is essentially vague and indeterminate, if not completely empty. It does not in fact logically support any of the conclusions which Stammler tries to derive from it, since quite different conclusions can be derived from his ideal with equal logical propriety. Let us take a concrete example. Two people mutually agree to live together in free love or being married voluntarily agree to separate and live with other parties. Stammler supports the view of most civilized countries in refusing to sanction a contract of this sort. Yet he does not *prove* that there is anything here contrary to the ideal of a community of free wills. Nor does this ideal enable us to determine whether a contract of

[15] See his *Theory of Justice* (1925) in the Legal Philosophy Series. For criticism see E. Kaufmann, *Kritik der neukantischen Rechtsphilosophie* (1921), and Wielikowski, *Die Neukantianer in der Rechtsphilosophie* (1914).

employment, dictated by economic necessity, should be enforced. Something more empirical and specific than abstract free will is necessary to arrive at a rational account as to what kinds of agreements should and what kinds should not be legally enforced. Again, in time of war we conscript men against their will to be killed or maimed for life. Can the right to do this be derived in strict logic from the ideal of a community of free-willing men? Stammler evades all real problems of this sort by circularly defining the free will as the rational will that is bent on rendering to each what is objectively just. This circular dependence of justice on free will and free will on justice makes the whole enterprise fruitless. Had Stammler used free will in its ordinary sense, as the freedom of the empirical wills of ordinary human beings, Stammler's theory might have been more a condemnation than a derivation of most of our criminal and public law, especially of our laws of taxation, tariff duties, etc. It is noteworthy that despite Stammler's pretensions to establish absolute principles, the only law he condemns as unjust is that of slavery. But what is slavery? Do not many labour contracts differ from slavery only in arbitrary legal form, rather than in the substance of things? His rule that "the contents of a given volition must not be arbitrarily made subject to another volition" is devoid of definite meaning, since the word "arbitrarily" is just a blank term that can be filled with question-begging interpretations, so that by means of it you can condemn or justify anything you please.

The radical indeterminism of the neo-Kantian formal ideal of justice can also be seen if we apply it to the conflicting interests of groups, for example, employers and employés. Here the ideal of free will agreement has certainly broken down and been productive of much injustice. Finally we need only ask whether the ideal of the community of free will is adequate for international law. Will Panama or Egypt, if independent, have a right to exercise free will and exclude the rest of the world from the use of the Panama or Suez Canals?

The fundamental indeterminism of Stammler's ideal has compelled some of his disciples [16] to speak of the ideal of "the place and the epoch" instead of the ideal valid for all time and place. Doubtless, the ideal prevailing in any given country at any time is much more definite. But that this always coincides with justice it is difficult to admit when we

[16] See Brütt, *Die Kunst der Rechtsanwendung* (1907).

remember that some opposing countries feel and think quite differently. The Austrians felt quite justified in regarding the Trentino as a region to be saved for German culture. Italy is certain that "sacred egoism" demands that the south Tyrolese be compelled to become Italians. Surely both cannot be right. Is it any different between peoples of different grades of culture? European nations professing Christian love feel themselves thereby justified in bombarding with air-bombs Mohammedan villages that refuse to be ruled by them; but the Mohammedans are not thereby convinced.

Nor do we get much farther if with the neo-Hegelians we set up the ideal of "the epoch of civilization." The term civilization is at best a vague one, and the division of the continuous stream of human history into epochs varies with the purpose of the historian. In any case, we get little help here in deciding the nature of justice. To assert that the ideal of justice varies with time does not help us to know what is and what is not just at any given time, for example, at the present. Indeed, no answer is possible at all on an Hegelian or on any other monistic determinism, which allows no real chasm between what is and what ought to be—a chasm which gives meaning to the human struggle and its tragic, though noble, defeat through the ages. The actual efforts of neo-Hegelians, like Kohler and Berolzheimer, to determine what is right by reference to the ideal of our own epoch are certainly not always and altogether fortunate—witness Kohler's remark on the retributive theory of punishment, or Berolzheimer's remark, in 1908, on the effect of strengthening the German fleet.[17] I am the last man in the world to maintain that the neo-Hegelians are the only ones to talk nonsense when they are certain that they have caught the idea of the epoch of our civilization. Kohler was undoubtedly a man of genius though erratic and uncritically opinionated. But his occasional insights have little support in the vague neo-Hegelian philosophy which he professes.

On a subject teeming with human significance rigorous logic is of the utmost importance. Such logic shows us a radical and incurable difficulty in all idealistic philosophies of law: either their ideal is, like the Kantian, purely formal and incapable of solving material issues, or, like the older theories of natural law and the post-Kantian idealism,

---

[17] Kohler, *Moderne Rechtsprobleme* (1913), and Berolzheimer, *System der Rechts- und Wirtschaftsphilosophie* (1907), end of Vol IV.

they uncritically assume certain material principles as self-evident. Critical reflection generally shows these self-evident ethical and jural propositions to be either question-begging, purely verbal plausibilities, or rhetorical justifications for preferences that are too spasmodic to serve as the basis of developed legal systems.

Can we abandon the effort to formulate a rational ideal that is to govern the facts of the law? Can the ideal of justice be derived from history or the empirical study of the facts themselves? Obviously not, if history or empirical study is restricted to the realm of existential facts, since our conclusion cannot contain an *ought* if all our premises are restricted to what *is*. But even if we begin with empirical judgments of what ought to be in concrete cases, we need some comprehensive ideal to organize our conflicting judgments into something like a coherent body. It is doubtless true that our ideal grows more definite as our experience expands and we get more opportunity to develop as well as to test our ideal by applying it. Still, any ideal that is to govern facts of conduct must be more simple, uniform and constant than these facts themselves. Otherwise it could not serve as a guide.

Dean Pound has used the postulates of civilization as a justification for the laws that secure the interests of personality, possession and transactions.[18] Certainly without such security our type of civilization is impossible. But these postulates do not undertake to settle questions of justice as to which of two heterogeneous and conflicting interests should prevail. An injunction is asked against striking employés. If the injunction is not granted, irreparable property injury will result. If it is granted, the workingmen will not be able to keep up the organization which protects their standard of living. Which of the two conflicting interests should justly prevail? The law has to balance interests but it has no scientific scale to measure the weight of conflicting interests, nor has it even any means of reducing them to some common denominator.

I conclude, therefore, that no ideal so far suggested is both formally necessary and materially adequate to determine definitely which of our actually conflicting interests should justly prevail. So long as this is the case, law must be, as it is, in large part a special technique for determining what would otherwise be uncertain and subject to conflict. It is

18 See his *Introduction to the Philosophy of Law* (1922).

socially necessary to have a rule of the road but it is morally indifferent whether it requires us to turn to the right or to the left.

This need for certainty and the inconvenience involved in changing men's rights enforce the human inertia which generally makes the law lag behind the best moral insight. As obedience to law rests on habit, a philosopher like Aristotle can plausibly argue that it is better that an unjust law should prevail than that law should become uncertain by change.

In the absence of an adequately determinate moral ideal many questions must be left to the discretion of judges and administrators. If the judge is both intelligent and well disposed, his exercise of the discretion is one of the ways whereby concrete justice is found. Yet the need for discretion indicates a limitation of legal justice. To be ruled by a judge is, to the extent that he is not bound by law, tyranny or despotism. It may often be intelligent and benevolent, but it is tyranny just the same. For political reasons it may be well to cultivate respect for judges; but from a philosophic point of view it is a crude superstition to suppose that any one can escape the limits of his intelligence and the bias of his limited experience by being elevated to an office. History does not show that the partisan bias of limited group experience can always be removed by legal training or by the criticism of the legal profession. Professional opinion is itself class opinion. This is not to disparage the judge or the legal profession. On the contrary! He surely is no friend of any profession who encourages its members to think that they are free from human limitations—ὕβρις invites the wrath of the gods.

(B) THE INTRACTABILITY OF HUMAN MATERIALS

The inevitable imperfections in the human beings that have to make, to enforce, and to obey the law constitute a second serious obstacle in the elaboration of legal ideals.

We may view the limitations of imperfect human nature (1) from the point of view of the legislator, (2) from the point of view of those who have to obey the law, and (3) from the point of view of those who have to enforce it or operate the legal machinery.

(1) INHERENT LIMITS OF LEGISLATIVE POWER. The obvious fact which no glorification of law can obscure is that it is made by human

beings subject to the limitations of human ignorance and of inadequate sympathy or good will.

(i) The ignorance of the legislator may relate to the end of the law which he helps to bring about. Moved by the demand for the redress of some grievance, the legislature enacts a statute. But changing the law is like making a change in the intricate plot of a highly organized drama. You cannot change one part without other parts being affected in unexpected ways. Legislatures thus seldom have an adequate idea of what they intend to bring about. The Napoleonic Code intended to guarantee that all the children shall have some part of the patrimony. Did its authors have any idea that they were erecting a check to the growth of population?

(ii) Assuming that the legislator knows what effect he wants to produce, he may be ignorant of the natural circumstances involved. All sorts of scientific facts have to be assumed in modern legislation. Our southern legislatures feel competent to pass on the truths of biologic evolution and one of our western states came near decreeing some absurd value for π. All our state legislatures feel competent to pass on the truths of history to be taught in the public schools. Modern states all contain heterogeneous elements. Hence laws applicable to a vast majority may be absurd for some groups or regions—e.g. trial by jury in those United States possessions inhabited by Negritos. Rural legislators do not know or sympathize with urban industrial life and city legislators do not always understand the conditions of farm life.

(iii) Of special importance is the imperfect power of the legislature to control the subsequent interpretation of its enactments. The legislature can express its intention only in general terms. It cannot foresee all actual cases. It cannot therefore completely control the interpretation and application of its statutes by courts and administrators bent on making the law serve other and wider purposes. The legislature generally looks to the removal of a specific abuse while the judge and the jurist must look upon any statute as a part of the whole legal system. Thus the first legislatures that wished to change the common law as to the property rights of married women were repeatedly defeated by courts that persisted in thinking of these statutes in terms of the traditional common law.

(2) INHERENT DIFFICULTIES IN FORCING OBEDIENCE TO THE LAW.
That in a democracy the law is the will of the people is a statement not
of a fact but of an aspiration. A great deal of the law is and necessarily
must be the work of legislatures, courts, and jurists, whose work is seldom fully known to the majority of the people. Indeed, as to rules
of conduct on which people are fairly unanimous there is no need for
the enactment of any law. Enacted law represents the will of some part
of the community and the rest obey either out of respect or by force of
habitual obedience to regular authority. Yet neither legislatures, courts,
jurists, nor all combined are omnipotent. There is no way of securing
perfect obedience where there are strong human motives for disobedi-
ence or evasion.

The failure of the law to secure obedience has been historically shown
in at least four fields of human life, viz. religion, personal morals. economics, and politics.

In the field of religion, the persistence despite persecution of the
Christian Churches in the early Roman Empire, of the Jews and Chris-
tion dissenters in Russia, England, and elsewhere, shows how persistently small minorities may defy and defeat the law.

The field of personal morals has always been a favourite interest of
the law, but legal failures in this field are proverbial. It is true that
sumptuary laws have often been generally obeyed, e.g. in the Middle
Ages. But this happened only so long as they conformed to the general
moral feeling of mediaeval society. When they cease to express a strong
moral consensus they cease to be effective. Take for instance the case of
the New York law which makes adultery a crime. Though many thousands of divorces have been granted for that offence there seem to
have been hardly any convictions for the crime. As a criminal law it
is a dead letter. Yet any proposal to repeal it would meet with widespread resentment. The majority of the people of New York State emphatically wish the statute book to express their disapproval of adultery.
Why, then, has there been no enforcement of it? The answer is to
be found in the inherent difficulty of enforcement. It is an unpleasant
and unedifying thing to bring into court. Moreover it is extremely
doubtful whether convictions would greatly reduce the number of actual
offences, or result in more good than harm.

The experience of our national prohibition law and the many evils which its enforcement involves, especially the violation of the law by the very agents of the government, are too flagrant to need anything but bare mention here.

It is interesting, however, to reflect that these evils are not merely contemporary. Long ago a most detached philosopher, Spinoza, wrote, "He who tries to fix and determine everything by law will inflame rather than correct the vices of the world." [19] He also observed:

"Many attempts have been made to frame sumptuary laws. But these attempts have never succeeded in their end. For all laws that can be violated without doing any one an injury are laughed at. Nay, so far are they from doing anything to control the desires and passions of men, that, on the contrary, they direct and incite men's thoughts the more towards those very objects; for we always strive for what is forbidden, and desire the things we are not allowed to have. And men of leisure are never deficient in the ingenuity needed to enable them to outwit laws framed to regulate things which cannot be entirely forbidden. . . . My conclusion, then, is that those vices which are commonly bred in a state of peace . . . can never be directly prevented but only indirectly. That is to say, we can only prevent them by constituting the state in such a way that most men will not indeed live with wisdom (for that cannot be secured simply by law), but will be led by those emotions from which the state will derive most advantage." [20]

Similar failure can be seen in the field of economics. The repeated failures of attempts to regulate wages by law, to prohibit mergers or trusts, to prevent railroads from having an economic interest in the products they carry, all illustrate how difficult it is to prevent evasion of the law by those who have a strong interest in doing so.

In the political sphere we need only mention the failure of the Fourteenth and Fifteenth Amendments to the United States Constitution to secure civil and political equality for the negro. Mr. Horwill [21] has recently called attention to the many ways in which Congress and the Executive evade the provisions of the Constitution, e.g. the failure to reapportion congressional representation after the census of 1920. The most interesting of these evasions is by what he calls "discreet

[19] *Tractatus Theol. Polit.*, Ch. 20.
[20] *Tractatus Polit.*, Ch. 10, 5-6.
[21] *The Usages of the American Constitution* (1925).

nomenclature," e.g. evading the duty of submitting a law for presidential approval by calling it a concurrent resolution.

There can be no doubt that the framers of the United States Constitution intended the Electoral College to serve as a barrier against the influence of political parties in the election of the president. This purpose has been defeated without changing the constitutional law, by the extra-legal device of the political convention.

(3) THE LIMITS OF LEGAL MACHINERY. Dean Pound has treated this topic with his usual thoroughness in an essay on "The Limits of Effective Legal Action" [22] and I may add only a few remarks. Legal machinery, we must remember, never operates apart from human beings, judges, juries, police officials, etc. The imperfect knowledge or intelligence of these human beings is bound to assert itself. It is therefore vain to expect that the legal machinery will work with a perfection that no other human institution does. We cannot expect results too fine for the discrimination of the ordinary juryman. A great deal of injustice cannot be prevented by law because the attempt to do so is bound to produce greater evil than it can cure. I think that the action for breach of promise to marry well illustrates this. Apart from the injury to public decency from the fact that this action is so often used purely for blackmail, it is a bad policy for the law to put a monetary value on the marriage promise and to seem in any way to force people into the marriage relation when the churches so strenuously insist that no matter what promises have passed between the two parties there shall be no marriage performed unless both parties are perfectly willing at the time of the ceremony.

It would be foolishness to contend that in the various fields mentioned, law has never been effective. Few injustices are absolutely beyond all human effort, if we are willing to make sufficient sacrifice to remove them. But it is folly to centre our attention on some result desirable in itself and ignore the fearful cost of incidental consequences required to attain it.

We conclude then that in view of the necessary limitations of any legal system and the many insuperable difficulties in the way of enforcing all sorts of moral considerations, it is a wicked stupidity that insists on "justice regardless of consequences." *Fiat justitia pereat mun-*

---

[22] *International Journal of Ethics*, Vol. XXVII (1917), p. 150.

dus is the device of the fanatic, too lazy to think out the consequences of his position. The more human wisdom is summed up in the saying, *summum ius, summa injuria.*

(c) THE ABSTRACTNESS OF LEGAL RULES [23]

The third limitation of legal idealism is due to the fact that legal justice must operate with abstract general rules. To guarantee equality before the law and eliminate favouritism or partiality, the law must operate with rules that are to apply to every one. No one is allowed to beg in the street whether he be poor or rich, strong or sick, young or old. Any one under twenty-one years has certain privileges in connection with contracts, no matter how developed intellectually or in the way of business experience. The injustice which this abstract uniformity works in particular cases has generally been recognized and many efforts are made within the legal system to correct it by some form of individualization, *equity*, or ἐπιείκεια.[24] This individualization itself, however, becomes subject to rule or else it remains lawless. The demand for abstract uniformity of legal rules is intensified by the fact that a developed legal system must assume a scientific form and that it can do so only as other sciences do, namely by the discovery of principles in terms of concepts so that multitudes of rules can be deduced from a few such principles. This abstract universality of legal rule is necessary to secure certainty. It ministers to a certain sense of justice which may be viewed as the rationalization of jealousy. But all rigid rules applied to life tend to suppress favourable as well as unfavourable variation. At times it becomes obvious that it is only the better element of a community that suffers from certain laws, while the worse elements evade or successfully defy them. Thus, while the law cannot admit the sovereignty of individual conscience without opening the door to anarchy, it must also recognize that individual conscience may be a much more delicate instrument for moral apprehension, so that to suppress it is to bar the way for more enlightened justice. This inherent difficulty of legal justice, like other difficulties of social life,

---

[23] I have treated this topic at greater length in "Rule vs. Discretion," *Journal of Philosophy*, Vol. XI (1914), p. 208.

[24] Aristotle, *Nicomachean Ethics*, V.

is not overcome by the Hegelian trick of making *Sittlichkeit* or social ethics supersede the morality of conscience.

An uncritical reliance on the abstract universality of legal justice is the crowning ethical defect of the so-called critical philosophy. It legalizes ethics without moralizing the law. It formulates its imperative in the abstract legalistic manner to the neglect of individual differences. Thus Kant's argument that it is always wrong to tell a lie is a striking example of legalistic ethics. In the end it is based on bad logic as well as on moral insensitiveness, as is Kant's horrible view of marriage as a mutual lease of the sexual organs.

Universality and individuality, justice, and the law, the ideal and the actual, are inseparable, yet never completely identifiable. Like being and becoming, unity and plurality, rest and motion, they are polar categories. Deny one and the other becomes meaningless. Yet the two must always remain opposed. Theoretically, the legal system may be viewed completely from either pole. You may even insist that there is little difference if any between a positivism like Gray's which allows for moral judgments upon the law, and an idealism that admits the inherent limitation of any ideal of justice that can be applied to human affairs. There is a sense in which the same system of legal rights and duties might be expressed in positivistic or in idealistic language. But not only is language itself a most important factor in human affairs— since all sorts of emotional differences arise from differences of expression—but so long as our knowledge remains incomplete it makes a difference from which end we view the legal system. The positivistic and the idealistic perspective cannot be identical on the level of human knowledge. Positivists fail in trying to separate law from all ideals, and Hegelians fail in trying to identify the ideal with some form of the actual, whether the Prussian or any other state.

The deeper and more ancient wisdom is to recognize that divine perfection is denied to human beings in legal as in other practical affairs. It is romantic foolishness to expect that man can by his own puny efforts make a heaven of earth. But to wear out our lives in the pursuit of worthy though imperfectly attainable ideals is the essence of human dignity.

Chapter Five

THE POSSIBILITY OF ETHICAL SCIENCE

IT IS tempting to suggest that the considerations as to law and justice discussed in the last chapter might with suitable adaptations
be applied to any other moral problem, and with that we might
leave the whole subject of ethics. But such a suggestion would overlook
the fact that as law and justice concern large numbers of individuals
living in relatively permanent groups, it has proved feasible for courts
and jurists to elaborate some more or less definite techniques to answer
certain of the questions involved. This, however, is not the case with
the more subtle and elusive problems of personal life. Can science be
applied to the whole art of living? It is easy to see that the problems
of law themselves involve assumptions as to the ultimate good of
human life, and ancient Semitic jurists suggested that he who would
deal with the law must meditate on life and death.

We cannot, therefore, avoid the question whether there can be a
science of ethics covering the whole field of human conduct. Let us,
before considering more objective difficulties inherent in the conception
of a moral science, review the more important of the human obstacles
in the way of adopting a scientific attitude towards the values of life.

The difficulties of social science which we have already noted are
intensified in the realm of morals. For moral judgments are deeply
rooted in our habitual emotional attitudes and in those of the community of which we are a part. It is thus most difficult to detach ourselves from the roots of our accustomed faith, to question what seems
obviously the right, and to devote the necessary patience and intellectual
sympathy to the understanding of opposing views that we almost in-
stinctively abhor and despise. This is true not only of the vast majority
to whom the ways of respectability are unquestionable and decisive, but
also of revolutionists in morals who move in groups that are inflexibly
proud of being "up to date," "emancipated," "forward-looking,"
"amoralist," etc. I remember as a child having great difficulty in

realizing that while the dome of heaven had its highest point directly over where I was, others living far off thought that the same was true for them. A similar realization in the moral realm is much more difficult.

Another difficulty in the way of attaining true views as to morals is the fear that these views will be perverted by the unintelligent or will have a bad effect on the young. Logically it might seem that if we believe new or heterodox views of morals to be true we should teach them to our children. But it is easier to change our theoretic views than our socially approved habitual attitudes. Thus many an agnostic sends his children to an orthodox Sunday school, and many confirmed Nietzschean amoralists are shocked when they hear their heterodox morality expressed before children. Some there are who justified this on the theory of vital lies, viz. that the young and the uneducated are not prepared for the truth and that we must keep them in check by convenient lies. Not many years ago a mother had no compunction about saying to her child: Don't tell lies or the bogeyman will get you! But older children are still taught the falsehood that the virtuous will be (financially) prosperous and that the wicked will always be punished either by society or by their own guilty conscience. This disregard of truth in moral education is as old as Plato. Nevertheless I suspect that it prevails largely among those who have not attained full confidence in the truth of their own views. There are many stages between entertaining a heterodox view as to morals, and actually living according to it. And the extent to which a man lives up to a new moral insight depends on his personal situation and courage rather than on the truth of these insights.

Closely connected with the foregoing obstacle is the irrationality of moral theory resulting from the effort to give justifying reasons for the institutions which happen to exist. It is obvious that if the maxim *What is is right* were true there would be nothing wrong in the world— not even with those who are always complaining of the evil in it—and all distinction between right and wrong would disappear as inapplicable or meaningless. Yet the fact that an institution exists gives arguments in its behalf an irrationally persuasive advantage over arguments against its value. We may illustrate this by a parable. Suppose that some magician came to us and offered us a magic carriage having great convenience, but demanded of us in return the sacrifice of thirty thousand lives every year. Most of us would be morally horrified by such an offer. Yet when the automobile is actually with us we can invent many ingenious arguments against the proposal to abolish it. Certainly an undue amount of moral philosophy is just an exercise in apologetics for what happen to be the prevailing moral institutions. Where the conclusion seems excellent we are not critical as to the supporting arguments.

Against the foregoing difficulties the moral philosopher must arm himself with the ethical neutrality of the scientist. Only by studying propositions about morals with the same detachment as propositions about electrons, caring more for the rules of the scientific game than for any particular result, can he hope to fulfill his function as a builder of sound ethical theory or science. How, indeed, can he promote the good life unless he first finds out what is the meaning of the good life and what are the conditions for attaining it?

In practice, the impetus to free scientific reflection on morality is greatly stimulated by familiarity with, and imaginatively living into, diverse moral systems. By taking note of moral variations, we may free ourselves from the absolute unreflective certainty which comes of not being able to imagine any possibility other than the one to which we are accustomed.

§ I. THE ILLUSIONS OF MORALITY

IF WITH the ideal of scientific detachment in mind we approach the task of developing a rational ethics, we find two conflicting ways before us—the way of those who believe in absolute principles and the way of those who think that the needs of life are cruelly crushed by such principles. To explain the persistence of the two parties it is safe to assume that each has some part of the truth in its possession; but the fact of conflict is also presumptive evidence that each party is in the grip of some illusion which prevents it from seeing the whole truth.

Moral rules are most often viewed as absolute. It does not occur to most people that there *can* be any genuine doubt about them. Men generally are surprised and painfully shocked at the suggestion that we need to search for new moral truth or to revise the old. For the most part the absoluteness of these accepted rules is supported by some authority regarded as beyond question, e.g. by some priest, sacred book, or prevailing respectability. When, however, these or other authorities are in fact questioned, any attempted justification must involve an appeal to some "scheme of things entire," of which these moral teachings form a consequence. Otherwise the moral teacher is in the position of the poor pedagogue who, when asked to explain or justify some questionable statement, stamps his foot and shouts, "I tell you so."

The sayings of great moral and religious teachers frequently find a magically responsive echo in our conscience. Yet our deepest moral feelings may seem to others no better than the superstitious taboos of primitive peoples appear to us. Indeed the morality of unreflective people does consist very largely of a series of taboos; you must not do so and so, and it is not proper to ask too insistently, "Why not?" Take one whom no one will lightly accuse of being unenlightened or irrational, to wit, Plato. The moral aversions which affect him most deeply are eating forbidden food (that is set aside for the gods) and incest.[1] Yet not only the former but also the latter aversion depends upon accidental or external traditions. To Biblical heroes like Abraham and David, there seems nothing wrong in any one's marrying his sister by the same father, provided there are different mothers; and among the Egyptians and others, a marriage between brothers and sisters was considered rather honourable, at least in royalty. Many have shared Hamlet's desperate horror of a man's marrying his brother's widow. Yet that was under certain conditions a pious command of the older Mosaic law.

When we are told that all civilized people are agreed about the immorality of murder, lying, theft, and adultery, we may well raise the doubt whether the agreement (of those who agree with us and are therefore called civilized) is not largely a linguistic phenomenon. We

[1] *Republic*, Bk. 9.

agree to use certain terms in a reprobative sense, but really differ as to what acts are to be so designated. There is certainly great diversity of opinion as to what acts we shall condemn as murder, lying, theft, and adultery.

Consider, for example, the commandment, *Thou shalt not kill.*

If this be viewed absolutely should it not apply to the killing of animals as well as of humans? Any one who has played with a dog (moralists are not reputed to be playful), or watched the gambols of lambs, knows how shocking can be the thought of killing them. The conventional argument that animals have not any reason like man need not be taken seriously. Are insane or idiotic men more rational than intelligent dogs? There seems to be no moral objection to killing a domestic pet to save it from suffering; why not justify euthanasia to relieve people of agonizing tortures?

Again, if the rule against killing be regarded as absolute, shall we not say that morally every heroic warrior is a murderer? If *Thou shalt not kill* be an absolute rule, how can it cease to be so because some one orders us to do it? "God will send the bill to you." Can we escape the difficulty by distinguishing between justifiable and unjustifiable wars, and say, for instance, that wars in defence of one's country justify the taking of life? Any such qualification obviously breaks down the absoluteness of our rule in making it depend on the somewhat shadowy distinction between offensive and defensive measures.

Furthermore, does the absoluteness of the rule, *Thou shalt not kill,* apply only to direct or short-range killing? We know perfectly well that unless more safety appliances are introduced into mines, railroads, and factories, tens of thousands of workers will surely be killed. Are those who have the power to make the changes and do not do so guilty of murder? If in economic competition I take away somebody's bread (to increase my own comfort or power), and he dies of undernourishment or of a disease to which undernourishment makes him liable, am I not killing him? If by monopolizing our fertile lands we confine the Chinese to a territory insufficient to keep them above the starvation line, are we or are we not guilty of killing them?

To common sense and to many moralists nothing seems morally so self-evident as the sacredness of human life. Yet there seems good reason to question the rule that life should always and everywhere be

431

increased and prolonged, and that to restrict birth or hasten death is always and everywhere evil.

The sacredness of life is sometimes supported on supernatural grounds, viz. that since it comes from on high we have no right to meddle with it. We must not lay human hands on the gates of life and death. But this cannot possibly be carried out consistently. Disease comes from the same source as life and death, yet few now follow those moralists who denounced efforts to cure diseases sent by God to punish sinners. Does any moralist condemn the martyr who throws away his life to testify to his faith? We characterize as base those who purchase life at the expense of freedom, honour, or convictions. Not life as a biologic fact, but the good life (involving some co-ordinated plan or pattern) is the object of enlightened endeavour.

We arrive at the same result by considering the false naturalistic conception of self-preservation as a law of nature that leads every one to seek always to preserve his own existence. In the chapter on biology we have had some indication of how misleading this phraseology is. But in the field of morals it is even less worthy of respect. Men generally have a positive preference or urge to live and want to postpone the pain of death. But they also want certain things for which they willingly shorten their lives by hard work, risks, etc. Of mere living existence we might soon get weary if it did not offer opportunity and hope of fulfilling some of the heart's particular desires.

These doubts do not diminish the horror of murder in the cases where we feel that horror. But they are sufficient to suggest that the absoluteness of our rule is generally saved only by refusing to think of many of its possible applications to life. The rule against murder does express a prevailing moral attitude in a number of clear, though not explicitly qualified cases. It claims extension to cases similar in principle. In such extensions, however, we have to introduce so many sorts of qualifications that the rule soon ceases to be categoric and becomes rather dialectical: to the extent that any action involves the destruction of life it is to be condemned—but other principles may supply countervailing considerations. As any definite course of action involves many elements, actual judgment upon it must depend upon some estimate of the relative weights of diverse conflicting moral rules that can be applied to it. What we call a situation involving a conflict of duties is really a case in which different results would follow if we attended to one or another of rival dialectical rules.

A great and noteworthy effort was made by Kant to prove all moral rules absolutely obligatory and to derive them all from one principle, the categoric imperative to so act that the maxim of our acts can be made a principle of universal legislation. To one who asks, "Why should I accept this categorical imperative as the rule of my conduct?" Kant offers no reason except to offer this principle as a formula for the unconditionally obligatory character of all moral rules, such as the absolute prohibition against lying. But why should I regard the latter as absolute? Why may not a lie to save a human being hovering between life and death be justified? There is no logical force at all in the claim that there is some absolute contradiction or inconsistency in telling a lie and wishing to be believed.[2] Nor is there any force in the argument that lying is morally bad because it cannot be made universal. The familiar argument, "If everybody did so and so . . ." applies just as well to baking bread, building houses, and the like. It is just as impossible for everybody to tell lies all the time as to bake bread all the time or to build houses all the time.

Empirically, of course, it is true that lying is subversive of that mutual confidence that is necessary to all social co-operation. And this justifies a general condemnation of lying—but not an absolute prohibition.

One of the consequences of the absolutistic conception of moral rules is the Stoic and Kantian contention that since the moral law demands that sin be punished, it is immoral to pardon any sinner. This appears in the contention that if we know the world is to be destroyed tomorrow we must see to it that the last murderer is executed, else we shall all perish with the blood of his victim on our heads. It shows itself in the orthodox theologic conception of a hell for most of God's creatures; for if God forgave sinners (without an expiating blood-sacrifice on His own part which these sinners must accept) He would transgress the moral law. It seems that we have here a glorified development of the

---

[2] People may not only be perfectly willing to take chances on others' deceiving them but may prefer that others should on certain occasions lie to them. Certain conventions and pleasant illusions are maintained that way; and diplomats are not the only ones who prefer a system of conventions where every one is privileged to half conceal when he half reveals his intentions.

primitive idea that honour demands the avenging of insults, and the greater the dignity of the one offended, the greater must be the vengeance.

A scientific ethics certainly cannot accept absolute moral rules of the character indicated by the foregoing examples.

(B) THE ILLUSIONS OF ANTINOMIANISM

The perception of the variations and inconsistencies of our moral judgments has, since the days of the Greeks, led people to entertain the view that morality is nothing but a matter of opinion or convention. There are many forms of this attitude, of which we may consider: (1) moral anarchism, (2) dogmatic immoralism, and (3) anti-rational empiricism.

(1) MORAL ANARCHISM. By moral anarchism I mean the view which denies that there are any moral rules at all, and insists that our moral judgments are mere opinions, having no support in the nature of things. In fact, however, no one of us believes that his own moral opinions are as bad or as absurd as those of others which fill us with repugnance or resentment. If, on the other hand, it is not true that every opinion is as good or as bad as any other, there must be some principle indicating the direction of preferable or more adequate judgment. We may not be always clearly aware of the principle involved in our actual judgments of approval or disapproval, and we may distrust abstract formulations of them, preferring to let tact or the feeling of the situation control us. But we cannot deny that such tact or intuition may involve serious error, and that such error might be corrected by fuller knowledge and reflection.

Scepticism is a natural reaction to the absurd claims of moral absolutism. It is justified in insisting that there is an arbitrary (in the sense of volitional) and indemonstrable assumption in every moral system, since we cannot have an *ought* in our conclusion unless there is an *ought* in one of our initial assumptions or premises. But from this it by no means follows that moral systems contain nothing but assumptions or that all assumptions are equally true or equally false.

(2) AMORALISM. By the term *amoralism* I mean the attempt to deny validity to the distinctively moral point of view, to wit, that from which

we judge that certain human acts ought or ought not to be. It is difficult to formulate without seeming self-contradiction a direct denial of any distinction between such seemingly different considerations as what *is* and what *ought to be*. However, there are many indirect denials of the validity of the moral point of view—by defining it in terms of the non-moral. The classical example of it is expressed by Plato's Thrasymachus when he defines justice as the interest of the stronger. More recently this has been expressed in the formula: Justice is the command of the sovereign, the interest of the dominant class, etc. If this means that we, the weaker or the subjects, *ought* to obey the expressed command of one who has the power to compel obedience, we have here a moral judgment, but of the kind that can rightly be called slave-morality. For it is slavish to *respect* brute power, however prudent it may be to obey it, and the free intelligent man refuses to let mere external power confuse his vision of what is *better*.

Those who define justice or *right* exclusively in terms of some sort of *might* generally, however, wish to insist that judgments of right are in fact determined by certain external forces. But that is not a question of the *meaning* of the moral judgment but rather of its *genesis*. It is doubtless true that modern ruling classes do have some power directly or indirectly to mould the moral judgments of the community. But it would be folly to deny the fact that men can and do distinguish between that which they see prevailing about them and that which they think ought to prevail. The facts of moral indignation or the persistent and bitter cry for justice are too vehement to be easily ignored.

It is, of course, true that our moral aspiration can never be realized in this world unless we find an effective machinery for it, and this involves recognition and acceptance of the necessary concatenation between available causes and desired effects. It is thus in a sense true that only by submitting to nature can we control it. But this does not deny the distinction between what is and what is desirable.

Hegel and others have argued that individual judgments as to morality or what *ought* to be must be subordinated to the actually existing social institutions, on the ground that the latter embody a fuller and less capricious world-reason. But while it is prudent for any individual to reflect and inform himself more fully before he condemns an existing social institution as immoral, an iniquity cannot cease to be judged an iniquity simply because it exists embodied in the Prussian or in any other state.

It is a significant indication of how far morality is popularly identified with conformity to the established order that Nietzsche calls himself and is called by others an immoralist. True, he attacks the moral value of Christianity, humility, and charity; but he himself is preaching a moral or categorical imperative: Act to obtain power regardless of ease and comfort. Despite his aversion for the Prussian state, this is a hard militaristic morality. Nietzsche's illusion that power is an absolute good is indicative of the uncritical character of his thought. It is rather obvious that power can be exercised only in society. If isolated in a cave or on a mountain man has no power over his fellows and is more dependent on external nature. And a study of social power shows that rulership always involves a heavy sacrifice of freedom on the part of the ruler. Warriors and rulers have to give up their enjoyment and their life in fighting for the protection of their subjects; and in time of peace they may be ruled by priests who are distinguished by slavish obedience to the rules of their orders. In general, rulers are successful to the extent that they recognize the superior power of mass inertia, custom, religion, etc., and do not put themselves in opposition to such forces. The expert ruler must practice the art of flattery and cajolery or else be, like an oriental despot, a slave to customary law and to the constant fear of losing his life. Stated more broadly, we may say that Nietzsche does not take into account the essentially social nature of man, i.e. his incompleteness by himself and his dependence on his fellow-man. The love of power is one element of our nature, but the love of ease and comfort makes most of us shun the arduous labour, responsibility, and risks that are inevitable in the exercise of power. On the whole, government or ruler-ship is possible because the vast majority find it easier to obey and thus be free from responsibility in all except some particular phase of life. The father may rule his children, who rule the mother, who rules the father. The scholar or artist may (and wisely so) care more for his learning or art than for the political governorship of his commonwealth.

There seem to be always moral protestants, who think that by merely breaking the traditional moral rules they will attain freedom and happiness. Alas for the irony of fate! In order to stand strong in each other's esteem and to make up for the disapproval of the multitude, these moral non-conformists must develop a code of their own. The Bohemians of the Quartier Latin or Greenwich Village have their own taboos no less rigid than those of the Philistines.

(3) ANTI-RATIONAL EMPIRICISM. Anti-rational empiricism in ethics generally sets up the claims of what is called "the concrete facts of the situation" against all abstract rules. It refuses to subordinate the actual needs of life to preconceived tags. It rejects all Procrustean rules into which all men must fit themselves regardless of their diverse characters and changing circumstances. All this is a natural reaction to the illusion of absolutism. But it is equally illusory to suppose that a humanly desirable life can be lived without rules to regulate it or without the recognition of invariant laws or relations on which these rules must be based. Changing conditions are not inconsistent with the possibility and serviceableness of a rule of conduct, any more than physical changes preclude the possibility of an invariant law of constant elements or proportions. Against the claim that there can be no moral rules because no two situations are absolutely alike, we may urge that there could be no sort of intelligence as to life, no tact, intuition, or empirical wisdom of any sort, if there were nothing about any situation applicable to another. No two physical situations are ever absolutely identical. Yet this does not preclude the possibility of abstract physical laws that give us control over nature undreamed of by other means. In action as in science, not all that exists is relevant; and neither the fullness of life nor the fullness of knowledge can be attained without scientific organization which ignores or eliminates the irrelevant. What is chaos but a universe in which there is no order or law ruling out certain possibilities. The raving maniac's mind, in which a piece of bread can become a burning volcano, or the ceiling a herd of elephants, points toward, though it falls short of, the absolute chaos in which all things are possible and anything may become anything else.

But if we cannot accept either absolution or antinomianism, whither shall we turn?

Our previous analyses in the light of the principle of polarity of the issues between absolutism and empiricistic relativism provide a vantage point from which to see the truth at the basis of both contentions. Concretely every issue of life involves a choice. The absolutist is right in insisting that every such choice logically involves a principle of decision, and the empiricist is right in insisting on the primacy of the feeling or perception of the demands in the actual case before us. If it is possible for us to be mistaken in our moral judgments, there must be some ground for the distinction between the true and the false in this as in other fields. Even if we deny that principles are psychologically primary sources of moral truth and view them as only the formulae which give us abstract characteristics of our actual judgments in moral affairs, the errors of the latter can be corrected only by considering our judgments in similar cases, and this means cases alike in principle. Principles express the essence or form of a whole class of individual cases. And they enable us to correct the individual judgment precisely because the recognition of what is essential in many cases helps us to distinguish the relevant from the irrelevant in any one case.

The moral rules whose absolutistic claims we have rejected are useful generalizations of human experience, like the cruder generalizations of popular physics, e.g. that all bodies fall. The existence of exceptions to such generalizations proves that we must either refine their statement so that the exceptions will be included in the rule (just as the law of gravitation accounts for some bodies not falling) or else formulate our principles dialectically, as what would logically prevail if other principles did not offer countervailing considerations.

The imperfection of our generalizations makes the relativist underestimate their importance or reject them outright, while the absolutist falls into the logical error of confusing generalizations of experience subject to exception with truly universal or necessary propositions. We must, therefore, accept empiricism as to the content of moral rules without abandoning logical absolutism in our scientific procedure.

§ II. THE APORIAS OF THE RATIONAL IDEAL

ETHICS, as we have seen, cannot be restricted to a natural history of what men actually do or even to a psychologic study of what in fact they believe they ought to do. An adequate science of ethics must, to be sure, include a good deal of such material. But it must not forget that it is a normative science, i.e. that it is a logical study of the validity of judgments of right or wrong, good or evil, implied in our expressed or tacit choices. Its primary interest is with the extent to which these judgments can be harmonized into a rational system.

This involves a distinction between pure and applied ethics, between the abstract theory of moral judgments and the art of living according to the best of these judgments. Interest in ethical science is thus not to be confused with such functions as those of the preacher and moral pedagogue who already know what is right or good in every situation and desire to help men to achieve it. The ethical philosopher need not necessarily assume the latter task—just as the physiologist need not go on to practise medicine—though it may be generally desirable that the moralist, like the physician, should be trained in the science at the basis of his art.

Is such a science of ethics possible?

The ideal of a rational ethics can be traced back to the Hellenic idea of deliberate wisdom concerning the management of life, parallel to the science of harmony at the basis of the art of music or the science of physics at the basis of the art of medicine. This finds its articulated expression in Aristotle's *Ethics*. But even in Plato we already have doubts as to whether the practice of virtue can be taught; and humanity at large has never accepted ethics as a science. Theologic or religious traditions have insisted that human life is not autonomous but must be subordinated to some supernatural objective, such as the will or greater glory of God, and this calls for more than human guidance. Modern romanticism too rejects the Hellenic ideal of the rational conduct of life, as bourgeois calculation devoid of inspiration, Dionysiac frenzy, ecstatic thrill, or the "kick" of great experience. With the rejection of the ideal of rational conduct, what room can there be for a rigorous science of the judgments involved in such conduct?

It is well to consider attentively the various persistent objections against this idea of ethics which reveal *aporias* or inherent difficulties.

(A) MORAL KNOWLEDGE AND THE MORAL WILL

The first objection is that this conception of a scientific ethics is not in harmony with the great stream of the world's moral thought. In the latter morality is not a matter of intelligence or deliberation but of strength of spirit to overcome the "fleshly screen" and to live up to

439

the moral truths clearly revealed by some divine teacher or other source of wisdom superior to ours.

This contention is sound as against moralists who believe that intelligence alone can change the course of our actual conduct. But Plato and Aristotle had already recognized the importance of habit and the influence of nature as well as of social institutions in effective moral education. Our power to change our moral nature or disposition, like our power to increase our stature, is limited. But with this granted we can still insist on the validity of the rational ideal. The moral precepts of the great religious teachers often conflict with each other—the teachings of Mohammed differ from those of Buddha, and neither are exactly those of Christ. And we may urge that what is common to them all has commended and maintained itself to the extent that it has stood the test of rational experience. In the round of our routine tasks we need strength and courage to live up to our ascertained truths rather than to be always searching for new insights. But when new and unforeseen situations arise, even the most pious soul may be perplexed and raise a cry of anguish for new light. In a changing heterogeneous society, in which conflicting moral and religious faiths are professed, the need for new insight can be met only in critical deliberation or scientific analysis of new situations in terms of relevant elements. And this shows itself also in the constant efforts to re-interpret the old authoritative teachings, so as to find in them answers to new moral problems. Examine the various published efforts to solve the problems of justice between different nations, or between diverse classes (such as employers and employés), which the various representatives of the different sects of Christianity, Judaism, or Islam have brought forth on the authority of their religions. Does not the element of authority in them become rather nominal when the authors come to grips with the realities of their problem?

(B) ETHICS AND DETERMINISM

Another objection is that theoretic ethical speculations must concern themselves with what an ideal creature would do in an abstract situation and that this enterprise can find no suitable material in the way men meet the tasks imposed upon them by their physical nature and social environment.

440

There is an element of justice in this as against moralists who achieve simplicity in their ethical rules by dismissing with the opprobrious term *temptation* those pressures which blot out abstract possibilities from the field of our actual choices. But such disregard of actual pressures is not required by a theoretic science.[3] It must indeed recognize the causal nexus in our conduct. But precisely because it is in that sense deterministic, it can inquire as to the significance of possible alternatives to the course which we actually take. Our analysis of history and of the nature of laws has prepared us to see this. We understand the significance of the Greek conquest of Persia or of Watt's perfection of the steam-engine if we ask what would have been the course of history without them. So, in our individual life, not only the prospective but even the retrospective analysis of the values of our acts helps to illumine their significance.

(c) THE PLURALITY OF MORAL ASSUMPTIONS

A third objection to the rational ideal may be put thus: Reason can enlighten us only as to the means necessary to bring about certain ends, but the choice of final ends is a matter of will. How then can there be a science where the arbitrary element is controlling?

This objection can be met only by frankly admitting the possible plurality of moral systems, bearing to the pure science of ethics a relation similar to that which various systems of geometry bear to pure mathematics. This runs counter not only to the absolutistic conception of moral rules but to the view that every act must be either right or wrong regardless of assumptions. But why not admit that the suicide of a given individual, say Cicero, may be right from one set of assumptions and wrong from another? I must condemn it if I assume that it contravenes the canon which the Eternal has set against self-slaughter. But if I do not make this or similar assumptions I may agree with the Stoics, who found no moral objection to suicide. If life has become an unendurable smoky chamber, all arguments as to why a man should stay in it must appeal to some actual desire in him. If he has no natural desires left, all "arguments" and the application of such epithets as *selfish, cowardly,* etc., are themselves arbitrary and irrational.

If I wish to preserve my health I must take account of the laws of

---

[3] See pp. 135 ff., supra.

physiology. But ought I always to preserve my health? That depends on some further assumptions. If I value the safety of my country, my family, or my own creative artistic activity, I may answer in the negative. And so, ethics may be viewed as dealing with hypothetical imperatives which condition a rationally coherent plan of life. Such a science directly enlightens us only as to necessary means, but in so doing it clarifies the choice of ends by showing what is involved in such choice. We can better decide what road to take if we know what we can reasonably expect on the way. While, therefore, rational science cannot give us absolute moral rules, it is, like mathematics, inherently applicable to the actual world.

The difficulties of such application, involving, as the latter does, a just appreciation of the diversities of power and inclination, cannot be underestimated; but the realization of such difficulties does not render a science of ethics impossible. It only helps to make it liberal and humane.

(D) THE COMMENSURABILITY OF VALUES

The foregoing analysis will help us to deal with the objection that ethical science is essentially inapplicable because human values are inherently incommensurable. Must not ethics as a science presuppose that we can always rationally determine which of two practically incompatible goods is the greater? And if this cannot be determined, is there any ground for ethical judgments? In every science we have fields in which all individuals are alike in a given respect. But though this is often asserted in the field of ethics, it seems absurd to ignore the fact that mere differences of time and space must control our moral judgments. No one in fact loves his neighbour's wife and children exactly as his own. Nor does it seem compatible with the conditions of human life that he should. Should we look after the helpless people in Thibet or Patagonia as after those of our own country, state, or city? Or are obligations to remotely future generations the same as to our immediate children?

It is fashionable to deny the force of the conflict between egoism and altruism by the assertion of an ultimate pre-established harmony between the interests of each and those of all. But even if we ignore the oppression and rapine against which men have cried out bitterly through the ages, we cannot lightly dismiss strife as a mere logical oversight. We live in a world of finite resources in which no two men can eat the same piece of bread or stand on the same ground. When the rulers of an overcrowded country threaten in the interest of sacred egoism to take the land that it needs from some other people, outsiders are apt to be scandalized. But these outsiders may approve of their own country's protecting the property of its citizens abroad when such property is used to exploit the economic resources of the weaker nation. The rule that each individual is to count for one will not help us to solve such ethical problems.

But the incommensurability of values holds not only in respect to different individuals but also in respect to different kinds of interests or values. Consider such a relatively simple question as how much of his opportunity for education and self-development as a great artist a dutiful son should give up in order to help his parents in their economic struggle. Are we to assume that the life of the young son, with his unusual promise, is equal or inferior in value to the life of one of his parents? Or shall we say that in abandoning his art for the greater comfort of his parents the resulting good (presumably by the example set to others) will outweigh the loss involved?

One ethical theory, hedonism, tries to overcome this difficulty by asserting that pleasure is the only human good and that conduct must be judged moral or good only in proportion to the balance of pleasure it ultimately produces. But while the common consciousness of mankind regards pleasure as a good, it rebels against regarding it as the only good. It associates morality rather with painful efforts to achieve things worth while which are not inherently pleasurable, though there may be pleasure in achieving them. Indeed the pursuit of pleasure is generally regarded as something reserved for our leisure moments when we are not engaged in the more serious and important life-activities. Few believe that the heroic fortitude of a Giordano Bruno or a John Huss going to the stake is good only because it was or is productive of more pleasure than pain. Consider also a musician gifted with the power of playing pleasing music or a cook with unusual skill in preparing food tastily. Surely the balance of pleasure involved in such an individual's activity is great. Yet what hedonist has considered this as the typical example of moral conduct? Many indeed will regard

443

such pleasing natural gifts, for example Mozart's, as having no moral value at all—morality being attributed only to conscious effort and striving. But no effort will produce as much pleasure as that which results from natural aptitude.

Some hedonists like J. S. Mill have tried to avoid this fatal weakness by distinguishing different classes of pleasure. But they have not been successful in establishing a hierarchy or scale to decide how much of a lower pleasure will outweigh a higher pleasure.

Consider the everyday question of estimating the degree of satisfaction of growing older and wiser as against the loss of zest and freshness which pass with youth. Are we really in a position to compare the two? Is there any common element in them which can be determined as being larger or smaller? Present satisfaction can be compared only with the memories of the past, and the latter are moulded as well as selected by our present interests. Statistical studies will not help us. Not only do different people have different experiences, but the same individual does not maintain a constant attitude to the different values which age brings or destroys. Sometimes we sigh for the springtime of our life. At other times we say, "Would you be young again? So would not I."

If hedonism has not satisfactorily established the commensurability of all values, neither have those moralists who maintain that moral effort or the good will is the only good. The latter point of view would make the material content of all such effort—natural life and all the things we strive for—devoid of any value. Doubtless moral effort is itself a good, but no more than pleasure can it be made the only good. To go on and assert that from the point of view of a science of ethics pleasure and moral effort are both necessary elements in the human good will not in itself solve the difficulties in the incommensurability of values.

(E) PRUDENCE AND THE PURSUIT OF THE UNATTAINABLE

The truth seems to be that there is not only a paradox in individualistic hedonism—that we are often happier when we successfully pursue some objective end without thinking of the pleasure of attainment—but that there is a wider paradox expressed by Santayana in the line, "It is not wisdom to be only wise."

We may exhibit this paradox in the form of a seeming antinomy:

444

Thesis: Wisdom means rationally managing or organizing our activities so as to achieve a maximum of attainable ends or goods of life. Without such wisdom our lives are foolishly wasted in the pursuit of ends which we cannot attain or of those which, when attained, produce only regret and unhappiness.

Antithesis: Men can achieve their noblest and their distinctively human character only by pursuing some great unattainable goal such as holiness or by subordinating themselves completely to some cause bigger than themselves. The pursuit of merely natural or worldly satisfactions will not of itself give human life sufficient scope, dignity, and real worth.

The thesis seems self-evident and in no need of supporting arguments. If the terms *wisdom* and *foolishness* are at all applicable to the conduct of life, there must be some rational ground which makes some choices wiser than others, and this demands a rational system of values or preferences. Few sensitive people can look back over their past life without a feeling of regret for lost opportunities or wasted years. "If I had only thought of . . ."

The upholders of the antithesis are, however, not without true insight. The Aristotelian or the Hellenic conception of a science or wisdom which shall illumine our natural desires tends to conceive of morality as mere prudence. Calculation is more readily applied to the tangible and more obvious phases of life, and we are thus induced to ignore the finer shades and larger vistas which give depth as well as significance to life. By ignoring the unattainable beyond us we cut ourselves off from that intense enthusiasm which brings out our utmost efforts and prevents our existence from becoming drab, our pleasures unimaginative, and our practical activities narrowly mechanical.

It is easy enough to find many ways out of this paradox of utilitarianism (in its widest sense, as synonymous with deliberative morality). But it is well not to minimize the real difficulty of a naturalistic ethics that does not fall into the vice of worldliness.

Not only is the rationalist who views life as an organization of means to ends apt to miss the subtler ends which other-worldly moralists emphasize, but he is also apt to pass over too lightly the immediate qualities of human activity which interest less rationalistic naturalists. The most obvious example of a sharp distinction between means and

445

ends is seen when men engage in hard labour to earn money. Yet all work no matter how unpleasant must have a certain incidental satisfaction as the successful expression of our energy overcoming obstacles. Without such satisfaction work could never get started or become habitual.[4] So the business man bends all his energies to the accumulation of a profit. But the driving force is not merely the reward of money but the incidental zest of the combat or game of making money together with the pride of victory. Myopic naturalism looks to the end or objective of love and misses the immediate delight of courtship, of mere communion between dear ones, of dreaming bright hopes for the future, etc.

A striking example of the failure to do justice to the immediate because of undue absorption in ends or purposes is the failure of some naturalists to appreciate the value of pure science as a most delightful expression of human energy apart from its useful applications.

Elaborate planning and efficient preparation may make more certain the attainment of certain ends, but it may also diminish spontaneity and make the whole enterprise less worth while. But it may not be unfair to claim that only a rationalistic naturalism can liberate us from false alternatives between means and ends. It does so by showing that logically the end or aim of any rational conduct is not something outside of our activity itself but a character or pattern of life itself. If the end is thus a whole which includes the necessary means, it is to be judged and justified (if at all) by the means which it involves.

§ III. RATIONAL METHOD IN ETHICS

Do not the foregoing aporias or difficulties bar the way to a science of ethics? The difficulties are real and they will remain real so long as we remain imperfect creatures in an unfinished world. But ethics as a science does not need to solve all our moral problems. It will serve its primary purpose as a science if it determines the degree of accuracy and probable truth that our moral judgments can attain at any state of our knowledge. This of course involves our possession of the logical ideal of perfect knowledge, but only on its formal side as de-

[4] Bücher, *Arbeit und Rhythmus.*

termining the direction of progress and as a standard of judging our actual attainments at any moment.

If we thus reconcile absolutism as to form or logical method and relativism as to material content, we can see the possibility of an ethics that is rigorously theoretical in its dialectical part and empirical in its material rules. The moral intuitions of which men have been most certain have turned out to be variable and fallible. Yet they are overthrown only by appeal to more enlightened conscience. There is no way in which we can recognize the error of a moral intuition except by showing its inconsistency with other moral perceptions. In this respect moral data are certainly no better than physical. Physical scientists are constantly correcting observations by theoretic considerations, but these theoretic considerations are such as enable us to organize our perceptions into a rational system; and the latter is a continuous process of self-correction.

Ethics, then, like any other science, must function with two elements, the dialectical and the empirical. Whether we are considering the ideal of the life of mankind or the ideal of an individual within a given community, we must assume certain empirical facts of human nature, e.g. that men generally have certain impulses, desires, sensibilities, etc., and that they will respond in certain ways to certain conditions. Without such empirical assumptions there would be no reason for preferring one mode of action to another. But we must also assume that certain things are generally more desirable than others, pleasure than pain, beauty than ugliness, honour than contumely, freedom of self-expression than slavery, etc. By generalizing such rules into hypotheses we can formulate their negatives, and then by comparing them with each other and with the facts of experience we may be led to draw distinctions and refine our rules so that they make for greater accuracy and are subject to fewer exceptions. Thus we may start with the idea that as the human desire for dominion over things is ineradicable some form of private property is essential. We may then compare the consequences of this rule with those of the opposite principle, viz. that if people are to live together the individual's use of things must be subject to rules in the interests of all. All sorts of empirical considerations will have to be taken into account before a generally satisfactory rule of property can thus be formulated. Similarly we may begin by recognizing the general desirability of monogamy so long as there is a general equality in the

447

number of men and women, so long as it is desirable for both father and mother to care for children, etc. We can then go on to note countervailing considerations, e.g. what happens when one of the parties becomes incapable because of crime, insanity, or the like, of fulfilling marital or parental duties. As the number of relevant factors that enter into any human problem is generally much larger than we are apt to think of at first sight, we must never forget the tentative or provisional character of the rules we formulate at any given time. At best we can only say: Taking the most important factors into account, these rules seem to fit most cases best.

We arrive at a similar result if we pursue a more abstractly dialectical procedure by taking one element at a time. We may thus argue that lying is always bad because it always works in the direction of undermining the basis of that mutual trust which is necessary for social life. We can then try to explain cases of justifiable lying as cases where the countervailing considerations are stronger. This will be found to involve some hierarchy of human values—a difficult yet inescapable task. In general dialectical rules as to a given factor, assuming all other factors to be indifferent, are seldom directly applicable. But like theoretic physical rules (e.g. the law of inertia) they help us to analyze actual cases and they are approximately applicable when divergences from them can be explained by some other rule.

The provisional and merely probable character of ethical judgments will remain a perpetual source of irritation to our craving for absolute certainty. But we need to learn to live in an uncertain world. Our action is seldom simple or due to a single motive. We can only say that so far as we understand a given act and so far as it runs counter to a certain accepted moral rule it is subject to moral condemnation. The possibility that we do not know all the facts or have not analyzed all the relevant factors is always present, and must mitigate our confident condemnation of others. Thus the honest and courageous recognition of the insuperable difficulties in the way of ethical judgments makes for a more viable set of moral rules.

The recognition of the validity of (unattainable) ethical ideals gives a direction to our efforts and prevents our conduct from sinking back into its animal origin. But the recognition that no actual temporal act can fully embody the ideal saves us from the idolatrous worship of some particular which, no matter how good, blocks the road to something better. We thus keep open the path of reform but keep from the fatal and desolating illusion that we can ever have or bring about a heaven on earth—an illusion which has been the source of much that is noble but also of that fierce fanaticism which has shut the gates of mercy on mankind. A sense of humour is the foe of fanaticism in that it is a manifestation of a sane sense of proportion. A wise Frenchman has well said: We must not throw to the dogs all that is not fit for the altars of the gods.

The spirit of scientific method is too sceptical, cautious, and sober to guarantee us many of the traditional hopes that like stars pierce the infinite darkness which envelops the human scene. What has science to offer us as sustaining as are these hopes which it dissolves?

Those who ask the last question are apt to forget how often these traditional stars of hope have proved will-o'-the-wisps, the pursuit of which has led mankind into hopeless swamps. But the readiness to sacrifice our fondest hopes in the single-minded pursuit of truth for its own sake is not without consoling compensations to a suffering humanity. We require not only hope but also strength of vision to look steadily into the darkness surrounding us. After all, the great religions relieve human misery not only by offering anodynes of hope and otherwise catering to our weakness but in greater measure by strengthening the spirit. So the great lesson of humility which science teaches us, that we can never be omnipotent and omniscient, is the same as the lesson of all great religions: man is not and never will be the god before whom he must bow down. If this seems depressing to those who think that the religious consciousness offers an actual union of the human and the divine, we may reply that in knowing our limitations we formally possess absolute knowledge; and to those who worship a suffering god it is surely not blasphemous to suggest that possibly this humility is a part of the divine consciousness. In any case,

"To bear all naked truth
And to envisage circumstance all calm,
That is the top of sovereignty."

END OF BOOK III

449

Epilogue

IN DISPRAISE OF LIFE, EXPERIENCE, AND REALITY

I N SPEAKING of the new philosophic movement which began with the present century, William James remarked: "It lacks logical rigour, but it has the tang of life." It is strikingly significant of the temper of our age that this was intended and has generally been taken as praise of the new philosophy. To any of the classical philosophers, to whom not life, but the good life was the object of rational effort, James's dictum would have sounded as a condemnation. For life devoid of logic is confused, unenlightened, and often brutish. Indeed the new philosophy itself maintains that it is precisely because unreflective life is so unsatisfactory that it gives rise to logic. Why then should the word *life* itself be a term of praise except to those who prefer the primitive and dislike intellectual effort?

I can imagine that a classical philosopher living long enough amongst us to penetrate some of our bewildering ways might conclude that our worship of mere life, rather than the good or rational life, reflects the temper of an acquisitive society, feverishly intent on mere accumulation, and mortally afraid to stop to discriminate between what is worth while and what is not. The same preference for terms of promiscuous all-inclusiveness, rather than for those that involve the discrimination essential to philosophic clarity, shows itself also in the use of the terms *experience* and *reality*. It is of course true that surface clarity can readily be obtained by ignoring fundamental difficulties, and that we cannot dispense with terms indicating the unlimited immensities of which our little formulated systems are but infinitesimal selections. But if the world contains many things and therefore distinctions between them, ignoring these distinctions is not the same as profundity. The honorific use of non-discriminating terms can only serve to darken counsel. That this has actually been the case in ethics and in theories of knowledge, in religion and in art, is the burden of this brief epilogue.

450

That the continuance of mere physical life is an absolute moral good seems to be axiomatic in current ethics. It serves as a basis for the un-qualified moral condemnation of all forms of suicide and euthanasia. Now I do not wish to question the biologic proposition that there are forces which make the organism continue to function after we have lost all specifically human goods, such as honour and reason. What I do wish to point out is that this setting up of mere life as an absolute moral good, apart from all its social conditions, is inconsistent with the moral approval of the hero or the martyr who throws away life for the sake of honour or conscience. It would be pathetically absurd to praise the abandoning of life by John Huss or Giordano Bruno on the ground that it increased or prolonged the total amount of life. Indiscriminate in-crease of population beyond any definite limit is of very doubtful moral value—despite the arguments of those who oppose all forms of birth control. We must not lose sight of the fact that life always carries with it not only the seeds of disease and inevitable death, but also the roots of all that is vicious and hideous in human conduct. We cannot, therefore, dispense with the classical problem of defining the good and discrim-inating it from the evil of life—a difficult and baffling problem, to be sure; but those who find it profitless are under no obligation to pursue moral philosophy.

The confusion of moral theory by the eulogistic use of the word *life* can be readily seen in the Nietzschean ethics—all the more instructive because Nietzsche himself starts from the classical perception of the inadequacy of ordinary utilitarianism in face of the moral values of heroism. The good life involves the sacrifice of ease and comfort, the receiving as well as the giving of hard blows. But just because Nietzsche is impatient of definition he falls into the easy error of sharply opposing the pursuit of life to the pursuit of knowledge—witness his essay on History. But the pursuit of knowledge is itself a form of life. This fact cannot be obscured by rhetorical contrasts between the life of the closet philosopher and the open-air or what is euphemistically called *real* life. To the eye of philosophic reflection the scholar or persistent thinker shows as much life or vitality as those who have to cover their naked restlessness by a gospel of strenuous but aimless perpetual motion—in no particular direction. This is not the occasion to sing the praises of the intellect and what it has done to humanize life. We may grant that the distaste for arduous intellectual tasks is natural, blameless, and in some cases even providential. But when such distaste sets itself up as a philosophy of life it is only ridiculous.

This brings us to our second point, the vitalistic theory of knowledge —or perhaps we should refer to it as the theory of a vitalistic intuition superior to knowledge—I mean the widespread notion that by mere living we get an insight superior to that of the intellect operative in mathematical and natural science. To prevent misunderstanding, let me say that I am not referring here to genuine mysticism which asserts that all intellection and language move in the mist of appearances and cannot reach the ineffable reality. Genuine mysticism always holds fast to the idea that the substance of reality is altogether beyond the power of language, and hence it does not use language to describe this reality. It holds that language can at best only indicate its own shortcomings and thus point the way beyond itself. When, however, as in the Bergsonian theory, the claim of the scientific intellect is set aside for an instinctive intuition, and when this is held to provide a superior explanation of empirical phenomena like the formation of the eye of the scallop, it seems to me that philosophy is then not far removed from glorified quackery where the philosopher's stone is expected to remove the effects of the evil eye or cure toothaches and other empirical ailments. We may grant that biology as a natural science does not carry us very far into the mystery of life. But it does not follow that our ignorance can be cured in any other way. The fallibility of scientific reasoning is best corrected only by definite experiments and the critical reasoning of science itself. When men despair of solving theoretic problems and appeal to undefined words like *life* they show themselves devoid of intellectual stamina. It is doubtless true that in the process of living our ideas develop, mature, and receive a solid amplitude through an enriched content. Time tests our judgments and eliminates clever, plausible sophistries. But it is also true that the older a lamb grows the more sheepish he gets. Nothing seems so solidly established by anthropology and history as that men will not learn from what has actually happened to them unless they have developed the power of reflection. The idea that experience alone will teach everybody is a thin optimistic illusion.

The use of the word *experience* without any ascertainable meaning is perhaps the outstanding scandal of recent philosophy. In its original sense, which it still retains in ordinary, intelligible discourse, and from which we cannot altogether liberate ourselves in philosophy, experience denotes conscious feeling or something which happens to us personally. Thus I make my meaning clear when I say, "I did not experience any pain during an operation," or, "I have never experienced what it is to be struck by lightning." I may also speak of not having experienced the panic of 1872 or the other side of the moon. The absence of such experience need not, however, prevent me from knowing a good deal about the operation, the lightning, the panic of 1872, and the other side of the moon—more indeed than about many of my own experiences. For experience in this personal and ordinary sense is but an infinitesimal portion of what is going on in the world of time and space, and even a small part of the world of ordinary human affairs. To identify the substance of the world with the fact of our experience of some part of it is to set up an anthropocentric universe, compared to which the mediaeval one is sane and respectable. For the mediaeval one rebuked the silly and arrogant pretensions of humanity by setting against it the great glory of God.

The absurdity of identifying the whole realm of nature with our little human experience of it is obscured in two ways—to wit, (1) by confusing the nature of possible experience, and (2) by stretching the word *experience* until it *excludes* nothing and therefore includes no definite meaning.

That things known are all objects of a conceivable possible experience to some possible being more or less like us need not be denied. But the object of a possible experience is a matter of intellectual consideration, not the object of actual personal experience. If, on the other hand, we stretch the meaning of the word *experience* and make it include everything that we can think about, e.g. the state of the earth before the advent of life, then there remains no difference between an object considered and an object experienced, and the proposition that knowledge rests on experience ceases to have significance. It is vain to define words so as to deny the fact that we know many things to be beyond our experience. In general, the term *experience* either means something personal and therefore limited, or it becomes so promiscuously all-inclusive that it ceases to have any intelligible negative. Without an alternative term to denote what is not experience it cannot have any pragmatic meaning. With characteristic sensitiveness to the difficulties of his own account, Professor Dewey has realized something of this dilemma in which the use of the term *experience* involves him. He has tried to defend it by the analogy of the use of the terms *zero* and *infinity*. But zero and infinity indicate at least definite directions. They indicate which of two definite terms is to the left or right of the other in a series. The term *experience*, however, in Professor Dewey's thought is equally applicable to everything that is an object of consideration. I cannot therefore see that it serves any definite intellectual function beyond carrying the faint aroma of praise.

In general, when familiar words are stretched and put to new uses, confusion is bound to result. For the meaning we attach to words is based on habits which arbitrary resolutions cannot readily change, and we invariably drag the old meaning into the new context.

An instructive instance of the confusing use of the term *experience* is the current phrase *religious experience*, used by those who regard it as a substitute for rational theology. Here again, I have no quarrel with any one who claims to have had the beatific vision of God or a special revelation of the truths of religion. One who makes such a claim puts himself beyond argument except when he asks others to believe what he believes. Then the doubt which Tennyson applied to his own vision certainly becomes relevant. Nor is my quarrel with those who assume the truths of their religion on the authority of an historic church or revelation fortified by the necessary truths of reason. The current fashion which talks about religious experience distrusts the great streams of historic tradition as it does the claims of systematic theology—witness James's *Varieties of Religious Experience*, in which none of the great historic religions receives any attention. He thinks he can establish "piecemeal supernaturalism" by the methods of natural science and the rules of empirical evidence. An elementary consideration, however, of the logic of induction shows the impossibility of proving the existence of miraculous or supernatural interventions on the basis of the postulate of the uniformity of nature involved in induction. Indeed, the naturalist can well maintain that as instances of mystic experiences have their parallel in the effects of drugs, starvation, etc., the naturalistic explanation of them is the only one that is scientifically worth investigating.

In any case, the spiritualistic hypothesis does not lend itself to the crucial test of affording us verifiable predictions. Not only a scientist but even a court of law would be derelict if it accepted as proved anything which rests on no better evidence than that offered by abnormal psychology for a finite, personal God and the immortality of the individual soul.

It is of course true that most people do not hold these beliefs as scientific hypotheses at all. Indeed, most people regard the cold, logical analysis of their religion with a horror like that which would be evoked by a funeral orator who proceeded to give a scientific examination of the character of the deceased. We come to mourn and praise our friend, not to hear him psychoanalyzed. But all this is irrelevant in moments of reflection or when our beliefs are challenged by the contrary beliefs of others. One may say: I hold these truths and the faith in them strengthens my life. But such assertions cannot keep out the lurking doubt that it is the psychologic attitude rather than the truth of what is assumed that produces the practical effects. The pragmatic glorification of belief contains the deep poison of scepticism as to what really exists, and this like a Nessus shirt will destroy any religious belief that puts it on. Religion may begin in ritual and conduct, but it inevitably goes on to reflective belief that must submit to the canons of logic. The popular and superficial contrast between religion and theology ignores the fact that where a diversity of religion exists it is impossible to stop a process of reflection as to which of two conflicting claims is true. In such a society, religious creed or theology (including the possibility of a negative or atheistic theory) becomes inescapable. Hazy talk about religious experience will not adequately meet the difficulties.

If terms that have no genuine negatives are to be condemned as devoid of significance, the word *reality* should head the list. I am not unmindful of the many attempts to define the unreal. But the question is: What corresponds to these definitions? The Hindoo mystic is deeply irritated when the wise Chinaman suggests that the realm of Maya or illusion does not really exist, or that it is not worth while worrying about it. The reality of illusion is the emphatic centre of the Hindoo's philosophy, and similarly, of all those who sharply contrast reality and appearance. The difficulty here is classic. What I am more especially concerned about, however, is to call attention to the fact that the word *reality* maintains itself as a term of praise rather than of description.

To be "in touch with reality" is our way of expressing what our less sophisticated brothers and sisters do by the phrase "in tune with the infinite." It is an expression which carries an agreeable afflatus without dependence on any definite meaning. Such edification is pleasing and would be harmless if it did not also cause intellectual confusion. This the eulogistic use of the word *reality* certainly does in the theory of art, especially in its realistic and expressionistic form.

Professor Neilson defined the realistic motive, in poetry and art generally, as the sense of fact. But whatever else art may involve, the process of selection is certainly essential to it in all its forms, useful as well as ornamental. Hence, the honourific use of a non-discriminating term like *reality* undoubtedly tends to justify the introduction of the inept and the ugly, which certainly cannot be denied to have real existence. But it is not only realism that is thus encouraged to escape or confuse the fundamental problem of what is relevant, fitting, or beautiful in representation and ornament. Expressionistic theories glorify the same lack of discrimination between the beautiful and the ugly. For expressionism is but a subjective realism. This becomes clear when we reflect that the real denotes, first, human affairs, then physical things, and now vivid impressions or emotions, so that abstractions are not real to us. The praise of reality, therefore, now has as its core the glorification of vivid impressions or violent expressions, regardless of fitness or coherence. This shows itself in an indiscriminate admiration for the breaking of all hitherto accepted rules of art—as if all rules were necessarily hindrances. But rules of art like the so-called rules of nature are at bottom only statements of what is relevant and what irrelevant to any given case. Hence it is doubtless true that new situations in art cannot always be profitably decided by old rules. But this again is a question of specific fitness, not to be disposed of by the violent assertion that the expression of inner reality is inconsistent with all rules.

It is doubtful, for instance, whether such a convention as the rules of the sonnet ever hindered a great poet from expressing himself, though it doubtless has aided many minor ones, perhaps unduly so.

To conclude, we cannot praise life without including in our praise moral and physical evil, corruption and death. As experience certainly includes error and illusion, we cannot praise it indiscriminately as a support of truth. Finally, as reality undoubtedly includes the useless and the ugly, its praise cannot but confuse the arts.

Instead of life we want the good life. Instead of accepting experience science discriminates between the experience of truth and the experience of illusion. Not all reality, but only a reality free from ugliness and confusing incoherence is the aim of art. Conduct, science, and art thus depend on rational discrimination. Rational philosophy tries to meet this need by defining the good, the true, and the beautiful. The essence of the romantic use of the terms *life, experience,* and *reality* is that it avoids this necessary task, and is therefore flattering to those to whom the use of reason is irksome. But the way to serenity and happiness through wisdom is more arduous and requires a purified vision into our hearts as well as courage to face the abysmal mystery of existence.

INDICES

INDEX OF SUBJECTS *

* Compiled by Dr. Ernest Nagel.

461

INDEX OF SUBJECTS

Index of Proper Names

Printed and bound by CPI Group (UK) Ltd, Croydon, CR0 4YY

21/10/2024

01777088-0014